INSTITUTE FOR RESEARCH IN
CLASSICAL PHILOSOPHY AND SCIENCE

Sources and Studies in the History and Philosophy of Classical Science

2

Edited by Alan C. Bowen and Francesca Rochberg-Halton

Science and Philosophy in Classical Greece

Edited with a Preface by

ALAN C. BOWEN

GARLAND PUBLISHING INC.

NEW YORK and LONDON

1991

Originally published in 1991 by the Institute for Research in Classical Philosophy and Science. Printed and distributed by Garland Publishing Inc., 136 Madison Avenue, New York, NY 10016

This book was typeset in TEX using the Computer Modern font family designed by Donald E. Knuth and Graeca designed by Phillip B. Payne, for the body text. Display type was set in the Times font family developed by Adobe Systems Incorporated.

Library of Congress Cataloging-in-Publication Data

Science and philosophy in classical Greece / edited, with a preface, by Alan C. Bowen.
p. cm. — (Sources and studies in the history and philosophy of classical science : 2)
Essays derived from a conference held by the Institute for Research in Classical Philosophy and Science, 1986.
Includes bibliographical references and indexes.
ISBN 0-8153-0214-2 (acid-free paper)
1. Science—Greece—History. 2. Science—Greece—Philosophy.
I. Bowen, Alan C. II. Institute for Research in Classical Philosophy and Science. III. Series.
Q127.G7S365 1991
590.38—dc20

91-37976
CIP

Printed in the United States of America
ISBN 0–8153–0214–2

Printed on acid-free paper

To my colleagues at the
University of Pittsburgh
in gratitude and appreciation

CONTENTS

PREFACE

This collection of previously unpublished essays derives from a conference, 'The Interaction of Science and Philosophy in Fifth and Fourth Century Greece', which was held by the Institute for Research in Classical Philosophy and Science in 1986. This conference was intended to be a first step in discovering ways to revitalize research in the history of ancient Greek philosophy and science. For it is a regrettable fact of academic life today that specialization often creates and enforces divisions between disciplines concerned with a common subject-matter or period, which few scholars even try to overcome. And this tendency is especially clear in the history of ancient Greek thought. Yet, the existence of competing stories told by the various academic groups—historians of philosophy, the exact sciences, biology, and medicine—who study this subject and do not communicate with one another is surely a challenge to begin afresh. Moreover, in the case of Greek thought in the fifth and fourth centuries, since scientists and philosophers were usually one and the same, meeting this challenge to begin anew should involve taking advantage of the results of specialization but without neglecting research in related disciplines. Still, to adjudicate among these stories in order to determine a single, unified account which is richer in its explanatory power is no easy task, since it demands a real appreciation of the questions asked in the various disciplines and sub-specialties, a mastery of the relevant information, and a critical grasp of how to use this evidence to establish results. In fact, given the expertise needed, it seems unlikely that many will undertake it themselves. So, what we proposed was to encourage collaboration. And to do this, we assembled leading historians of ancient Greek philosophy, the various exact sciences, the life sciences, and medicine to focus on three topics: (a) how Greek philosophers and scientists defined science and demarcated the particular sciences during the fifth and fourth centuries BC, (b) the role played by observation in theory as well as by theory in observation, and (c) whether philosophical debates about the ontology and character of scientific explanation occasioned any changes in what the Greeks later regarded as science, and were in fact instrumental in the emergence of new sciences.

This nexus of issues will interest all students of Greek culture. For, to the best one can determine given the evidence available, it was during the fifth and fourth centuries that the Greeks first began to elaborate their idea of the differences between the scientific and philosophical enterprises and of what is knowable through each. This period was, therefore, a critical point in the history of Western thought, defining as it did two intellectual activities and setting terms for their interaction or mutual influence then and later. Of course, there have been other such points in the subsequent history of philosophy and science. But this one has a special charm, since it is the earliest on record, and since the figures involved—Archytas, Plato, Aristotle, Eudoxus, Euclid, and so on—continue to have influence on the course of Western intellectual history. Moreover, on the principle that the challenge of new information is not to tailor it to pre-existing views but to reconsider all the evidence at hand, it was (and still is) a particularly good time to raise these issues again. For, not only has there been during the last decade a substantial increase in the publication of scientific documents from the Near East, documents which raise far-reaching questions about the nature of Greek science and its relation to that of other cultures, there has also been a dramatic increase in the output of research in the history of Greek biological science and its philosophical underpinnings.

The meeting itself made notable progress towards our goals, thanks indeed to the efforts of my Co-Chair, James G. Lennox, and to the assistance of James Allis, Stephen C. Wagner, Arlene Woodward, and Barbara Woolf. But, as one might expect, there are numerous differences among the conference planned, the conference itself, and this volume of essays. Of the two most important differences, the first is that this collection captures only the traces of the excitement of the meeting and nothing of the free-wheeling, thought-provoking discussions among the participants throughout the proceedings. Yet, it is my sincere hope that this volume will still serve to draw the reader into the conversation begun in those few days, a conversation the Institute aims to support and promote.

The second difference, as the contributors to this volume make clear, is that the question of the interaction of Greek science and philosophy has to be refined, and that there are many subsidiary issues which have to be dealt with in order to address the three topics originally proposed. Thus, for instance, there is a fundamental question about how to read and interpret technical, scientific documents, the answer to which will affect our determining what information, if any, they afford about the interaction of science and philosophy.

This volume of essays, then, is a record of a series of inquiries into a basic problem in the study of ancient Greek science and philosophy. At no point are the answers final. What is important in this collection is the interdisciplinary effort to indicate new directions for research by addressing the general question of how ancient science and philosophy interacted once they were first differentiated, and by applying to narrowly defined instances the latest techniques of modern philology, philosophy, historiography, and literary studies.

In conclusion, I should like to record my gratitude to the sponsors of the conference: the National Endowment for the Humanities; the Pennsylvania Humanities Council; the Research and Development Fund at the University of Pittsburgh; as well as the Departments of Classics, Philosophy, History and Philosophy of Science, and the Center for Philosophy of Science, at the University of Pittsburgh. Moreover, it is with great pleasure that I thank all the contributors to this volume and acknowledge the invaluable assistance of William R. Bowen, Marjorie Cars, and Stephen C. Wagner in preparing it for publication.

Pittsburgh, Pennsylvania

CONTRIBUTORS

ANDREW D. BARKER

Senior Lecturer in Philosophy at the University of Warwick (Coventry, England). Fellow of the Institute for Research in Classical Philosophy and Science (Princeton, NJ). Author of *Greek Musical Writings* (2 vols: 1984–89) which offers translations of Greek texts on the musical arts (vol. 1) and harmonic and acoustic theory (vol. 2), with extensive commentary; as well as of articles on Greek philosophy and science, especially musical theory, and on music as a practical art in the ancient world.

J. L. BERGGREN

Professor of Mathematics at Simon Fraser University (Burnaby, Canada). Fellow of the Institute for Research in Classical Philosophy and Science (Princeton, NJ). Author of *Episodes in the Mathematics of Medieval Islam* (1986) and numerous papers on the mathematical sciences in ancient Greece and Islam in the Middle Ages. Currently preparing a translation of Euclid's *Phænomena* with historical introduction and commentary.

ALAN C. BOWEN

Director of the Institute for Research in Classical Philosophy and Science (Princeton, NJ); Adjunct Associate Professor in the Departments of Classics and of History and Philosophy of Science at the University of Pittsburgh. Fellow of the Center for Philosophy of Science at the University of Pittsburgh. Recipient of a Fellowship from the American Council of Learned Societies (1983–84) and Research Grants from the National Endowment for the Humanities (1986, 1988–91). Author of *The Ancient Greek Science of Music: A Preface to Platonic Dialectic* (University of Toronto Press: in press), and of articles on the history and philosophy of the exact sciences in antiquity. Co-Editor of *Ancient Philosophy*. Currently writing a book on early Hellenistic astronomy.

D. H. FOWLER

Reader in Mathematics at the University of Warwick (Coventry, England). Fellow of the Institute for Research in Classical Philosophy and Science (Princeton, NJ). Author of *The Mathematics of Plato's Academy: A New Reconstruction* (1987) and of articles on the history of mathematics. Now writing on further ramifications of his proposals concerning Greek mathematics, and on the history of continued fractions.

CHARLES H. KAHN

Professor of Philosophy at the University of Pennsylvania (Philadelphia, PA). Recipient of Research Fellowships from the American Council of Learned Societies (1963–64, 1985) and from the National Endowment for the Humanities (1974–75, 1990–91). Member of the Managing Committee of the American School of Classical Studies in Athens since 1964. Author of three books, including *Anaximander and the Origins of Greek Cosmology* (1960) and *The Art and Thought of Heraclitus* (1979), as well as of numerous articles on ancient Greek philosophy and on linguistic theory. Currently completing a book on Plato and the creation of the Socratic dialogues.

WILBUR R. KNORR

Professor in the Program in History of Science at Stanford University, with a joint appointment in the Departments of Classics and Philosophy. Author of over 40 articles and four books, including *The Ancient Tradition of Geometric Problems* (1986) and *Textual Studies in Ancient and Medieval Geometry* (1989). Specialist in ancient and medieval mathematics, with particular emphasis on geometry and the geometric sciences. Current projects include studies of the medieval Latin geometric commentators, ancient optics, and Plato's cosmology.

JAMES G. LENNOX

Associate Professor of History and Philosophy of Science at the University of Pittsburgh. Former Junior Fellow of the Center for Hellenic Studies; Life Fellow, Clare Hall, Cambridge; currently Fellow of the Center for Philosophy of Science at the University of Pittsburgh; Fellow of the Institute for Research in Classical Philosophy and Science (Princeton, NJ). Author of many articles on the philosophy of Plato and Aristotle; co-editor of *Philosophical Issues in Aristotle's Biology* (1987).

G. E. R. LLOYD

Professor of Ancient Philosophy and Science at Cambridge University; Master of Darwin College. Fellow of the Japan Society for the Promotion of Science, and of the British Academy; Senior Fellow of the Institute for

Research in Classical Philosophy and Science (Princeton, NJ). Sather Professor at the University of California at Berkeley (1983–84), and recipient of the Sarton Medal from the History of Science Society (1987). Author and editor of some ten books on early Greek science and philosophy including, most recently, *The Revolutions of Wisdom: Studies in the Claims and Practice of Ancient Greek Science* (1987).

ALEXANDER P. D. MOURELATOS

Professor of Philosophy at The University of Texas at Austin, where he founded and directs the Joint Classics-Philosophy Graduate Program in Ancient Philosophy. Past member of the Institute for Advanced Study (Princeton, NJ); recipient of a John Simon Guggenheim Memorial Foundation Fellowship (1988–89). Author of studies on Plato's science, Presocratic philosophy, and philosophical linguistics.

IAN MUELLER

Professor of Philosophy at the University of Chicago. Recipient of Fellowships from the American Council of Learned Societies (1972–72, 1991–92), the National Endowment for the Humanities (1984–85), and the John Guggenheim Foundation (1991–92). Former Fellow of the Center for Hellenic Studies (1977–78) and of the Fondation les Treilles in Salernes, France (1982). Author of *Philosophy of Mathematics and Deductive Structure in Euclid's* Elements (1981), and editor of a collection of essays on Greek mathematics, Περὶ τῶν μαθημάτων (1991). Currently working on a study of the role of mathematics in later Greek philosophy, and on translations of Alexander's commentary on Aristotle's *Prior Analytics* and Simplicius' commentary on Aristotle's *De caelo.*

JOSEPH OWENS, CSsR

Professor Emeritus of the University of Toronto and Emeritus Fellow of the Pontifical Institute of Mediaeval Studies. Fellow of the Royal Society of Canada and member of the Catholic Commission on Intellectual and Cultural Affairs. Author of *The Doctrine of Being in the Aristotelian Metaphysics* (1978, 3rd edn) and other scholarly books and papers on Aristotle and Thomas Aquinas.

ROBERT G. TURNBULL

Professor Emeritus in the Department of Philosophy at Ohio State University (Columbus, OH). President, Central Division, American Philosophical Association (APA) 1977–78; Chair, Board of Officers, APA 1989–present. Director of a Summer Institute in Greek Philosophy and Science at Colorado College, 1970. Author of numerous articles in scholarly journals and books; editor (with P. K. Machamer) of *Motion and Time, Space and Matter* (1976) and *Studies in Perception* (1978).

Science and Philosophy in Classical Greece

1

Some Remarks on the Origins of Greek Science and Philosophy

CHARLES H. KAHN

This is not the occasion for new and surprising theses concerning the Presocratics, and that is just as well, for I have no new and surprising theses to present. Instead I shall defend some old theses and try to put them in a perspective that may be useful here as a background for the more specialized papers to be presented in this volume.

Philosophy in the strict sense is pretty clearly a Greek invention; but at first sight Greek science, and above all Greek astronomy, seems to be a borrowing from the Orient, like sculpture, architecture and the alphabet. Does this mean that the older view is wrong, I mean the view presented by Tannery and Burnet, who wrote before we learned so much about Babylonian astronomy and mathematics, and who portray Greek science and natural philosophy as coming into the world *together*, one and indivisible, first in Ionia and then in southern Italy and Sicily, in the sixth and early fifth centuries BC? I want to argue that the old view is right after all, and that once we have absorbed the discoveries of Neugebauer and other explorers of Mesopotamian science, we can see that Greek science is essentially a new creation, inseparable after all from the origins of Greek philosophy in the earliest phase of these two disciplines. In short, I want to defend the traditional view that Greek astronomy and natural philosophy (and the beginnings of geography and history too) first developed in Miletus in the middle of the sixth century BC and then spread like an epidemic throughout the Greek world, first by contagion to the neighboring cities of Samos, Colophon, Ephesus and Clazomenae, then to Ionian colonies in the northern Aegean (Abdera and Apollonia); and soon, travelling with refugees and immigrants to the far west, to Croton and Metapontum, to Elea and Acragas. So within the two generations that separate Anaximander from Parmenides, Ionian science had been carried across the Greek world, paralleling the diffusion of the alphabet some two centuries earlier.

Now the utility of the alphabet is obvious; but in the case of what the Greeks called περὶ φύσεως ἱστορία (the investigation of nature), it is not immediately clear why this should have been so widely attractive so soon. The new science had its utility, no doubt, in so far as it included map-making (probably derived from the East) and observational astronomy (certainly derived from the East). But I think it was above all the intellectual power of a new, naturalistic or rational world-view which captured the imagination of an amazingly curious, open-minded people, beginning with a handful of pioneers along the Anatolian coast and in neighboring islands, but spreading swiftly throughout those bustling Greek cities scattered across half the Mediterranean. We can form some notion of the motivation and diversity of these first two generations (from about 550 to 490 BC) in the glimpses we get of three very striking and very different personalities—Pythagoras, Xenophanes, and Heraclitus—who contributed both to the renown and to the rapid physical diffusion of Ionian natural philosophy.

I do not propose to retell this familiar story. Instead I want to concentrate on the two features which best mark the radical break with earlier world-views, both in Greece and in the Orient, and which illustrate the close links between exact science and philosophical speculation in this earliest period. I think that a clear grasp of these two features will protect us against three seductive errors which can distort our understanding of the origins of Greek science and philosophy. The first error is to see Presocratic natural philosophy as a continuous development from mythopoetic thought in Homer and Hesiod, without a revolutionary break. The second error is to see Greek astronomy (and/or mathematics) as essentially a continuation of Mesopotamian science, without radical innovation. The third error is the view championed by D. R. Dicks [1966], which treats the development of Greek observational astronomy as if it were completely independent of the speculative theories of the early natural philosophers. I have argued the case against Dicks in my response [Kahn 1970], and I refer you to this article for detailed documentation. Here I summarize some of my conclusions.

I take it for granted that the new science that arose in Ionia in the sixth century was heavily dependent upon Babylonian astronomy, much in the way that the creation of the alphabet was dependent upon Phoenician sources and the creation of Greek sculpture and architecture was dependent upon models from Egypt. (This is all part of the 'orientalizing period' of Greek culture, in the broadest sense.) Herodotus tells us of some essential borrowings in the case of Greek astronomy: 'The Greeks learned of the πόλος and the gnomon and the twelve parts of the day from the

Babylonians' [*Hist.* ii 109]. For once he is right (although his guess, in the preceding sentence, that geometry was discovered in Egypt, has not been confirmed). For example, the identity of the Morning Star and the Evening Star, which had been known in Mesopotamia for many centuries, is first attested in Greece for Parmenides [Diogenes Laertius, *Vitae* ix 23 = Diels and Kranz 1951–1952, i 224.29–31]. (The absence of Greek evidence before Parmenides is surely an accident of our meager documentation for this early period: no doubt the information passed to Italy through Ionia.) How much Babylonian lore was available in Ionia in the sixth century we cannot know. But the Milesians added something for which there seems to be no Mesopotamian precedent. This is a geometric model for the heavens, a clear-cut scheme of concentric circles and other figures by which the observed motion and changes of the heavenly bodies were to be explained. The first and crudest of these models is attested for Anaximander: a series of circles set at numerically definite distances from a disk-shaped earth in the center. The model was quickly transformed and improved by his successors. Within two generations we get the classical scheme of a celestial sphere to which the fixed stars are attached so that their observed motion is explained by the daily rotation of the sphere. (When exactly the stellar sphere was introduced is not clear from our shabby evidence, but not later than the poem of Parmenides, *ca.* 500 BC.) Somewhat later, but before the time of Plato, the flat, discoid earth is replaced by a spherical model for the earth as well. It is this kind of geometric model (but without the spherical earth) that permits Anaxagoras to come up with a correct optical explanation of lunar eclipse by the middle of the fifth century BC. Now the important thing is not that the early models were so crude—that is only to be expected. What was important was that a geometric model for celestial motions had been proposed, with explanatory intent.[1] At the technical

[1] By a model here I mean a good deal more than a cosmic picture of the sort that one might find in Hesiod's *Theogony*, and more also than the picture of the heavens that served in Babylonian astronomy for plotting the movement of the Sun, Moon, and planets relative to the fixed stars, since such a picture serves only to describe but not to explain the observed phenomena. Scientific astronomy in the Greek sense begins with the attempt to give such an explanation by means of a definite structure conceived in terms of precise geometrical figures and relative sizes and distances. Such a model made possible the correct explanation of the Moon's light by the time of Parmenides and the correct explanation of lunar eclipse by the time of Anaxagoras. Since we have no astronomical texts from this period, we cannot know when the various technical advances were made that are incorporated in the later theory of spherics. The ascription of five zones to Parmenides by Strabo [*Geog.* i 94 = Diels and Kranz 1951–1952, i 225] on the authority of Posidonius is probably unreliable. But the pseudo-Platonic *Erastae* 132a [= Diels and Kranz 1951–1952, i 393] implies that schoolboys in the middle

level, it is this model that essentially defines the new philosophical view of the natural world as a κόσμος, a system governed by regularity and order. And it is this same model that brings into existence scientific astronomy in a new sense: a structured theory capable of explaining (or trying to explain) the observed phenomena of the heavens. In this sense the cosmology of Anaximander and Parmenides is closer in principle to that of Ptolemy and Copernicus than it is to Hesiod or to any of their predecessors—unless one finds a geometric model in Babylon.

The second great innovation of Greek science is the notion of mathematical proof. On this point I can quote Neugebauer [1963, 530]: 'the discoveries of the Old Babylonian period had long since become common mathematical knowledge all over the ancient Near East'. What the Greeks added was 'a fundamentally new aspect..., namely the idea of mathematical proof. It is only then that mathematics in the modern sense came into existence.'

What Neugebauer does not see, but what seems obvious to me, is that the idea of proof plays the same role here in the creation of Greek mathematics as the kinematic models for the heavens plays in the creation of astronomical theory. When did this fundamental innovation in mathematics begin? Neugebauer tends to date it relatively late, in the fourth-century work of Theaetetus and Eudoxus. But the reports on Hippocrates of Chios take us back earlier, to the last half of the fifth century. Hippocrates is said to have been the first author of *Elements*, that is, a presentation of geometry in deductive form; and, in a long extract from Eudemus' history of geometry, we can see him operating with the 'method of hypothesis' or explicitly recognized premisses. Before Hippocrates we have no detailed documentation, so I shall not claim this achievement for my hero, Anaximander. Of course the tradition recorded by Eudemus actually assigns the earliest geometric proofs to Anaximander's predecessor, Thales of Miletus [Friedlein 1873, 157.10–13, 250.20–251.2, 299.1–5, 352.13–18 = Diels and

---

of the fifth century, in the time of Anaxagoras and Oenopides, were supposed to be familiar with a structure showing the obliquity of the ecliptic relative to the celestial equator:

> the boys seemed to be arguing about Anaxagoras or about Oenopides. For they appeared to be drawing circles and imitating certain inclinations by their hands <relative to one another>.

Once these two circles are drawn on a celestial sphere, a partial system of zones is given. Whoever wrote the *Erastae* thought that Anaxagoras and Oenopides were doing this kind of astronomy. I see no reason to believe that we are better informed than the author of the *Erastae* on the development of scientific theory in the fifth century.

Kranz 1951–1952, i 79.8–19]. Thales is more a figure of legend than of history, but this image of the sage who predicts an eclipse, uses geometry to measure the height of the pyramids, and also begins cosmological speculation, nicely reflects what I take to be the fundamental historical fact: that observational astronomy, speculative cosmology, and mathematical research were developing together within those small circles of intellectual activity that carried the new science from Ionia to Magna Graecia and beyond.

We cannot reconstruct the early history of mathematical proof in Greece as we can reconstruct (to a certain extent) the development of a celestial model. But we can see the two enterprises interacting or coinciding in the work of Oenopides of Chios (after Anaxagoras and before Hippocrates), who did major work in astronomy but studied certain problems in geometry 'because he thought they were useful for astronomy' [Friedlein 1873, 283.4–10 = Diels and Kranz 1951–1952, i 395.10–14] and specifically for measuring the obliquity of the ecliptic. Oenopides also contributed to speculative cosmology by explaining the Milky Way as the path previously marked out by the Sun's annual course, before it settled in the ecliptic [Achilles, *Isag.* 24 = Diels and Kranz 1951–1952, i 394.29–32]. And we can see the same interaction between observational astronomy, technical work in geometry, and philosophical cosmology (as well as map-making) in the case of Democritus, whose contribution to the mathematics of the cone and the pyramid is recognized by Archimedes [Heath 1921, i 180].

Although we cannot reconstruct the early development of mathematical proof in Greece, we can perhaps see this development reflected in the examples of philosophical arguments that happen to be preserved. The oldest and most elaborate of these arguments has reached us intact simply because it was embedded in the hexameters of Parmenides' poem. (For this one argument preserved from the beginning of the fifth century there must have been dozens if not hundreds of arguments contrived by the mathematicians but lost, because they were in prose or not even written down.) Parmenides begins with a clear statement of his premiss or first proposition, presented in the choice between a pair of contradictories, *it is* or *it is not*; and he gives reasons for rejecting the second alternative. He then proceeds to derive a number of attributes of Being from the single premiss *that it is*. So it must be one, unique, dense, symmetrical, immobile, ungenerated and imperishable. The argument for the attribute *ungenerated* is preserved in full. It is an indirect argument proceeding from a trilemma of assumptions: if what-is has come into being, then it must have come to be (a) from Not-Being, (b) from Being, or (c) from nothing at all. All three assumptions are shown to be incompatible with the basic premiss *that it is* and so they must be rejected. Hence, by a series of *reductio* arguments which eliminate the alternatives, the thesis *that it is ungenerated* is established. This argument clearly parallels the form of an indirect proof in geometry. And we can recognize the same general pattern of argument over and over again in the preserved fragments of Zeno, Melissus, Anaxagoras, and Diogenes, where it is used either to defend or destroy a thesis. Either we have p or not-p. My opponents assert p. But look what absurd (or false or contradictory) results follow from p. Therefore not-p. Or else, I assert p. For just imagine the opposite, not-p. But if not-p were the case, things would be quite different (or impossible or ridiculous). Therefore p must be the case.[2]

Incidentally, I think we can also see the influence of a mathematical mode of argument in the way Anaxagoras and other thinkers such as Diogenes first construct their ἀρχή or starting-point ('all things together', plus νοῦς ready to start things rotating, in the case of Anaxagoras), and then show how the world-order develops *naturally* and *inevitably* ('by necessity') out of these initial conditions. The physical ἀρχή occupies the place of premiss or *hypothesis*; the development of the cosmic order is analogous to the derivation of theorems.

In the case of these philosophical arguments, preserved almost by chance for the whole length of the fifth century, it must remain anyone's guess how far they presuppose, or how far they prepare the way for, the use of formally similar arguments in geometry. Some scholars (notably Szabó) have argued that the development of proof by the mathematicians is essentially dependent upon the (supposedly) earlier deductive exploits of the Eleatic philosophers. In the absence of any good textual evidence for mathematical proof earlier than Hippocrates in the late fifth century, it is impossible to *refute* Szabó's thesis; but I think there is nothing to be said in its favor [cf. Berka 1980, Knorr 1981a, Bowen 1984]. Hippocrates' proof is so elaborate that it clearly presupposes a considerable tradition of some technical sophistication; and it is a sheer accident of our documentation that we cannot trace this tradition back to its origins. My own hunch is that the development of more or less rigorous proof in mathematics and in philosophical argument went hand in hand, but that the geometrical application is likely to have led the way from the beginning, even before Parmenides. In these matters it is normal for philosophy to borrow from mathematics, just as we can see Plato in the *Meno* taking over the method of *hypothesis* from geometry. But in the beginning the philosophers and the mathematicians will often have been the same people, as the tradition

[2] Compare Geoffrey Lloyd's remarks [1979, 25 and 71–78] on the use of *modus tollens* and *reductio* arguments in fifth-century philosophical and medical texts.

tells us of Thales and Pythagoras, and as we can see later in the case of Democritus (and of a sophist like Hippias). The real contribution of philosophy was not in the specific techniques of proof but in the very idea of proving geometric propositions: taking some things for granted, either as obviously true or temporarily assumed, in order to establish what follows from them. This seems to me just as deep and philosophical an innovation as the introduction of geometric models for the heavens.

There is a third fundamentally new idea which we can detect in the fifth century and which has been elegantly documented by Geoffrey Lloyd, the concept of nature as a uniform system implying a regularity of cause and effect. This is the doctrine asserted in *Airs, Waters, Places* 22: 'each affliction (πάθος) has its own nature and none of them occurs without a natural cause (φύσις).' As Lloyd points out [1979, 33], the origins of such a view can be glimpsed in Anaximander's fragment on cosmic justice; and a dogmatic generalization is found in the somewhat questionable 'fragment' of Leucippus [Aëtius, *De plac.* i 25.4 = Diels and Kranz 1951–1952, ii 81.3–6]; but for a full articulation we must turn to the Hippocratic treatises of the late fifth century. This again seems to me simply an accident of our documentation: the older Hippocratic treatises are the *only* non-fragmentary scientific/philosophic texts that have reached us from the fifth century. Their close connection with Ionian science seems to me clear; but I leave this topic to Geoffrey Lloyd.

What I am proposing, then, is the traditional view of the origins of Greek science and philosophy in the emergence of a closely connected bundle of diverse but interrelated activities in the sixth and fifth centuries BC, before the systematic specialization and separation of the disciplines that becomes more characteristic of scientific work in the fourth century and later. There may well have been astronomers and mathematicians in the fifth century who were not also natural philosophers, but the more typical case is that of Oenopides and Democritus who worked in all three fields. Socrates is probably the first philosopher whose conception of his calling is essentially independent of work in astronomy and cosmology; and, according to the biographical sketch of the *Phaedo*, that was not true even of Socrates in his youth.[3] The close connection between philosophy, science, and mathematics is just as characteristic of Greek philosophy in its first century and a half as it is of the initial period of modern philosophy in the 17th century.

[3] The only possible precedent for Socrates that comes to mind is Protagoras, but his rejection of a realist view of truth seem unthinkable without the traditions of Eleatic ontology and Ionian cosmology. Socrates may well have been fascinated by natural philosophy in his youth, but there is no trace of this in his own positive conception of philosophy.

And it is this essential interconnection which is exemplified in the two key ideas which I have emphasized: geometric models for the heavens and the development of deductive proof. If we bear these two ideas in mind, we will not be tempted by either of the three errors to which I referred in the beginning: to see Presocratic cosmology as a continuous development from Hesiodic mythopoetry; to see Greek science as a mere borrowing from the East; or to see the development of scientific astronomy as essentially independent of speculative cosmology, as Dicks and Neugebauer would have us do. It is really painful to see a great scholar like Neugebauer saying, in his masterful *History of Ancient Mathematical Astronomy* [1975, 572], that there is 'no need for considering Greek philosophy as an early stage in the development of science. Its role seems to me only comparable to the influence on science of the Babylonian creation myth or of Manichean cosmology'. I think it would be difficult to find a more profoundly mistaken view in any serious book ever written on our topic. I can only put it down to an extraordinarily narrow construal of science and to a morbid dislike of speculative theory on Neugebauer's part.

More interesting, and perhaps more prevalent, than the line taken by Dicks and Neugebauer is the related error of exaggerating the continuity between Ionian science and its poetic antecedents. Of course, the early natural philosophers were men of their time and place; their language and much of their conceptual equipment were inherited from the Greek past. But their debt can be overestimated and their originality masked by reading back into Homer and Hesiod some of the most characteristic ideas of the Milesians and their followers. There is a Cambridge tradition for this, going back to Cornford and still visible in Kirk and Raven (even in the second edition of 1983) and in some of Guthrie's work.[4]

Thus Kirk and Raven claim that in 'the naive view of the world' in Homer 'the sky is a solid hemisphere like a bowl' [Kirk, Raven, and Schofield 1983, 9]. If there had been such a clear geometric model before the Milesians, the invention of the stellar sphere would have marked a relatively trivial advance in a continuous tradition. But in fact there is no trace of the notion of a celestial hemisphere or bowl in Homer or in any poet earlier than Parmenides; in so far as οὐρανός in Homer has any definite shape, it is that of a flat roof or a steep incline rising to the zenith [Kahn 1985, 138–140]. But the very notion of a clear geometric model composed of circles and spheres (as distinct from an anthropomorphic structure like a house or a tent) is alien to mythopoetic thought as we find it in Homer and

[4] But this tendency is not limited to Cambridge. Wade-Gery of Oxford once wrote an essay [1949, 81] in which he described Hesiod as 'the first Presocratic'. And compare, for example, Solmsen 1950.

Hesiod. Even more seductive is Cornford's misreading of χάος in Hesiod as a somewhat distorted version of the Polynesian myth of the separation of heaven and earth. Thus we find in Kirk and Raven: 'For Hesiod's source, at all events, the first stage in the formation of a differentiated world was the production of a vast gap between sky and earth' [Kirk, Raven, and Schofield 1983, 41]. Cornford found an echo of this supposed source in some verses from Euripides [*Frag.* 484]: 'Heaven and earth were once one form; but then when they separated from one another, they gave birth and brought all things to light'. But of course Euripides is quoting not Hesiod but Empedocles or Anaxagoras or some other natural philosopher. To project this view back into pre-Milesian mythopoetry is once more to make the revolutionary novelty of Milesian cosmology invisible. It is also to make hash of Hesiod, for whom χάος, the primeval gap, came into being *first of all.*

This is not the occasion for a more sympathetic reading of the *Theogony*, along the lines of Paula Philippson's study [1936] or of Norman Brown's sensitive interpretation [1953]. I want only to remark that Cornford's reductive approach to Hesiod, reading not the text of the poem but looking through it to find the more primitive 'source', has the effect not only of disguising the radical novelty of Ionian cosmology but also of doing an injustice to Hesiod's own speculative achievement. He set out to imagine what there could have been first of all *in the beginning*, before anything had taken shape. Different mythic poets conceive this beginning in different ways. Hesiod's ploy was to imagine an enormous gap, a vacant, yawning hole with no sides. In the genealogical language of mythopoetry, the negative character of this primordial chasm is revealed by its offspring, infernal darkness (Erebos) and black Night. What we have then is a kind of black hole, into which things could only fall and be lost. The first positive item to appear is 'broadbosomed Earth, a safe seat for all things forever'. Earth is a safe seat because it prevents things from falling into the dark chasm below. Hence everything positive and solid will now be produced from Gaia. Her first product is the starry Heaven 'equal to herself, to cover her all around'. So the world now has an upper floor, and Gaia now has a mate. This story has a beautiful coherence of its own, which is to all appearances Hesiod's own creation. And if some of the succeeding misadventures of Gaia and Ouranos *do* have a non-Greek source, this has nothing to do with a primeval Polynesian 'clinging together', and also nothing to do with the cosmologists' quite different attempt to understand how a differentiated universe could emerge *in a natural way* from the initial hypothesis of an undifferentiated mass, whether this mass is described as ἄπειρον or as 'boundless air' or as 'all things together'.

In conclusion, let me say one word about the demarcation between science and philosophy, I can be brief, because in the period before Socrates there is no such demarcation. The investigation of nature (περὶ φύσεως ἱστορία) comprises both. Looking back from our point of view, we might see the prefiguration of a distinction between philosophy and science in the division between the two parts of Parmenides' poem: the Way of Truth, which presents a metaphysical account of Being, and the Way of Opinion, which describes the genesis of the natural world. When we come to Plato's *Timaeus*, we can see the transformation of this dichotomy into something resembling our distinction between a philosophical account of reality (in the doctrine of Forms) and Plato's own version of Ionian natural philosophy or physics, as a 'likely account'. But that lies outside my *propos*.

I should add that I have considered only the internal history of Greek science and philosophy in its earliest phase. The external history—the social, economic, and political conditions of this momentous innovation—would require another paper. But I would have nothing substantial to add beyond the very instructive parallel between the emergence of rational thought about the physical universe in the sixth century and the contemporaneous development in Greek political life—the parallel that was (to my knowledge) first usefully drawn in J.-P. Vernant's *Les origines de la pensée grecque* [1962], and then convincingly developed in the last chapter of Lloyd's more recent study [1979]. I can add only one final question. If one accepts the thesis developed by Jasper Griffin [1977], as I am inclined to do, then the author of the *Iliad* must be seen as having made a systematic attempt to eliminate, suppress or play down all of the more strikingly miraculous and monstrous elements in the older epic tradition. That means that the Greek tendency to think of the circumstances of human life in 'naturalistic', non-magical terms can be seen at work already in the late eighth century. In this perspective it is Homer and not Hesiod who might properly rank as the first Presocratic. Now can we trace back to the time of the *Iliad* that political and sociological parallel which works so well for the sixth century? If not, is this too an accident of our documentation? Or is this evidence for a more Weberian view of the essential autonomy of intellectual history?

2

Plato's Science—His View and Ours of His

ALEXANDER P. D. MOURELATOS

I propose to canvass the essentials of Plato's conception of science. The two 'views' alluded to in the title of this paper correspond to these two questions: What is it that *we*, given our conception of science, find either especially congenial or especially uncongenial in Plato's treatment of science and of the sciences? What does *Plato himself* choose to emphasize in the various contexts in which he does what is recognizable to us as science, or talks about either particular sciences or the general topic of scientific inquiry?

1. The judgmental approach to Plato's science

Those elements in Plato's conception of science that are jarringly out of line with modern presuppositions are almost tediously familiar; so I shall simply allude to them in general terms. In the famous methodological passage of the *Phaedo* [95e–99e], Plato expresses a decided preference for teleological explanation. Later, in the *Timaeus*, he reiterates that preference and displays it in the actual explanations he offers. In both the *Phaedo* and the *Timaeus* Plato also implies that there is little point in offering an *explanans* that falls short of providing a fully satisfying rational insight. In other words, an explanation that is worth giving must either itself be, or be a proximate step toward yielding, what we would call 'ultimate' explanation. The impatient optimism Plato shows in this respect may well strike modern scientists as puerile. Well known, too, and much deplored is Plato's depreciation of the testimony of the senses in the search for truth and his specific disparagement of experiments and of the amassing of observational data.

But there is also much that modern readers have found congenial, or even strikingly prophetic of twentieth-century doctrines and trends. Remarkably, the very same dialogue that earns the greatest scorn from Plato's modern detractors, the *Timaeus*, is the one that has stimulated the more appreciative readings too.

The *Timaeus* might be said to constitute Plato's encyclopedia of the sciences. It could also be said to constitute Plato's statement of a certain doctrine that holds considerable appeal in our day: the unity of science. If we set out to record discernible early contributions to what eventually came to be distinct branches of science, we are likely to cite passages of the *Timaeus* not only under the headings of number-theory, geometry, stereometry, astronomy, and harmonics—these passages are well known and conspicuous—but also under such headings as mechanics (notably fluid mechanics and the theory of projectiles, 77c–81e), acoustics [80a–b], optics [45b–46c], physical chemistry and mineralogy [53c–61c], the physiology of perception [61c–68d], general human physiology [69d–81e], as well as human pathology [81e–86a], and psychopathology [86b–90d]. It matters not at all that many of these passages contain not statements of Plato's own theories but rather Plato's informed reportage of theories propounded by his scientifically minded predecessors and contemporaries. The details in the individual articles of the encyclopedia may be borrowed, but the vision and organization of the whole is authentically Plato's. The system is as thoroughly mechanistic in its substance as it is teleological in its regulative principles. Within the limits set by the teleology—the *desideratum* of always realizing the best of relevant possibilities—the system both of the heavens and of terrestrial phenomena involves only two factors: geometrical structure and motion. This is a far cry from the sort of mechanism Plato decries when he attacks materialist theories in *Laws* x 889b–c;[1] it is rather a highly abstract doctrine of the geometry of motion, something akin to the pure mechanism propounded by Descartes.

Modern readers have also been struck by what appears as a prefiguring in the *Timaeus* of the standard hypothetico-deductive model of scientific explanation [cf. Vlastos 1975, ch. 2–3]. We should remind ourselves here of the two cases that work best for Plato. At 36b–d he constructs a geometric model of the seasonal spiral-like movement of the Sun between the tropics. He theorizes that the spiral is the resultant of two regular and uniform motions: the diurnal westward rotation of the whole sphere of the

[1] The universe develops through chance interactions of reified opposites.

heavens; and an eastward annual movement along a great celestial circle drawn between the two tropics, the circle of the ecliptic. The construction is elegant, and the fit with the empirical data known to Plato is nearly perfect [cf. Vlastos 1975, 54–57].

At 51b–61c he uses a structural-mathematical model to account for both intrinsic and interactive properties of the four elements. The theory is that each element has a distinct but regular molecular structure, each type of molecule having the geometry of one of the regular solids, the dodecahedron excluded. A fair number of facts concerning the elements are captured by Plato's construction, but the fit between theory and data is far looser here than it is in the account of solar motion. According to Gregory Vlastos [1975, 85], the looseness should be blamed on the type of data concerning terrestrial phenomena that Plato had at his disposal: 'ordinary, uncontrolled, unrefined, unanalyzed observation—things which everybody was supposed to know and no one was expected to investigate'.

Even Plato's disparagement of empirical knowledge has a bright side when judged in the light of twentieth-century developments in epistemology and the philosophy of science. Thus, Plato appears squarely in agreement with the view that the status of knowledge about the physical world is at best that of a conjecture, an εἰκὼς μῦθος (probable account).[2] Also very much in the spirit of recent philosophy of science is Plato's recognition of the tug-of-war relationship between theory and data. We say today that a theory, controlled as it is by the relevant empirical data, is nonetheless under multiple *a priori* constraints—those of the regulative principles of science and of the dominant paradigm. Moreover, the pull of the theory on the data is often strong enough to regiment and mold the data—we speak of observation as being theory-laden. Plato tells us that in cosmology or natural philosophy we must first take note of the contribution 'reason' makes to the universe, which calls for an abstract *a priori* understanding of structures and of their optimal combinations and transformations; but then we must also take note of the factor of brute facticity or raw givenness, what Plato calls the 'wandering cause' and 'necessity'. In the dialogue's ontological and quasi-mythical terms, cosmic Reason 'exercises sway' over Necessity as it 'persuades' her to assume intelligible structure to the greatest extent possible [*Tim.* 48a]. The epistemological corollary is that the

[2] The point is made well by G. E. R. Lloyd [1983b, 11–30] in a succinct and beautifully balanced discussion of Plato's conception of science. See especially 1983b, 22: '[Plato's] reluctance to claim any more than a certain probability is readily understandable, indeed laudable when we reflect on the excessive dogmatism shown in this general area of inquiry not only by most of Plato's predecessors but also by most of his successors.'

realm of Necessity is inherently so indeterminate as not to constitute a legitimate object of understanding [51a–b, 52b]; it is intelligible only to the extent that theoretical structures have been projected on to it [49e–50a].

Then, too, there is that feature of the physics of the *Timaeus* that has delighted Whitehead, Friedländer, and many a modern physicist: the atomism of Plato's *Timaeus*, which posits not unsplittable elementary corpuscles but abstract generative formulae for the systematic articulation of space, is closer to the spirit of our wave-theory of matter or of quantum mechanics than is the classical atomism of Dalton, or its ancient predecessor, the atomism of Democritus and Epicurus.[3]

2. A survey of Plato's 'philosophy of science'

Let us now try to shed our modern ideology in the hope we may come as close as possible to Plato's own view of science. There is, of course, one preconception we are unable to shed: our selection of texts will necessarily be guided by what *we* recognize, in an appropriately broad sense, as 'science'. The semantic focus of the English term 'science' is clearly in its count-noun use (a), the one that allows us to speak of a distinct body of knowledge, e.g., geometry or chemistry, as 'a science'. Radiating out from this focus are: (b) the collective-noun use, and (c) the abstract-noun use, which makes it possible for us to speak of the thinking that characterizes the sciences. The use of ἐπιστήμη includes these three patterns, but there are some major complications. The Greek term also carries (d) the state-noun senses of 'knowledge' and 'understanding', and corresponds also to two uses of the English term 'skill': (e) the abstract-noun use (as in 'He shows skill'), and (f) the count-noun use (as in 'Horse-riding is a skill'). Plato shifts quite comfortably from one to the other of these six patterns of use of ἐπιστήμη. Indeed, because of the absence of the indefinite article in Greek, it is only in uses of ἐπιστήμη in the plural that the two count-noun uses, (a) and (f), are immediately recognizable without recourse to the context. Since the theme of the present collection of studies is history and philosophy of science, not epistemology or learning theory, it is reasonable that we should select those texts in which the sense of ἐπιστήμη that is thematically prominent is that of 'branch or body of knowledge'.

A good text with which to begin is not the *Timaeus* but the digression in the *Philebus* concerning precision and purity in the arts and sciences.

[3] See Whitehead 1933, 126: 'The modern wave-theory of the atom sides with Plato rather than with Democritus: Newtonian dynamics sides with Democritus against Plato.' Cf. Friedländer 1958, ch. 14.

Mathematics and the 'professions': Philebus 55c–59c. The genus under scrutiny is initially referred to as that of τῆς περὶ τὰ μαθήματα ἐπιστήμης. But as the scheme unfolds, the term ἐπιστήμη is used interchangeably with the term τέχνη (art)—a significant pattern of equivalence that reflects the established sense of 'skill' for ἐπιστήμη.[4] At 55d the genus is split into two subgenera: the first comprises those arts and sciences that provide 'services to the public' or 'work/practice for hire' (δημιουργικόν), which Plato also calls 'practical arts' (χειροτεχνικαῖς, 55d), and which we might call 'professional occupations' or simply 'professions'; the second comprises those pursued for the sake of education and culture (περὶ παιδείαν καὶ τροφήν). Plato then makes the observation that if one were to subtract from each of the professions the component that involves calculation, measurement, and weighing, the remainder could well seem trivial (φαῦλον, 55e). The sciences of arithmetic, metrics, and statics are, accordingly, called 'leaders' (ἡγεμονικάς, 55d) among all the professions. This list of three leaders is not, however, meant to be exhaustive: Plato will shortly refer to λογιστική (the science of computation) as distinct from ἀριθμητική (properly the theory of numbers) and as likewise a leader [56e]; indeed, he eventually speaks of a whole train of sciences that are of the same genre as arithmetic and metrics—what we may call, with no fear of anachronism, the 'mathematical sciences'.

After the initial division between professions and educational or cultural subjects, the professions are in turn subdivided into those that make maximum use of counting, measuring, and weighing, and those that make limited use of them, relying mainly either on experience and practice (ἐμπειρίᾳ καί τινι τριβῇ, 55d) or on—what may well be meant as the rival 'leader'—στοχαστική (the art of taking aim), or 'the art of trial and error' [55d: cf. 56a]. The art Plato cites as representative of the first group of professions is τεκτονική (the art of building), which includes shipbuilding, architecture, and carpentry. The prime example for the second group is music (performing), other examples being medicine, agriculture, the art of navigation, and military science. Characteristic of the professions in the first group is that they show greater reliance on knowledge—or should the translation be, 'have more to do with science' (ἐπιστήμης μᾶλλον ἐχόμενον, 55d)?—treat of things in their purest state (ὡς καθαρώτατα νομίζειν, 55d), involve more of what is clear and certain (σαφές... βέβαιον: cf. 56a), and have a greater share of precision (ἀκριβείας μετισχούσας, 56c: cf. 56b).

[4] The terms ἐπιστήμη and τέχνη are used interchangeably not only in this passage but often in Plato. The translation 'craft' for τέχνη is possible only in certain contexts, not standardly. See Roochnik 1986, 295–310.

Though Plato speaks of a division, his use of comparatives and superlatives suggests that we are dealing with a continuum or spectrum, with the art of building at the one end and music at the other. Clearly, the distinction does not correspond to that between 'exact' and 'inexact' sciences, as we use these terms; but it will be convenient for us to refer to the first and second groups as the 'exact professions' and 'inexact professions', respectively.

Dichotomous division is not pursued further in this passage. Instead, Plato focuses directly on the mathematical leaders. Their place in the scheme is unclear. They are expressly said to be leaders of *all* arts and sciences (ἑκάστων αὐτῶν, 55d; πασῶν, 55e); so they cannot be said to be leaders of the exact professions only. Nevertheless, the wording at 56c suggests that they are to be included among the exact professions, albeit standing far above the rest, being superlatively exact (τούτων δέ [the reference is presumably to the exact professions] ταύτας ἀκριβεστάτας εἶναι τέχνας). Still, the rubrics, χειροτεχνική and δημιουργικόν (practical art, work/practice for hire), of 55d certainly do not fit them. The suggestion ready to hand from the initial dichotomy is that the mathematical sciences are not professions at all; they are pursued for the sake of education. But, then, how are we to explain their intricate and extensive involvement with the universe of the professions?

Plato rules that the question is to be approached as one not of genus-species classification but of homonymy (ὁμώνυμον, 57b: cf. 57b–e). He inquires whether a distinction analogous to that between exact and inexact professions may not be drawn *within each* of the mathematical sciences, and he reaches the conclusion that there are 'two arts of arithmetic and two arts of measurement, and a whole train of arts of this genre (ταύταις ἄλλαι τοιαῦται συνεπόμεναι συχναί) that have this twin character but have come to share a single name' [57d].[5] The distinction is, say, between the arithmetic

[5] The use of συνεπόμεναι here calls to mind the use of συνεπομένας at 56c. There the participle is used to express the relation that holds between the individual professions and their paradigm profession (τεκτονική in the case of the exact professions, μουσική in that of the inexact ones). One might, therefore, suppose that the participle at 57d is meant to refer to the relation the individual professions have to the mathematical sciences, in the latter's capacity as leaders. If that is the relevant meaning of συνεπόμεναι here, we have the implication that the duality extends through the whole spectrum of the professions—the inexact ones included, since they too are properly led by the mathematical sciences. The mathematical component present in each of the professions would, accordingly, constitute in each case a distinct mathematical science that has the duality postulated. An attractive feature of this interpretation is that it accounts for the omission of the science of harmonics from Plato's list of mathematical sciences and for the rather surprising choice of music as the paradigm of a trial-and-error

employed within the professions, which involves units of various sizes, and only approximately equal at that (two head of cattle, two military camps), and the arithmetic studied by theorists (τῶν φιλοσοφούντων), which posits constant and exactly equal units [56d–e]. Plato alludes to a corresponding distinction that obtains between the art of calculation employed within the profession of building or of business and the theoretical study of geometry or of numerical operations (τῆς κατὰ φιλοσοφίαν γεωμετρίας τε καὶ λογισμῶν καταμελετωμένων, 56e–57a). It is safe to refer to this distinction as that between applied and theoretical or pure mathematics.

The extent of the difference between applied and pure mathematics receives considerable emphasis:

> We have come to note a degree of difference in clarity between sciences that is amazing.... Even after saying that these [*scil.* the mathematics employed in the professions] differ greatly from the other professions, the sciences employed in the enterprise of true theorists (αἱ περὶ τὴν τῶν ὄντως φιλοσοφούντων ὁρμήν) differ to an astonishing degree in precision and truth concerning measures and numbers from their respective professional versions. [*Phil.* 57c–d][6]

These words, however, are spoken not by Socrates but by Protarchus. When Socrates asks at 57e, 'Shall we then say that these [theoretical mathematics] are the most precise of sciences?'[7] Protarchus does not hesitate to answer, 'Very much so.' But Socrates goes on to make the characteristically Platonic point that still superior is the art of dialectic [57e–59d]. To mark its supreme status, Socrates bestows on it the 'finest' and 'most honorable' of names, νοῦς (understanding) and φρόνησις (thought) [59c–d]. Thus, while exalting dialectic Plato also suggests that the title of ἐπιστήμη for theoretical mathematics is secure.

In the course of his exaltation of dialectic Socrates makes the point that the theoretical study of nature (περὶ φύσεως... ζητεῖν... περὶ τὸν κόσμον τόνδε, 59a) ranks with the ordinary professions, inasmuch as it deals not

art. The suggestion would be that there are also two arts of music: the hit-and-miss affair practiced by performers, and the theoretical science that is also called 'harmonics'. But it is more straightforward to assume that the phrase ταύταις ἄλλαι τοιαῦται συνεπόμεναι συχναί at 57d functions merely as an 'etc.' that covers recognized mathematical sciences other than the two, arithmetic and metrics, that are explicitly mentioned.

[6] τούτων δ' αὐτῶν clearly refers back to αὗται, which can only be the mathematics used in the professions. The genitive, therefore, should be taken as governed by διαφέρουσιν, not as a partitive genitive.

[7] I discuss shortly below a certain pregnant ambiguity in this question.

with timeless truths but with things subject to change. Not only the title νοῦς but equally the title ἐπιστήμη is withheld from it.

That names such as φρόνησις, νοῦς, and ἐπιστήμη are, in an important sense, honorific and subject to recall is a theme that runs throughout this passage in the *Philebus*. In the earlier stages of the division it is the term τέχνη that has this character. Thus, at 55e–56a Plato observes that it is only 'the many' who use τέχνη of that component of the arts that does not involve mathematics, and at 56c he shows some reticence about using the term τέχνη of the genus that encompasses the whole spectrum of the professions, including the inexact ones (θῶμεν τοίνυν διχῇ τὰς λεγομένας τέχνας, 'Let us divide in two the so-called professions'). The implication is that, as the mathematical component becomes smaller, the title of τέχνη becomes progressively dubious. There is also a poignant ambiguity of syntax in the two passages that place mathematics among the exact professions (ταύτας ἀκριβεστάτας εἶναι τέχνας, 56c; ἐπιστήμας ἀκριβεῖς μάλιστ' εἶναι, 57e): Plato may be saying that these rank as exact τέχναι or ἐπιστῆμαι; or he may be saying that these are τέχναι and ἐπιστῆμαι in the most exact sense of these terms. Clearly what we have in the *Philebus* is not a purely descriptive classification but an axiology of the arts and sciences.

The scheme of Statesman *258e–260b.* The initial distinction at *Philebus* 55d between professions and educational-cultural subjects appears to correspond, at first blush, to the distinction drawn in *Statesman* 258e between πρακτικὴ ἐπιστήμη and γνωστικὴ ἐπιστήμη. But, as further divisions are drawn in the latter dialogue, we find at 260b that γνωστικὴ ἐπιστήμη has two species, one purely κριτικόν (judgmental) and one ἐπιτακτικόν (directive, executive, or managerial). The science of calculation falls entirely within the judgmental species; but architecture, which involves the issuance of directions and plans as well as of judgments, falls in the executive species [259e–260a]. As lower-level divisions are articulated, we find under the heading of the executive species—and thus under the general heading of γνωστικὴ ἐπιστήμη—not only the art of the statesman but also various forms of husbandry. Obviously, the γνωστικὴ ἐπιστήμη of the *Statesman* does not correspond to the 'enterprise of true theorists' referred to in the *Philebus*. The relevant sense of ἐπιστήμη here must be that of 'skill', and the initial distinction is not one between applied and theoretical science but rather one between purely practical skills—which are not discussed further—and cognitive skills. Thus, all the distinctions drawn in the *Statesman* are within the domain of what in the *Philebus* would count strictly as professions. But the rationale of the distinctions in the

Statesman is quite different from that of pointing up the contrast between exact and inexact professions.

Plato's citing architecture as a skill that combines judgment and execution does, however, firm up the point that was made in the *Philebus*, where architecture appeared among the 'exact' professions. Even though the dialectical method encourages dichotomous division, the truth is that the professions properly constitute a spectrum: those involving only practical skills at one end; those that involve a heavier component of cognitive and intellectual skills at the other.

The sublimated sciences of Republic *vii.* The doctrine of the *Philebus*, that once the mathematical component is subtracted from any of the arts and sciences the remainder is 'trivial', appears also in the *Republic*, where we are told that 'every art and science' must necessarily involve number and calculation [522d]. The remark occurs at the beginning of the long section in book 7 that lays out the program of higher education of the guardians: first the five mathematical subjects of arithmetic, geometry, stereometry, astronomy, and harmonics; and then the 'coping stone' and 'main song', for which the mathematical studies are only a 'prelude', dialectic. *Philebus* 55c–59d quite obviously repeats many themes from this famous passage of the *Republic*. The duality of arithmetic, geometry, and astronomy appears in the dialogical play between the practical version of each that Glaucon appreciates—the military commander's counting of troops, measuring of fields, and knowledge of seasons—and the austerely theoretical version Socrates is promoting. And, as later in the *Philebus*, the contrast between the mathematical sciences and dialectic is made to seem bigger than that between the two homonymous versions of each of the five mathematical sciences.

There are also, however, some intriguing differences from the account in the *Philebus*. Astronomy and harmonics, neither of which is mentioned in that later dialogue, are given in the *Republic* an additional twist of theoretical sublimation. Applied harmonics, surely, is none other than music in the ordinary sense—that prime example of an inexact profession. Significantly, Plato does not have Glaucon dwell on the usefulness of music, since this was fully covered in book 4. Instead, Glaucon, who by this stage in the discussion has caught the drift of Socrates' interest in the mathematical subjects, immediately volunteers his censure of the aural approach to harmonics. In distinguishing musical tones and measuring intervals, these aural harmonists hew to the practice of musical performers. Just as the latter do their tuning by trial-and-error, the aural harmonists seek to determine the smallest detectable interval through intent listening after repeated tunings and retunings of strings. Socrates dismisses that approach with the rather repugnant elaboration of a metaphor of strings being tortured, and proceeds to draw a distinction between two versions of properly *theoretical* harmonics: the science of the Pythagoreans who seek the ratios that correspond to aural concords, and the purely mathematical science practiced by Platonically minded harmonists who would 'ascend to problems to investigate which numbers are consonant and which not, and why the ones are so and the others not' [531c]. The wording suggests a program for some sort of purely mathematical unified theory of selected ratios. The phrase 'consonant numbers' (σύμφωνοι ἀριθμοί) has been convincingly interpreted by Andrew Barker [1978b, 342] as 'those numbers which, from their place in an intelligible system, we shall call "consonant" on the *analogy* with heard sounds'.[8]

When astronomy is discussed in the *Republic*, the applied version of the subject is represented—as in the case of arithmetic and geometry—in Glaucon's jejune endorsement of its usefulness.[9] Here, too, Socrates draws a further distinction, one within the domain of *other-than-applied* astronomy: standard contemporary astronomy (ὡς νῦν ἀστρονομεῖται, 530c) *versus* 'real astronomy' (cf. τῷ ὄντι ἀστρονομικόν, 530a; ὄντως ἀστρονομίας, 530b). What Socrates considers standard astronomy here is not the application of astronomical data in the construction of calendars or in agriculture; for it is expressly said to have no utility (ἀχρήστου, 530c), an important detail in the text that has often been overlooked. As Socrates' conventional image of the gaping stargazer (ἄνω κεχηνώς, 529b) suggests, it is a form of θεωρία (natural-historical or disinterested inquiry). It aims to chart the heavens by seasons, and to ascertain the paths and periods of revolution of the planets in their easterly motion through the zodiac. The definition of astronomy given in *Gorgias* 451c envisages clearly this standard version: 'the account (λόγοι) concerning the revolutions of the stars and of the Sun and Moon, specifically, how they relate to one another with respect to speed'.

Real astronomy is *toto caelo* different. A purely programmatic subject, it is not discussed under its proper name in any other part of the Platonic

[8] I regret having missed this important article in my earlier investigations of this subject.

[9] The applied version of astronomy that Glaucon appreciates must have been part of the Athenian public's perception of μετεωρολογία. In *Symp.* 188a–b the pompous medical doctor Eryximachus, who comes across clearly as a spokesman for Ionian science, conceives of 'the science of the revolutions of the stars and of the seasons of their respective years' as applied to the forecasting of violent weather and of epidemics.

corpus. It is defined in strikingly revisionary terms as a science of 'solids in revolution' or of 'the revolution of that which has depth' [528a, 528d]. The elaboration of these definitions Plato gives in 529c–d is most plausibly read as envisaging a science of general and pure kinematics. As with Plato's sublimated version of harmonics, real astronomy uses a 'problem' approach; it treats the motions traced by the stars and planets not as evidence but merely as suggestive of problems in the analysis and synthesis of motion.[10]

The sciences generally and the 'love of wisdom'. We have touched on the major texts in which Plato gives us, what we would recognize as, his philosophy of science. Let me now ask a general question. Within the large Platonic context of the 'love of wisdom', what is the contribution made by the sciences? I shall proceed by raising three subordinate questions. First, what is the *contribution made by the arts and sciences* at the broadest range of the spectrum? In answering this sub-question Plato would almost certainly have recourse to the familiar Socratic theme of the arts and sciences as paradigms of human rationality: each of the arts and sciences has a product or object as its focal concern; for each there is a τέλος and an implied teleological framework that is necessarily coherent; each is teachable; each affords a fairly accessible and transparent distinction between experts and non-experts; each affords compelling procedures for settling disputes that arise in the course of its practice. These several aspects of the paradigmatic function the arts and sciences have for Plato's Socrates are now understood quite well [see, e.g., Irwin 1977, 73–75; Roochnik 1986, 303–310; Brumbaugh 1976]. I choose here to expand briefly on the first aspect, which is remarkable for its presence at all stages and contexts of Plato's thought. Even the most menial and least exact of the arts has the virtue of focusing intelligent endeavor on a fairly well-defined field: the cobbler will concern himself with the making of shoes; the flutist with making music on pipes; the doctor with healing the sick. This disciplined and collected 'about-ness' of the arts and sciences, their engaged intentionality, is an authentic manifestation of philosophic ἔρως. The relevant contrast is with the scatterbrain mentality of dabblers, meddlers, and impressionable acolytes; the sophists and their pupils; imitative artists and their audiences; demagogic politicians and their gullible and sycophantic followers.

[10] Here is an example that has an obvious astronomical analogue: When a sphere revolves uniformly on an axis, what is the curve traced by a point that moves along a great circle the plane of which is inclined to the sphere's axis? Answer: a spiral of alternately descending and ascending coils. I have argued at length for this interpretation of 'real astronomy' in Mourelatos 1980, 33–73 and 1981, 1–32.

The strict code of *Laws* viii 846d–e, which bars an artisan from practicing two professions, is not only a continuation of the Socratic theme of the rationality of specialization, it also points up the affinity between this theme and the Platonic conception of social justice—in the reinterpretation worked out in the *Republic* of the traditional precept, 'doing one's own and not being a busy-body'.

Mathematics and the 'love of wisdom'. The second sub-question is *What contribution do the* mathematical *sciences make*? Plato would cite four distinct contributions. As we saw in the passage from the *Philebus*, mathematics constitutes the core and essence of both exact and inexact professions. Mathematics is the relevant paradigm in both the Platonic and the modern (Kuhnian) sense: it is the ideal Form toward which all arts and sciences aim; and it is the source and repository of the acknowledged techniques and procedures that mark the difference between the true professional and the charlatan. Second—here the relevant text is *Republic* vii—mathematics promotes the περιαγωγή and μεταστροφή, that momentous turn-around, away from preoccupation with the world of sensibles and toward the contemplation of purely theoretical entities. Third—the relevant text is again *Republic* vii—mathematics is the appropriate προπαιδεία (preparatory training) for dialectic. Let me dwell briefly on this last contribution.

Human feelings run high when it comes to discussions of justice or happiness; but the topics of number and shape lend themselves to dispassionate discussion [cf. *Euthyphro* 7b]. Progress toward real insight in mathematics can be quick, which fortifies us against crises of μισολογία (hatred or mistrust of argument) and sensitizes us to the important distinction between genuine and purely eristic argument. This facet of the propaedeutic function of mathematics is well understood. But there is also another, which is sometimes overlooked: it involves the series of the five mathematical sciences, with astronomy and harmonics understood in accordance with the revisionary definitions of *Republic* vii. This series of five subjects paves the way to that 'synoptic view' of reality that is characteristic of the dialectician [*Resp.* 537c]. The five have an obvious affinity one to another (οἰκειότητος ... ἀλλήλων τῶν μαθημάτων, 537c). Indeed, they seem to constitute a systematic circle. The truths of the theory of numbers can be studied in isolation from those of the other sciences. But our intellectual horizon progressively expands as we move to plane geometry, which incorporates all truths of the theory of numbers, and then to stereometry, which incorporates all of plane geometry. Our horizon is expanded further when, with real astronomy, the geometrical sense of shape is generalized into that of curves and trajectories in two and three dimensions. Then, finally, when Plato's version of harmonics seeks to understand the principle of unity of the ratios that underlie concordant motions, the circle is closed: we return to the theory of numbers, now focusing on a pre-eminently unified part of the theory of ratios [see above, and Barker 1978b]. Putting together this theme of the systematic unity of the five mathematical sciences of *Republic* vii with the theme of the *Philebus*, that the core and essence of *all* the arts and sciences is mathematics, we come to appreciate how large a part of the work that leads up to the dialectician's crowning achievement of synoptic vision is done at the stage of mathematical study.

Turning now to the fourth contribution made by mathematics, we begin to make contact with features of Plato's conception that were cited at the start of this paper as holding a special appeal for modern readers. As the *Timaeus* shows, the mathematical sciences provide the appropriate *explanantia* in cosmological speculation. The point can be made concerning the mathematical sciences generally; but it holds with greater force when mathematics is conceived of in the terms of *Republic* vii.

Mathematics dispenses appropriate *explanantia* in two ways.[11] First, the only explanation of specific phenomena that promises not to beg further questions is one in terms of mathematical structure. Why does earth (the element) tend to be cohesive, stable, unreactive, and heavy? Answer: because earth molecules have the geometry of a cube. Accordingly, when contact is made with a plane surface, all six faces of the cube offer either a secure square base or at least the possibility of a smooth slide, with no tumbling. Moreover, as cubes make contact with one another, their six square faces make for firm and compact stacking. Why, by contrast, is fire expansive, highly mobile, destructive, and light? Answer: because its molecules have the geometry of a tetrahedron. The four triangular faces of this solid make for a structure that is liable to tipping and tumbling, and the pointed vertices and sharp corners that join the faces insinuate themselves more easily into other structures, thus promoting breakup. In other words, mathematics offers ultimately satisfying *formal causes* in our study of particular natural phenomena.

What is more, mathematics provides the appropriate context toward answering those general and metaphysical questions for which only a *teleological* answer would be appropriate. Suppose we ask, Why should the elements have the structure of regular solids? or Why should the real (as distinct from the phenomenal) motions of the heavenly bodies be uniform

[11] In this paragraph and in the one that follows I condense an argument developed at length in Mourelatos 1981, 24–30.

and circular? or Why should there be seven planets, their periods related to one another in accordance with the specifications Plato gives in the *Timaeus*? Plato's answer in each case is that this arrangement is 'the best'. The answer may strike us today as quite arbitrary, a mere aesthetic prejudice. Plato's rejoinder might well be that the original question was not raised *in vacuo*. It is inherent to the 'generative' logic of mathematics that within each of the mathematical sciences certain structures are promoted as paramount and pre-eminent: in arithmetic, the smaller integers and the contrast between odd and even numbers; in plane geometry, straight lines, triangles, simple polygons, perfect arcs, and circles; in solid geometry, the sphere and the regular solids; in real astronomy, rotary motions; in harmonics, the three means, arithmetic, geometric, and harmonic. Thus a certain axiology of structure is built into mathematics. The asking of questions as to which structures are 'best' is necessarily placed against that background, and the answer forthcoming expresses a preference educated by mathematical insight.

Mathematics as the essence of and paradigm for all the arts and sciences, as the discipline of the 'turn-around' toward the world of Forms, as the preparatory training for dialectic (in the cultivation both of habits of argument and of the synoptic view), and as the dispenser of appropriate explanatory structures in cosmological inquiry—these are four major contributions the mathematical sciences make to the Platonic quest for wisdom.

Cosmology's contribution: relevance of the Timaeus. The contribution last mentioned naturally prompts us to ask whether there is not also a *contribution made by cosmology as such.* This would be the third in my main series of subordinate questions concerning Plato's conception of the role played by the sciences.

Here the distinction between cosmology-in-general and Plato's use of it in the *Timaeus* is crucial. The verdict with respect to cosmology-in-general is clear from the *Philebus*. Though it often is a disinterested inquiry (this is presumably the force of the concessive εἴ τε καί at *Phil.* 59a), it ranks, as we saw, with the practical professions. For, like them, it is concerned not with timeless truths but with various events and processes; its mode of cognition is that of δόξα (belief, conjecture); the entities it deals with fall squarely within the realm of δόξα [*Phil.* 58e–59b]. If it contributes to the grand Platonic quest, it does so only in the Socratic way, by serving as one of the examples of well-focused intelligent activity. It is astonishing that Plato never makes comments of the sort we commonly make in discussing the contributions of Presocratic natural philosophers to the development of science. We recognize that, regardless as to whether the Presocratics saw the world in dynamic or static terms, their inquiry into φύσις was itself a momentous 'turn-around'; that it was, in the language of Plato's Cave, a release from the bondage of irrationality and superstition; that in distinguishing between appearance and reality the natural philosophers were close to the spirit of Plato's corresponding distinction; and that—as Charles Kahn has again reminded us [see ch. 1 above]—mathematics and περὶ φύσεως ἱστορία developed hand-in-hand in the sixth and fifth centuries. What Plato keeps seeing, instead, are the perils and the distractions posed by natural philosophy. Thus, in the *Phaedo* [96c] natural philosophy threatens to 'blind' Socrates, robbing him of his common sense; and in *Laws* x 891b–892c it is said to promote the fundamental error that soul and τέχνη are posterior in the order of things to 'nature'.

Plato acknowledges the epistemological limitations as applicable to his own cosmology in the *Timaeus*. But the metaphysical certainties that frame Plato's otherwise 'probable account'—the principles that the universe is a structure of the highest beauty and value, and that it reflects the workings of intelligent soul at many levels of activity—permit Plato to claim for his cosmology a redeeming ideological function, one that no naturalistic cosmology could perform. The explanatory mathematical structures of the *Timaeus* exhibit the ideal order that cosmic intelligence has built into the world. As we contemplate that order, we come to be in possession of patterns we can use in ordering our immediate moral universe, our lives on earth. This essentially heuristic, edifying, and inspirational function of Platonic cosmology is alluded to throughout the dialogue and is given rhetorical emphasis in the dialogue's climactic paragraph, which enjoins us to tune the motions of our soul in accordance with the 'harmonious structures and revolutions' of the universe [*Tim.* 90c–d].[12]

So in answer to our third question concerning Plato's conception of the contributions the sciences make, we must say that cosmology *as such* makes no special contribution. If, however, a cosmology is framed by a Platonic metaphysics, and only on that condition, it has a powerfully edifying function. The compliment Plato seems to pay cosmology in the *Timaeus* is a thinly disguised act of self-praise for his metaphysics, after all.

---

[12] There is good reason to think that, for Plato, human souls have their own proper motions not metaphorically but in a fairly literal sense of κίνησις: Kung 1985, 17–27.

3. Platonic natural philosophy and modern science

The two approaches to Plato's science that I have pursued have produced significantly different observations. The encyclopedia of the sciences in the *Timaeus* is a mirage of anachronistic reading. With the exception of the mathematical sciences, none of the sciences listed in what I earlier called the encyclopedia get any recognition from Plato as distinct fields of inquiry. His own scheme is simply: the arts and sciences in the ordinary sense; the mathematical sciences; and, as something of an appendix, global natural philosophy, provided it is invested with a Platonic ideology. The Platonic thesis of the unity of science is actually more sweeping than the *Timaeus* would suggest, and more sweeping than its modern counterpart: it is a thesis that all of the arts and sciences have a common mathematical core. Plato never seems to let go of the conviction that empirical inquiry is a fundamentally flawed and unsatisfactory undertaking. The bare logical schema of hypothetico-deductive science can, no doubt, be found in the *Timaeus*. But so far as Plato is concerned, its function is to furnish illustrations, from familiar experience, either of certain metaphysical principles or of mathematical propositions, the truth of which, in either case, is known *a priori*.

The bearing of 'hard facts'. When I referred earlier to the looseness of fit between Plato's geometric-kinematic theory of matter and the relevant data of observation, I cited the explanation that no one had done for terrestrial phenomena the sort of assembly and analysis of 'hard facts' that fifth-century astronomers like Meton and Euctemon had done for celestial phenomena [see Vlastos 1975, 85]. And yet I am not convinced that, had there been such a body of data for terrestrial phenomena, Plato would have searched them for evidence against his theory. Even in the case of astronomy, once Plato turns to the motion of the Moon and the other five planets, he shows little inclination toward canvassing 'hard facts'. At *Timaeus* 38d–e Plato's astronomical spokesman begs to be excused from giving details concerning the positions in the cosmos assigned by God to the three slowest planets—Mars, Jupiter, and Saturn[13]—and the reasons for those particular assignments. That Timaeus defers this discussion is not in itself significant: his plea, that it would require too long a digression,

[13] τὰ δ' ἄλλα οἷ δὴ καὶ δι' ἅς αἰτίας ἱδρύσατο. It is clear that ἄλλα excludes the Moon, the Sun, and the Sun's two companions, Venus and Mercury. The positions of all four of these bodies were explained at *Tim.* 38c–d.

makes perfect sense in the context of the dialogue and should be taken at face value [cf. Lloyd 1983b, 21]. What is significant is that he should suppose that a scientific account of the positions in the cosmos of the three slowest planets *can* be given. Empirical data that would have bearing on questions of the distance of the outer planets from the earth, or even of their order in space, must have seemed as inaccessible to Plato as they seemed down to the invention of the telescope [Mourelatos 1987, 93–96]. For all of antiquity, including Ptolemy, such questions with respect to all five planets (not just with respect to the outer three) could be the subject only for *a priori* constructions or numerology [see Van Helden 1985, 15–27, esp. 21 and 26]. It is speculation in the latter vein, not a canvassing of hard facts, that Timaeus defers.

It is also significant that Plato betrays not the slightest concern that the data obtained through observation might put in jeopardy his rationalist conviction that the periods of the planets are systematically related to one another in intelligible ratios (ἐν λόγῳ δὲ φερομένους, 36d).[14] Remarkably, he seems to have made this part of his theory virtually safe from refutation by tying it to the doctrine of the Great Year, the time it takes for all planets and the sphere of the fixed stars to complete simultaneously an integral number of revolutions [*Tim.* 39c–d]. The intelligible ratios Plato speaks of must apply either directly to the planetary periods or to the number of revolutions each planet completes in a Great Year. The first possibility could hardly have seemed worth pursuing. The best approximations that may have been known to Plato do not seem to fall into any intelligible pattern: $^1/_{366}$ year for the revolution of the sphere of the fixed stars (one sidereal day);[15] $^1/_{13}$ year for the Moon;[16] 1 year for Sun, Venus, and Mercury; 2

[14] Only six ratios would be involved, since there are only four distinct numerical figures for planetary periods (the Sun, Venus, and Mercury are said to have the same period: cf. *Tim.* 38d). I cannot agree with A. E. Taylor [1928, 216] that Plato's claim that the planetary orbits are related by ratios means simply that 'the fraction $\frac{\text{period of } x}{\text{period of } y}$ is always a rational fraction'. In Platonic contexts in which the theme of harmonious structure is prominent, terms such as λόγος and συμμετρία are likely to refer to proportions that are systematically significant, proportions generated in accordance with some intelligible principle of progression: see Mourelatos 1980, 39–41, 54–56: cf. 1987, 88–90, 96–101.

[15] At *Laws* 828a–b Plato gives the figure of 365 days for the solar year; and his account of the motion of the Sun in the *Timaeus* implies that he understood that the solar day is slightly longer (we know it is approximately 4 minutes longer) than the sidereal day.

[16] The relevant figure would be that of the sidereal month, which is nearly two days shorter than the synodic month (the familiar month of the lunar phases). That Plato is aware of the distinction between the sidereal and the synodic month

years for Mars; 12 for Jupiter; 30 for Saturn.[17] The λόγος Plato was hoping for was presumably expressed more perspicuously in the formula for the Great Year; it was a λόγος that applied somehow to the whole numbers that represent the integral number of revolutions each of the eight bodies performs in a Great Year—in other words to the smallest possible integers that can be substituted for the variables in this grand equation:

$$\begin{aligned} a \text{ sidereal days} &= b \text{ sidereal months} \\ &= c \text{ years of Sun/Venus/Mercury} \\ &= d \text{ Martian years} \\ &= e \text{ Jovian years} \\ &= f \text{ Saturnian years.} \end{aligned}$$

The λόγος should, of course, be the same, whether our unit is solar years or any other of the five units implied in this equation. The importance of the Great Year formulation is, obviously, that—in the absence of clocks and instruments of precise celestial measurement—it constitutes the ideal algorithm for establishing the relevant λόγος. The figures concerning the periods are, after all, rough approximations, subject to correction for observational error; and cycles, such as that of the Great Year, would offer, in principle, the most impressive check on received figures. In fact, the concept of the Great Year is utterly useless for purposes of verifying values for the planetary periods. Each of the figures would have to be verified in smaller cycles against each of the other five; or, ultimately, the two exactly similar celestial events that mark the beginning and the end of the Great Year proper—a certain configuration of stars and planets, and the next occurrence of that same configuration many aeons later—would have to be observed and recorded, and all the intervening revolutions would have to be accurately counted and their figures duly recorded. Given the enormity and chimerical nature of either of these research projects,[18] it was not unreasonable of Plato to have discounted the possibility that the empirical

is shown by the wording at *Tim.* 39c: 'The month is completed when the Moon, having made the full round of its own proper orbit (περιελθοῦσα τὸν ἑαυτῆς κύκλον), should catch up with the Sun (ἥλιον ἐπικαταλάβῃ).' Note that the text says τὸν ἑαυτῆς rather than τὸν αὐτῆς, and that the syntax (aorist participle modifying the subject of the finite aorist) strongly suggests that two distinct events are involved.

[17] Concerning early ancient knowledge of the periods of the three slower planets, see Neugebauer 1975, 681, 688.

[18] Computations of the Platonic Great Year in the ancient and medieval tradition vary wildly, and, in any event, run in the thousands of solar years: see Taylor 1928, 216–220.

data might some day give the lie to his idealizing postulate of a λόγος that governs the periods.

Vlastos [1975, 91] is right on the mark when he speaks of Plato's 'lordly insouciance for the empirical verification of his elaborate and ingenious physical theory'. But this attitude of Plato arises not from the paucity of relevant 'hard facts' in *certain areas* of inquiry but from Plato's rationalist conviction of the tediousness and marginal relevance of the empirical data in *all* science. In the dialogue between Reason and Necessity, it is Reason that has the last say.

Cosmology and myth. I should like to close with a coda about Plato's use of the term μῦθος to describe the genre of his cosmology. To be sure, μῦθος does not necessarily mean 'myth'; it can mean 'story' and 'account'. Yet there is an unmistakable playfulness to the *Timaeus*. And this playfulness becomes progressively more noticeable as we move from the domain of Reason to that of Necessity. In the final pages, which discuss the differentiation of sexes and lower animals, the style is almost comic—a return to the Aristophanic idiom of the *Symposium*. Indeed, at *Tim.* 59d the εἰκὼς μῦθος (likely story) told in natural philosophy is explicitly called a 'quiet and thoughtful form of play' (μέτριον... παιδιὰν καὶ φρόνιμον). Modern accounts that claim to find in the *Timaeus* a prefiguring of our hypothetico-deductive conception of science fail to hear the chuckles of play and irony in Timaeus' voice. Given that in Platonic natural philosophy the explanatory function is completely subordinate to the edifying, inspirational, and ideological function, there is little point for faithfulness to the details of natural history [cf. Mourelatos 1987, 96–102]. It suffices to say, 'It could be like this, or like this, or yet like this.' What matters is that we grasp the principal truths: that the universe is good and beautiful, that it shows the workings of intelligence at all levels of its organization, and that it is articulated in harmonious structures.[19]

With a little transposition we could easily make the preceding comments apply to the Great Speech of the *Protagoras*: 'Civilization may have started more or less like so.' What matters is not the details but the insight the

[19] In so far as Timaeus' cosmological account promotes our grasp of these principal truths, it is not only 'no less likely' than any competing account, or 'as likely as possible' (μηδενὸς ἧττον εἰκότος, *Tim.* 29c; μάλιστα εἰκότος, 44c–d; μάλιστα εἰκός, 67d), it is even 'more likely' (μηδενὸς ἧττον εἰκότα, μᾶλλον δέ, 48d).

story elicits into certain truths concerning the nature of man and the origins of the conception of justice. The sense in which the *Timaeus* is a μῦθος is not so different, after all, from the sense in which Protagoras' account of the origins of civilization is a myth.[20]

[20] I thank Charles Kahn for detailed critical comments on the version that was presented at the 1986 conference at Pittsburgh. I also thank Alan Bowen for some very thoughtful corrections. An abridged version was presented in February 1989 in French translation at the Maison des sciences de l'homme in Paris, under the joint auspices of L'École des hautes études en sciences sociales, the Centre nationale de la recherche scientifique, and the University of Lille III. I thank my French hosts for their kind invitation, which afforded me that second occasion of stimulating discussion of the argument of this paper.

3

The Aristotelian Conception of the Pure and Applied Sciences

JOSEPH OWENS CSsR

Much has been written on Aristotle's conception of the pure and applied sciences. His tripartite division of science into theoretical science, practical science, and productive science is well known.[1] The surface features of this division may suggest an equation of the Stagirite's concept of theoretical

[1] The tripartite division occurs four times in the Aristotelian corpus: *Top.* 145a14–18, 157a10–11; *Meta.* 1025b18–26, 1064a10–19. The passages in the *Topics* merely refer to it so as to illustrate other points, while those in the *Metaphysics* give its rationale. On other occasions Aristotle speaks in terms of a bipartite division, in which theoretical science is always one member. The other member is practical science at *Meta.* 993b19–23 [cf. *De an.* 407a23–25, 433a14–15, and *Polit.* 1333a16–25]; whereas it is productive science at *Meta.* 982b11–28 and 1075a1–3, *De cael.* 306a16–17, and *Eth. Eud.* 1216b10–19. The last text here is interesting in so far as it notes how theoretical science may nevertheless be useful, and as it treats an instance of practical science as a productive science:

> This approach holds good in the theoretical sciences: nothing belongs to astronomy or natural science or geometry except knowing and apprehending the nature of the objects which fall under these sciences; though incidentally they may well be useful to us for many of the things we need. Of the productive sciences, however, the end is distinct from the science... health is the end of medicine, good social order... the end of political science. [trans. Woods 1982]

Aristotle's bipartite division of the sciences need not cause much surprise against its own background. The really basic partition was between sciences whose purpose was knowledge for its own sake and those whose purpose was something else (conduct or product), and between those whose starting-points were in the things known and those whose starting-points (choice or plans) were in the agent or producer. Plato calls the crafts practical sciences [*Polit.* 258d–e] as well as productive [*Soph.* 219b–c, 265a–266a], and at *Soph.* 266d the expression 'pro-science with today's notion of pure science, leaving his practical and productive sciences to the realm of applied science. Closer acquaintance with the rationale of the Aristotelian grouping, however, quickly raises doubts about the supposed correspondence. Indeed, Aristotle's division calls for scrutiny against a background wider than the currently accepted notions of pure and applied science, and our *prima facie* impression that there is such a correspondence needs to be modified in the light of a careful examination of the problems involved.

The first difficulty lies in attaining correct understanding of the Aristotelian notions of what a science is. As an English word, 'science' carries the general meaning of the Latin, '*scientia*', in this context, which translates the Greek ἐπιστήμη as used here by Aristotle.[2] For him ἐπιστήμη meant an organized body of knowledge, quite as it does today. But our modern notion of science requires considerable adjusting if we are to get an accurate account of Aristotle's tripartite division. Under the Aristotelian caption of theoretical science come metaphysics and the philosophy of nature, disciplines that today are not commonly regarded as scientific. Moreover, sciences such as astronomy, harmonics, optics, and mechanics are viewed by the Stagirite as 'the more physical of the branches of mathematics' [*Phys.* 194a7–8: Hardie and Gaye 1930, *ad loc.*][3] in the sense that they are theoretical sciences in themselves and not just applications of mathematics to concrete, material domains. Next, practical science for him finds its starting-points or principles in the correct habituation of the moral agent, and its conclusions in the actions that issue from those principles [*Eth. Nic.* 1095a3–6, 1103b6–25, 1147a18–28: cf. *Meta.* 1025b18–24, 1064a10–16]; it is viewed as a science of a different type from the theoretical, and not as essentially the application of theoretical principles to conduct. Rather, practical science involves intrinsically a correct appetitive habituation. Finally, the productive sciences consist in correct habituation for producing things, as carpentry for making houses. The only productive sciences on which Aristotle himself has left treatises are poetry and rhetoric, subjects that today

ductive practice' is used. Sharp contrast between the two terms, 'practical' and 'productive', did not always have to be observed.

[2] Cf. texts cited in n1 above, and others listed in Bonitz' *Index Aristotelicus* [1870, 279b38–280a4]. For Aristotle, ἐπιστήμη can also signify (a) intellectual knowledge in contrast either to sensation [1870, 278b58–579a1] or to opinion [1870, 279a4–10], (b) reasoned knowledge in contrast to intuition [*Eth. Nic.* 1140b31–35]. The etymology of '*scientia*' is not certain [see Ernout and Meillet 1951, s.v. *scio*]; on its subsequent history, see n6, below.

[3] On the science of mechanics in this context, see *An. post.* 78b37.

would hardly be regarded as applied science, while accomplishments like carpentry would be considered crafts and not sciences at all.

These observations indicate at once that our notion of science must be revised if we are to grasp the Aristotelian conception of what science is and how the sciences are divided. In general, Aristotle tends to approach science by treating it as a habit that qualifies a human person. Thus, he categorizes [*Cat.* 8b29] the sciences under habits, a subdivision of quality, and claims that science is also [*Cat.* 11a20–31; *Top.* 145a15–18] something relative in so far as it bears essentially upon a subject-matter. For him any science is fundamentally a habit of a human person. This helps explain why he thinks that moral and technical habituation are involved intrinsically in their respective types of science, the practical and the productive. The intelligible content of any science will remain the same when the science is viewed as a habituation, since for Aristotle [*De an.* 415a14–23] the specification of the human faculties comes from the objects of these faculties through the acts by which the objects are apprehended, and habits are formed by repeated acts. But approaching science as a personal habituation opens out in directions different from those indicated by approaching it as an objective body of knowledge; and it requires changing the way our notion of science is specified in its various branches. And to make the differences even greater, Aristotle regards a science as a body of knowledge given essentially in terms of a thing's causes, where 'cause' is understood in a much wider sense than is usual today.

This adjustment in focus should not be too much to ask of those considering the classification of the sciences in Aristotle. In recent times what we call science has in fact shown considerable flexibility. Not so long ago the natural scientists resented the use of the term for the social sciences. Today nobody raises an eyebrow at the mention of political science, behavioral science or human science. Experts in those fields do not hesitate to call themselves social scientists, and even the term 'human scientist' has been used.[4] In the past, the English word 'science', like the German '*Wissenschaft*', has been applied to philosophy.[5] But no philosopher or theologian today would think of calling himself a scientist, no matter how scientific he may consider his procedure to be; nor would he bring it under the heading of Science (capitalized or personified). The term 'scientist' dates only from 1834, and was quickly appropriated to the natural and life sciences. But,

---

[4] E.g., Bernard Lonergan [1972, 3, 23, 210]. The terms 'metaphysical science' and 'divine metaphysical science' have been used recently by an aerospace scientist, Gordon N. Patterson [1985, 3, 58–65].

[5] E.g., by Edmund Husserl [1965, 71–147]. Cf. Lauer's comments [1965, 5, 8–19, 25–27].

traditionally, the term 'science', translating the Latin *scientia*, has been used of philosophical disciplines such as metaphysics, philosophy of nature, moral philosophy, and classical logic.[6] Now one should not be too hasty in concluding that there is no reason for this usage in the notion of science itself. Words have their magic. Their use at a particular time can easily cast a blinding spell over one's ability to draw out the potentialities latent in a concept. To study Aristotle with genuine empathy one needs at least to leave open the possibility that the very notion of science itself may in virtue of its own implications be extended as far as he and centuries of Western tradition have seen it range. Without this empathetic understanding, much of what Aristotle wrote on the division of the sciences will appear badly confused and even nonsensical or self-contradictory. The possibility that our notion of science has a wider range than current use allows needs to be entertained seriously.

The classification of the sciences as pure and applied stems from the nineteenth century. It has proven serviceable for practical purposes, as a glance at the large number of titles listed in a library-catalogue under applied mathematics or applied science will indicate. Yet this distinction is not easily described theoretically. Pure sciences are usually described as sciences that deal with generalities which may be applied to particular subject-matters. For Aristotle, all human knowledge originates in particular sensible things.[7] From Descartes on, however, ideas have been regarded

[6] See Mariétan 1901; McRae 1961; Weisheipl 1965, 54–90. On the shift in the usage of 'science' during the late eighteenth and early nineteenth century from its primary signification of a human attribute to that of an objective body of knowledge, see Williams 1958, xii–xvii.

[7] 'The nineteenth century invented the terms "pure" and "applied" mathematics ... a terminology which is far from being adequate and satisfactory' [Lanczos 1964, 1]: cf. '... we have now the "pure analyst," who pursues his ideas in a world of purely theoretical constructions, and the "numerical analyst," who translates the process of analysis into machine operations' [1964, v]. The earliest notice listed in the *Oxford English Dictionary* [*s.v.* pure II d] of the distinction 'between pure science, which has to do only with ideas, and the application of its laws to use of life' is from Johnson's *Rambler* in 1750, though the contrast between pure and mixed mathematics is noted from a century earlier. Implicit in the sense of 'pure', as understood in this context today, is pursuit of the discipline just in itself on account of its own intrinsic appeal, as opposed to pursuit for utility or some other extrinsic purpose: 'On the other hand, the pure mathematician studies mathematics in its own right and finds great aesthetic appeal in its logical structure and abstract systems' [Jackowski and Sbraga 1970, 1–2]. Yet one may doubt the original independence suggested by 'in its own right' [see Lanczos 1964, 1], and applied branches such as astronomy and computer science may have their own intrinsic appeal. Moreover, disciplines like theoretical physics and theoretical mechanics appear to function as applied mathematics in their spe-

as the first and immediate objects of the mind, with the result that they may be considered apart by themselves in their generality prior to their application to particular subjects. Kant's distinction of pure reason from practical reason, and then of both from empirical knowledge, has further deepened this cleft. However, the distinction of pure and applied sciences has in fact proven so successful that there is no question now of looking upon Aristotle's tripartite division as something that could be substituted for it or used alongside it. As Tennyson wrote:

> Why take the style of those heroic times?
> For nature brings not back the mastodon,
> Nor we those times.
> [*The Epic* 35–37]

The purpose of looking at Aristotle's tripartite division now is not rivalry or confrontation. Rather, it is to see what philosophical insight and understanding Aristotle can give in regard to the nature and the functioning of the sciences, in the way that different philosophies can disclose deeper and inspiring conceptions that may be missed by views to which one has become accustomed. In this spirit, then, let us examine Aristotle's tripartite division to see how it compares with the current classification of the sciences into the pure and applied.

The first question, obviously, will be whether Aristotle's notion of theoretical science coincides with the modern notion of pure science. In the ancient Greek context what was theoretical did not imply any contrast to what was real or actual. It did not have any connotation of the hypothetical or the uncertain, and it did not bear upon something worked out solely in the mind and now waiting to be tested by observation or experiment for its truth. Rather, 'theoretical' entailed the contemplation or study of something already existent and lying before the mind's eye for examination. It meant something that just in itself revealed its own truth and in a thoroughly objective fashion.

The objects that came in this way under the mind's theoretical gaze were of three ultimate kinds: they were either metaphysical or physical or mathematical. Hence, the three broadest divisions of theoretical science

---

cial areas, while application to practical use is seen in applied physics, applied psychology, applied cybernetics, applied radiology, applied geography, and so on. These considerations suggest difficulties in our distinction of the pure and applied sciences, in spite of the practical success the division enjoys in current use. The two terms will continue to be used, 'even if they have been coined wrongly by conjuring up associations that are not warranted philosophically' [Lanczos 1964, 1–2].

were metaphysics, philosophy of nature, and mathematics. Of these three the first, metaphysics—called by Aristotle the primary philosophy—was the highest and most exact. It was the highest, since it was understood by the Stagirite as the explanation of things in terms of their causes, and metaphysics explained things through the causes that were first and highest [*Meta.* 981a24–982a1: cf. 982b2–10]. Further, metaphysics was clearly for Aristotle the most 'exact' of the sciences, because it explained things through the most precise or least complicated of causes, such as being and actuality [*Meta.* 982a12–14, a25–28].[8] In both these ways metaphysics was science in the most excellent grade. Moreover, as the first or primary instance of science, metaphysics was be the paradigm to be imitated as far as possible by the other sciences, in accord with the focal reference that located in the primary instance the nature expressed in all others.[9]

Does a discipline so conceived count as a pure science today? Certainly it was regarded as the most general of disciplines, for it treated universally of all things under their aspect of being. In addition, it was looked upon as a type of knowledge pursued purely for the sake of knowledge, even though its conclusions about intellectual activity and God and soul were put to significant use by ethics. One of its explicitly listed tasks [*Meta.* 1005a19–b34] was to protect the other sciences against assaults from scepticism and extreme relativism, by defending such general principles as the first principle of demonstration (which was later called the principle of contradiction). But it kept this watch entirely from the outside, like a police unit that patrolled the streets without entering into the family life of the citizens and in fact that was seldom if ever called upon to exercise its authority. On these counts, metaphysics can well qualify for listing as a pure science, remaining aloof as it does from observed particularities and keeping its hands unsullied through any immediate contact with the practical or the productive orders. Indeed, it appears from this standpoint as an object so pure and thin to the modern mind that one may have difficulty in seeing any real science in it at all!

Yet some hedging seems required. There is a difference in motivation. Because it is pursued for its own sake, Aristotle ranks theoretical science

---

[8] Thus, mathematical science is not the most exact, even though Aristotle's example here is taken from mathematics (arithmetic is said to be more exact than geometry, because it is less complicated). Cf. *Meta.* 995a8–11, where mathematics is mentioned as an instance of accuracy in discourse.

[9] For Aristotle a term like 'health' signified a nature found in a primary instance as such, i.e., in health as the disposition of a vital organism, and in other instances by reference to that primary instance, i.e., in cooked food as a cause of health. This was felicitously termed focal meaning by G. E. L. Owen [1960, 169].

higher than sciences meant for other purposes. But today's pure science, no matter how attractive it may be in itself, is in most quarters valued for the mediated contribution it can actually or possibly make to practical living. Otherwise it would tend to be regarded as lacking relevance. Further, the universality of focal reference in Aristotle's metaphysics, though unlimited in extent, is based upon a definite type of being, the divine. Accordingly, by virtue of its object, metaphysics is for Aristotle a theology.[10] It is specified by separate, that is, supersensible substance. Still, granted these reservations, there seems no reason to demur at looking upon Aristotelian metaphysics as a pure science, if it is to be classed at all as a science—the phrase 'applied metaphysics' would in fact seem incongruous and difficult to endow with meaning.

Next, philosophy of nature for Aristotle explains sensible things through their substantial principles or causes, namely, matter and form, and thereby accounts for their extension in space and their multiplication in singulars of the same species. With the two further causes, namely, the efficient and final, the philosophy of nature also explains in its own way generation and perishing, change, and time. As in the case of metaphysics it is a theoretical science, pursued for its own sake even though it proves helpful for metaphysical study and for ethical matters concerned with passions and self-control [see *Meta.* 1071b6–10, *Eth. Nic.* 1102a18–b11]. Consequently, the philosophy of nature can be classified as a pure science just as metaphysics. There will be similar reservations in regard to its bearing upon a particular kind of things, namely, real sensible beings, and in regard to its pursuit for the knowledge it gives in itself apart from its relevance to the control of nature. If no supersensible substance existed, it would be the highest science [*Meta.* 1026a27–29, 1064b9–11].

Finally, the third general type of Aristotelian theoretical science is mathematics. Mathematics deals with corporeal things in abstraction from their sensible qualities. It explains them in terms of quantity, both continuous and discrete. Once abstracted from the restraints of sensible qualities, the objects of mathematics can be expanded indefinitely through dimensions beyond the third and fourth. Irrational numbers may also have their place, and the way is left open for non-Euclidean geometries and the other developments that took place long after Aristotle's time. In any case, mathematics, as Aristotle saw it, is to be pursued as a pure science in abstraction from the sensible conditions of everyday life; there need be no special hesitation in classifying his conception of mathematics in general

---

[10] See *Meta.* 1026a19 (where primary philosophy is called theological) and 1064b3 (where it is the 'theological' type of the theoretical sciences).

as pure science. Indeed, mathematics, as Aristotle accounted for it, is so pure in its detachment from sensible qualification that it is able to sustain epistemologically all possible advances of mathematics, even in areas of which Aristotle himself did not have the least inkling.

But also recognized by Aristotle as mathematical sciences were astronomy, harmonics, optics, and mechanics. As noted above, the first three of these were explicitly called by him 'the more physical of the branches of mathematics'. Mechanics can readily be seen as coming under this type too, as also today's far-flung mathematized natural and life sciences, and by extension the behavioral sciences. In such sciences the mathematics would today be regarded as applied.[11] But from Aristotle's viewpoint, these sciences could be pursued purely for the sake of the knowledge they gave, even though this knowledge was restricted to particular areas and could be put to practical use; it was in itself theoretical. The sciences that pursued such knowledge were in consequence regarded as theoretical sciences. Aristotle [*An. post.* 78b39–79a16] also speaks of non-mathematical sciences of nature, such as navigational astronomy and acoustical harmonics. Apparently, he looked upon sciences of this kind as coming under the philosophy of nature, on the grounds that from the knowledge of a thing's specific form all the thing's natural developments could be deduced somewhat as knowledge of the blueprint shows how a house is to be constructed.[12] Of this type would be the extensive research carried out in Aristotle's biological works.

But, in regard to the present point of investigation, the situation is sufficiently clear. The sciences that deal with the physical universe, whether Aristotle classified them under mathematics or under the philosophy of nature, are in his conception regarded as one and all frankly theoretical. None of them can be looked upon as either practical or productive science. Their type is markedly different from either of those two divisions.

What, then, is there left for Aristotle's practical science to bear on? It deals with human conduct, but in a way very different from the theoretical procedure of our social sciences. It finds its principles not in what is already there before it, but in the human choice that originates conduct [*Meta.* 1025b8–24, 1064a10–16]. It is focused upon not what is going on or taking

---

[11] E.g., as at *An. post* 78b35–79a16. For treatment of these Greek sciences expressly as applied mathematics in contrast to 'pure mathematical subjects', see Heath 1921, i 17–18; and 1949, 58–61. On mathematics in today's life sciences, see Defares and Sneddon 1964; for the social sciences, see Bishir and Drewes 1970.

[12] 'For in house-building too it is more the case that *these* things take place because the form of the house is *such*,.... And this is the way we should speak of everything that is composed naturally' [*De part. an.* 640a15–b4: Balme 1972, *ad loc.*]. Cf. Balme 1972, 86–87 and Aristotle, *De an.* 402b6–403a2.

place, but upon what should be done through human choice and the right reason that is meant to guide that choice. Practical science is concerned with something yet to be done, an object that is not already determined by nature. More specifically, it bears upon something that is to be chosen.

While truth in the theoretical order consists in the conformity of one's judgement with reality, truth in the practical order consists for Aristotle in conformity of one's judgement with correct moral habituation.[13] This correct habituation requires training and education from the earliest years [see *Eth. Nic.* 1103a14–1107a2, 1179b29–1180b28]. In conformity with such moral habituation, one judges immediately that some things are right and that others are wrong, and in that way one acquires the premises of practical science [*Eth. Nic.* 1095a2–8, 1095b4–13]. Hence, Aristotle's own claim, so strange and unacceptable to many today, is that only a morally good and mature person is capable of undertaking the study of ethics as a science.[14] The conclusion of practical reasoning, moreover, is the action performed. It is not a detached proposition uttered by the mind [*Eth. Nic.* 1147a25–28: cf. 1095a5–6]: it is something done in human action. The whole purpose of practical science is to bring about good conduct [*Eth. Nic.* 1103b26–31, 1179a35–b4: cf. 1102a7–12].

This conception of practical science is obviously different from the ordinary understanding of science today. Perhaps Aristotle's presentation of truth in the practical order is too brief and this may be why it has been overlooked or forgotten.[15] But in the three works on ethics there has

---

[13] '... truth in agreement with right desire' [*Eth. Nic.* 1139a30–31: Ross 1915 *ad loc.*].

[14] See *Eth. Nic.* 1095a2–8, 1095b4–13, 1103a14–1107a2, 1179b29–1180b28. There can of course be a *theoretical* study of what good persons do [*Eth. Nic.* 1169b33–1170a3], and an evaluation of the actions of others in analogy with the way one relates one's own actions to one's own chosen ultimate goal. But to be truly scientific, from Aristotle's viewpoint, the reasoning must be based upon the *true* principles, and these the man who is not morally good just does not have. The evil man would be merely repeating by rote what he has learned from the morally good man, like the drunkard reciting the verses of Empedocles [*Eth. Nic.* 1147b9–14]. With the premises accepted in this way on faith or authority, the immoral man can construct his reasoning on analogy with his own habituation. But his reasoning does not thereby become truly scientific. His failure to appreciate τὸ καλόν would be like tone-deafness in a person writing about the sound of music.

[15] As Adler states:

> That there is such a widespread ignorance of Aristotle's introduction of a twofold conception of truth, sharply distinguishing between the truth of theoretical (or descriptive) statements on the one hand, and the truth of practical (or normative) statements on the other hand, can, perhaps, be explained, though hardly excused, by the fact that his treatment of this

been handed down a moral philosophy that is acknowledged to have perennial worth.[16] Yet Aristotle's practical science does not fit into the modern conceptions of the sciences. It is not pure science, since it is embedded in action, and since in its totality it bears upon individual needs. It is not applied science: instead of being an application of theoretical norms to practical acts, it starts as a science from the individual acts of right and wrong and reasons from there to its general conceptions. To speak of applied ethics would in this case seem to be a mistake.[17] Practical science can and has to make use of the findings of theoretical science [see *Eth. Nic.* 1102a16–b11, *Eth. Eud.* 1216b10–19], but it does not itself *consist* in an application of these findings. Basically, perhaps, the notion of a genuine science of conduct that is a unitary habituation of will, appetite, and passion as well as of intellect, is what estranges Aristotle's practical science from contemporary empathy and acceptance as a science. There is certainly nothing corresponding to it in the modern division of pure and applied sciences.

Finally, there are for Aristotle the productive sciences. As with practical science, they too have their starting-points within the producer. But unlike those of practical science, these starting-points are fixed plans or designs; they do not originate in choice. The productive science works the design into some material, such as bricks and stones and lumber in the case of a house, pigments or bronze in the case of a painting or statue, words and images in the case of a poem or speech. Like practical science it makes abundant use of theoretical knowledge. But it is not itself that theoretical knowledge applied externally. It is a habituation in its own right, a habituation that resides in unitary fashion in mind, nerves, and muscles.[18] It is intrinsically a different type of knowledge from the theoretical. It involves knowing how to do something rather than knowing what something is. Thus, one may know motor-mechanics and the rules of the

crucial matter is contained in a single paragraph in book 6, chapter 2 of the *Ethics* (1139a21–b31).

[16] John Herman Randall [1960, 248] portrays how 'Aristotle's practical philosophy' can be generalized to fit '*any* cultural heritage'.

[17] On this topic in general, see MacIntyre 1984, 498–513.

[18] Plato had written 'But the science possessed by the arts relating to carpentering and to handicraft in general is inherent in their application ...' [*Polit.* 258d–e: trans. H. N. Fowler 1925]. Gilbert Ryle [1949, 27] distinguished 'know' from 'know how' as the knowledge 'of this or that truth' from 'the ability to do certain sorts of things'. With Aristotle the productive *science* is the ability, and not just a theoretical knowledge applied by the art. The 'science' of boxing would be a case in point.

road, and yet not know how to drive. No matter how much theoretical knowledge one has about wood and nails and tools, one is not a carpenter if one can hit a nail on the head only occasionally. The habituation that enables one to build a house is the productive science. As practical science is the habituation that brings one to conduct oneself properly, so productive science is what equips one to produce things expertly, whether houses, automobiles, poems or music. Using, but not giving, knowledge of what things are, practical and productive sciences consist in knowing in habitual fashion how to behave and how to make things. In this perspective the theoretical, the practical, and the productive are three different types of science.

The preceding remarks make plain the difficulties encountered in efforts to fit the Aristotelian sciences into the modern categories of pure and applied science. To the question, What is Aristotle's classification of the pure and applied sciences?, the straightforward answer is that Aristotle has no such classification. But *we* can take the sciences the way he divides them, and ask how they fit under the modern classification. From this viewpoint metaphysics may, with certain reservations, be regarded as a pure science. Philosophy of nature and mathematics in their general conclusions would also come under pure science, though, when their general principles are studied in particular areas, the procedure would be akin to that of applied science. Aristotle's practical science does not seem to fit at all under the modern classifications, since it is neither pure nor applied science. And today his productive sciences would be looked upon not as sciences at all, but as crafts and fine art. Moreover, sciences set up to cover their activities would be regarded from his standpoint as theoretical, since they would be studying what was already existent instead of bearing on what is yet to be produced. Indeed, all the modern pure and applied sciences would count for Aristotle as theoretical knowledge. Logic, which was not classified by the Stagirite but left as a preparation for them all, would be no exception [see *Meta.* 1005b2–5, 1059b14–19]: mathematical logic would obviously enough come under mathematics; and traditional logic, though concerned with the structure of human thought, does not bear upon behaving or producing and would be akin to theoretical knowledge. But it was left by Aristotle as something outside the classification of the sciences and functioned as a preparation for them all.

The modern classification of the sciences in terms of pure and applied has proven serviceable. Nobody would wish to substitute Aristotle's tripartite framework as far as curricular and ready reference purposes are concerned. The multiplication of the natural, life, and social sciences would throw the tripartite division badly out of balance for presenting a general picture of the situation today. But one need hardly look upon the modern division of pure and applied sciences as a contemporary sublimation of everything worthwhile in the Aristotelian conception, meekly acquiescing 'that a truth looks freshest in the fashion of the day' [Tennyson, *The Epic* 31–32]. No, Aristotle presents a radically different philosophical approach from ours. The Stagirite's viewpoint affords a panoramic understanding of the various phases of human knowledge that is lacking in the modern approach based on the classification of pure and applied sciences. It allows room for bringing the supersensible under the scrutiny of genuine science, for penetrating to the substance of material things, for safeguarding human conduct from inclusion in the grip of a necessitarian interpretation of reality, and for maintaining the human dignity that is expressed in the arts and crafts.

None of this Aristotelian understanding of scientific knowledge need be expected, one may note, to help the individual modern sciences strictly within the areas of their own work. What the Aristotelian explanation gives, rather, is a marvelously elevating philosophical view of the enterprise as a whole. It satisfies a deep longing for a well-rounded conception of the role the sciences play in human culture and human personality, and of the indispensable aid they give in bringing about the good life for which human nature is fashioned.

4

Platonic and Aristotelian Science

ROBERT G. TURNBULL

My intention in this paper is to show that both Plato and Aristotle describe and explain procedures of inquiry which can appropriately be called scientific, and that both offer plausible rationales for those procedures. By calling the procedures scientific I intend more than the claim that they are somehow precursors of later developments that really must be called scientific. Since they make use of observation and theory (in their own ways) and derive conclusions which explain phenomena and can be rationally corrected, I think that those procedures are genuinely scientific in their own right. I must caution, however, that in making these claims I am not asserting that the results of using these procedures can now be defended as plausible science.

Plato's procedure of collection and division and Aristotle's procedure for arriving at definitory middles are the basic procedures I have in mind in making the above claims. Their respective descriptions and explanations of those procedures are, however, imbedded in rich philosophical contexts, so that the effort to explain and defend the claims made above is rather complicated and controversial. I shall try to reduce the complication and narrow the field of controversy as much as possible in the exposition that follows. I think that it is proper to begin with an account of concepts for both Plato and Aristotle.

1. Concepts

Plato. As everyone knows, Plato holds that there are eternal and unchanging Forms, and that the Forms constitute and provide objective norms of all intelligibility. Since—especially in the earlier dialogues—Plato often uses visual metaphors in speaking of our awareness of the Forms, a number of interpreters have assumed that what he has in mind is some sort of non-visual *seeing* of the Forms, rather like Russell's 'acquaintance' with sense-data. I think this assumption is mistaken even for such dialogues as *Phaedo*, and I am confident that it is mistaken for the later dialogues. From the earliest introduction of the Forms in the dialogues, Plato is at pains to contrast sensation or perception and dialectic, insisting that our awareness of the Forms is non-sensuous.

The effort of the Socratic dialogues is, in the main, to arrive at some sort of definition—whether or not definitions are thought to delineate the structure of the Forms—and to criticize or defend proffered definitions by some sort of reasoning or dialectic. Though there is in this effort reference to applications of a proposed definition, where these may involve perception or memory of individuals characterized in some appropriate way, the effort is not plausibly understood as a heuristic preliminary to non-conceptual and non-visual staring at a Form.

In later dialogues, notably as I shall shortly show, *Cratylus* and *Parmenides*, Plato explicitly claims that we must have some sort of conceptual ability in order to be aware of or have knowledge of the Forms. But even in earlier dialogues, the rationale of the argument presumes that it is possible for us to learn and that such learning involves a change *in us*. The change in Polus (in *Gorgias*) in recognizing that it is better to suffer than to do injustice or the change in Glaucon and Adeimantus (in *Republic*) in coming to recognize justice as each doing his/her own is what we should ordinarily call *conceptual* change—whether this is to be thought of as catastrophic change or as recognition of the inner workings of a concept imperfectly articulated.

In *Cratylus* Socrates claims that there are Name-Forms which are *of* the Forms of which they are Name-Forms.[1] I understand the 'of' here as πρός τι, so that Name-Forms stand to the Forms they are of rather as double stands to half or slave to master. (In particular, I wish *not* to suggest that the 'of' expresses what we should call intentionality.) Plato is at pains to make it clear that Name-Forms are not themselves bits of language. His mythical name-giver, however, must have them in mind in inventing a language suitable for expressing the distinctions among the several Name-Forms. And Plato makes it clear that different natural languages may

[1] For a detailed defense of the line of interpretation of Name-Forms given in this paper, see Gold 1978.

be suitable for such expression and equally clearly suggests that a child is fortunate to live in a society having a language suitable for such expression.

A person who acquires a suitable language has, of course, a set of linguistic abilities tied in a variety of ways to perceptual episodes, actions, procedures, and so on. And this person may have those abilities without having any idea that the language which is the vehicle for those abilities is expressive of the Name-Forms. As I read Plato, it takes reflection upon terms and their uses to bring about recognition of the Forms and further reflection to achieve recognition of Name-Forms. But let that matter stand for the moment.

What I wish to contend is that Plato uses the notion of Name-Forms with the implication that in this, as in other cases of Forms, we may speak of having shares of or participating in such Name-Forms. And I wish to contend that having a share of a Name-Form is having a concept or conceptual ability, an ability the articulation of which is, of course, linguistic. Thus, to have a share of the Name-Form which is *of*, say, The Triangle Itself is to be able to recognize something seen *as* a triangle; and, on reflection, to be able to define The Triangle Itself and, thus, to have an articulate awareness of that form. Indicative, therefore, of having the concept, triangle, is the ability to pick out triangles or triangular objects and say what they are. And indicative of having reflected properly is the ability to give a definition and defend it.

In *Parmenides* Plato speaks of Knowledge-Forms in a context that requires our thinking of a given Knowledge-Form as (in the πρός τι sense) *of* some commensurate Form.[2] Thus, knowledge and the several kinds of knowledge are of appropriate Forms or clusters of Forms. Again, as in the case of the Name-Forms, to have a share of a Knowledge-Form is to have a conceptual or 'knowing' ability.

In *Timaeus* Plato speaks of the soul as being 'made of' Being, Same, and Different (the most pervasive Forms); and in *Theaetetus* 184c–185d,[3] he lists Being, Same, Different, and Number among the resources of the soul in perceiving various things *as* something or other, *as* the same as like or similar things, *as* different, and so on. I think that, if one were to look for a contemporary materialist equivalent for what Plato states in these

[2] *Parm.* 134a. I shall be making a parallel point about 'of' in the πρός τι sense concerning Aristotle, *An. post.* ii 19.

[3] Needless to say, I am thinking of Being, Not-Being, Likeness, Unlikeness, Same, Different, and Number as the conceptual means employed by the soul when, as Theaetetus is made to say [185d7–10] 'I think that the principle is no special organ as in the case of the others, but that the soul, by means of itself, discerns the commons in everything.'

places, one would find it in the claim that in any sort of perception there is not merely stimulation but a great deal of processing going on. Thus, evolution has so modified the primitive nervous system of our ancestors that we 'see' a world of things and their modifications, movements, similarities, differences, and so on. This processing is, of course, pre-linguistic or non-linguistic, and, in this materialist way, explains our ready acquisition of linguistic abilities. Plato, of course, has no such theoretical resources and thus, to explain our conceptual abilities, turns to what he takes to be the nature and character of soul.

Earlier Plato, as in *Phaedo*, has Socrates depend heavily upon the sufferance of an interlocutor in inquiry—usually inquiry that promises to lead to a correct definition. Here, I believe, Plato relies on standard linguistic usage and Socrates' interlocutor's recognition of it when pressed. Later Plato, recognizing the multiplicity and interrelation of the forms, introduces collection and division (best described, I believe, in the 'Prometheus' passage at *Philebus* 16c–17a[4]). With this and clear recognition of one/many patterns in genera/species orderings of the Forms, one has a procedure for a disciplined, one-person search for definitions. For any given term, one looks for a 'one', i.e., a genus in which it might plausibly be placed. Having determined such a genus, one then 'divides', i.e., tries to determine what species fall immediately under that genus, using some *prima facie* appropriate *differentia*. In turn, one takes the species which seems appropriate for the term to be defined and divides in the same manner once again, using a *differentia* congruent with that first used. And so on until reaching the term to be defined. What emerges from this process is a definition by multiple genera and *differentiae* by means of which one can explain some necessary feature or other of the term or Form inquired about.

Aristotle. Aristotle introduces the term 'universal' (τὸ καθόλου) for something which is 'in the soul', which 'holds for many', and which ultimately depends for its 'coming-to-be' on sense perception and memory. Given Aristotle's doctrine of soul, a universal has to be some sort of acquired ability of a human being, an ability which, among other things, enables a person to be aware of something as being the sort of thing it is. Given this description, it seems clear enough that Aristotelian universals are concepts.

[4] Socrates says of the procedure described in the 'Prometheus' passage: 'It is a path which is not very difficult to point out but exceedingly difficult to use. For by its means have been brought to light all of the discoveries of science (τέχνη)' [*Phil.* 16c1–4]. See also Moravcsik 1979.

In *An. post.* ii 19, Aristotle offers his well-known account of the coming-to-be of the first universal in the soul.[5] He says that there will be no such coming-to-be unless there is (a) the capability of sense perception, (b) the capabil'ty of memory, and (c) 'repeated memory of the same'. I understand 'memory' here to mean or include the ability to hold a past perception of something sufficiently long to fit it to a new perception of that (or a similar) thing. Given these conditions, human beings at least, being endowed with the capability of (experience) ἐμπειρία, undergo the coming-to-be of the universal in the soul (or, perhaps, then become capable of ἐμπειρία). Aristotle proceeds to give us a famous image, that of the army in retreat. As I understand the application of that image, the initial condition of one who has no universals but is exercising perception (in the narrow sense of αἴσθησις, that is, perception that is not conceptually appropriated and thus rather like Kant's 'manifold of sense') is like the very confused situation of an army in headlong retreat and total confusion. The coming-to-be of the first universal in the soul is like having a single soldier turn around and take his stand. With increasing numbers of universals in the soul and with the formation of generic as well as specific universals, more soldiers take their stands, and the ordering of the army by squads, platoons, companies, and so on, emerges. And Aristotle proceeds to explain very briefly the ascending order of universals or concepts in the soul. Before saying anything about that, however, I should like to turn to another matter, one raised by Aristotle in this same chapter.

It is a *cliché* of standard Aristotelian interpretation that knowledge is of the universal and that perception is of the individual. And, indeed, Aristotle says as much. But a common way of understanding that merits criticism and rejection. For 'the universal' is, in this way, understood as claiming that, if someone knows something, what one is aware of and knows is a universal. But what this would seem to require is that the *object* of a knowing state of a person is either something 'in the soul' or some sort of Platonic Form. The first alternative is absurd; the second impossible (since Aristotle emphatically rejects Platonic Forms). As noted above concerning Plato, the trouble with such reading of Aristotle lies in treating the genitive case as signifying some sort of intentionality.

The clarifying but very brief passage is *An. post.* 100a15–b1, which reads as follows:

> When one among the indiscriminables has made a stand, the first universal is in the soul: for, though one perceives the individual, the

[5] For a more detailed account, with translation of this passage, see Turnbull 1976, 28–56.

> perception is of the universal; it is, for example, of man, not of a man, Callias.

I think the proper sense of this passage is:

> Before the earliest universal is present in the soul, there is no discrimination of what is presented to sense. With a universal in the soul, a person can recognize as standing out from the confusion one individual as belonging to a kind. Thus, one can, in perception, recognize an individual as being a man.

Unless perception is conceptually informed, there is nothing that produces the discrimination necessary for recognizing an individual. In this interpretation the genitive of *man* is used to qualify the kind of awareness one may have if one has the concept *man*—in the illustration, an of-man awareness of the object of perception (while bare αἴσθησις is confused of-particular awareness).

Attic Greek does not have separate terms distinguishing perception from mere sensation, and thus the single term, αἴσθησις, slides between the two. Some niceties aside, the English terms treat perception as requiring sensation plus some sort of conceptual appropriation and sensation as not requiring such appropriation. Sensations are indeed individual and can hardly be taken as 'holding for many'. The conceptual element in perception, of course, holds for many and makes possible the awareness of something as being of a kind.

With this understanding of *An. post.* ii 19, the universal in the soul is not an *objectum* but rather an *instrumentum a quo*, that is, a means by which things can be recognized as being of a kind. This accords with the medieval understanding of first intentional consciousness and with the medieval idea of second intentional consciousness as awareness of the 'contents' of concepts, to say nothing of the medieval understanding of concepts as *instrumenta a quibus*.

But now return to the army in retreat with the soldiers taking their stands, first individually and then as ordered groups of individuals. With the idea of universals as concepts, one can understand this as the conceptual analogue of the predicational scheme of *Categories*. Thus, even as the concepts, man, mammal, animal, and living thing, can be predicated of precisely the same thing (say, Callias) and illustrate a species/genus ordering, so one may have of-man, of-mammal, of-animal, and of-living thing conceptual awarenesses of the same thing. And, of course, in the process of acquiring higher-level concepts, one may also link kinds of things together and not merely 'bare' individuals.

In describing this process of acquiring higher-level universals or concepts, Aristotle uses ἐπαγωγή, a term commonly translated as 'induction'.[6] This translation is quite misleading, especially since it invites the ghosts of the modern 'problem of induction'. A more perspicuous translation is 'assemblage', for what Aristotle describes is the assemblage of kinds under a genus, of genera under a higher genus, and so on. Strictly speaking, of course, ἐπαγωγή is the assemblage of conceptual means under higher conceptual means: as we shall shortly see, Aristotle thinks we can by rather clear-cut means modify and improve our concepts (though we could hardly improve objective species and genera). And, of course, objective species and genera would come perilously close to being Platonic Forms.

Physics i 1 is a major key to understanding Aristotle's ideas about the possible improvement of our concepts [see Turnbull 1976]. It is one of the passages where Aristotle distinguishes between what is clear and lucid to us and what is clear and lucid in nature. Improvement of our concepts is, of course, movement from the former to the latter. In the same chapter he says of our primitive concepts (if you please, first universals in the soul) that they are close to sense and very confused indeed. And he illustrates the movement from what is clear to us to what is clear in nature with the movement from a confused (and sensible) concept of (a) triangle to a concept articulately incorporating the definition of (a) triangle.

The accounts of *An. post.* ii 19 and *Phys.* i 1 get us into the neighborhood of Plato's procedure of collection and division. And, indeed, Aristotle is well aware in *Posterior Analytics* that the procedure he is clarifying and defending is very close to Plato's. I turn, therefore, from this rather quick account of concepts in Plato and in Aristotle to the attempts of both to improve our concepts and thus our ability to offer explanations, explanations which I think can plausibly be thought of as scientific.

2. The patterns of explanation

Aristotle is at some pains in *Posterior Analytics* to distinguish his procedure for arriving at acceptable explanatory 'middles' from Plato's procedure of collection and division. I shall get to that in the next section of the paper. Here I should like to emphasize the similarity of the procedures of Aristotle and Plato.

The pattern of explanation both subscribe to is that of explanation by means of definitions. Thus, should one wish explanation of X's being F,

---

[6] For a discussion of uses of this term in Aristotle's *Analytics*, see McKirahan 1983.

one will find it in the definition of X, a definition which, in the cases where there is scientific explanation, will *include* F. From the definition, one can see, e.g., the necessity of a (or this) triangle's having three sides. And one may also be able to explain how it is that an animal *may be* (i.e., in Aristotle's terms, is δεκτικός) well or ill.

Plato's insistence in the 'Prometheus' passage at *Phil.* 16c–17a that in division one determine exactly how many something is before letting the ἄπειρον intrude is an insistence that one determine exactly how many intervening genera and species there are between the object of inquiry and the 'one' chosen as a genus under which the object falls. Given that it has been determined just how many the object of inquiry is, it has *ipso facto* been determined what the definition is—indeed, the definitions of all or most of the intervening genera and species. Thus, Socrates says that the old procedure (presumably, e.g., that of *Phaedo*) of going directly from the form to the world of becoming does not really explain anything. What he has in mind is, of course, the old way of explaining Helen's being beautiful by saying that she has a share of the Beautiful Itself. But the new procedure of *Philebus* (and the collection and division dialogues) provides an explanation of X's being F by means of the definition of X (and that definition's including F).

It is worth underlining that this mode of explanation (explaining why X is necessarily or καθ' αὑτό F), though applicable in universal propositions, does not make individual propositions claiming necessity depend upon universal propositions. Given that the definition of X includes F, 'This X is F' has as much or as little necessity as 'All Xs are F'. The burden of necessity is carried by the subject term and its definition, quite aside from quantification. If one includes propensities or potentialities in the definition (whether directly so or as a result of higher generic classification), then, again regardless of quantification, one may claim that this X is *possibly* F, i.e. that it is δεκτικός.[7]

The senses of 'necessity' and 'possibility' so explained are quite different from those invoked in contemporary modal logic. And I believe it is a mistake to interpret Aristotle's use of such modal terms simply using the resources of contemporary logic. I think it is also a mistake to explain Aristotle's syllogistic by adversion to Boolean logic with its functional notation. Without attempting full-scale demonstration of these claims but

[7] As this paragraph suggests, I am prepared to argue that for 'possible' Aristotle uses δεκτικός (receptive). Thus, an animal is possibly (δεκτικόν) well or ill; but it is impossible for a stone to be either. See *Cat.* 13b13–19, where it is noted that only if something is δεκτικός can one or the other of a pair of contraries hold for it—otherwise neither can.

attempting some explanation of them, I shall quickly note what I take to be the proper form of the standard Aristotelian syllogism in Barbara, in so doing, hoping to make clear the divergence from Boolean syllogistic.[8]

The standard Aristotelian syllogism in Barbara has the form,

$$C \text{ holds for } B \text{ holds for } A.$$

Thus, for example, 'three-sided' holds for plane figure bounded by three straight lines, and 'plane figure bounded by three straight lines' holds for triangle. The middle term, a definitional formula, explains A's being C (or C's holding for A). Less obvious invocation of definitory middles could explain, e.g., why 'having a backbone' holds for a mammal. And, of course, 'if receptive of illness' holds for man (or any other animal), then a man is possibly ill (though a stone would presumably not have such a possibility). It goes without saying that Barbara is Aristotle's standard form for the demonstrative syllogism, i.e., the syllogism used in scientific explanation.

Though Plato does not use Aristotle's terms (in the 'Prometheus' passage or elsewhere), I think that Aristotle, in *Posterior Analytics*, is in his own way detailing Plato's explanatory pattern in *Philebus*. And I do not believe that Aristotle's explicit rejection of Platonic Forms in the *Analytics* makes any difference for my argument in this paper. Give both Plato and Aristotle doctrines of concepts (which make possible awareness of things as what they are or as qualified in various ways), and Aristotle can readily claim that there is no need for Platonic separated Forms in explanation.

At bottom, Plato needs the world of forms to give legitimacy to the process of collection and division with its exercise of conceptual ability. And, at bottom, Aristotle needs the *De anima* identification of form in the soul and form in the thing (and, perhaps, the remarkable ability of νοῦς ποιητικός) for the same reason.[9] One may, therefore, object that Aristotle's quick rejection of Platonic Forms in the *Analytics* is illegitimate. But what is at issue is the conduct of inquiry, not the invoking of the frame or frames which make it and its results fully legitimate.

---

[8] See Corcoran 1972, for a formal treatment of important divergences.

[9] What I have in mind is, of course, the grounding of concepts *in rerum natura*. Plato, if I am right about *Cratylus* and *Parmenides*, finds in collection and division a means of linking concepts with the world of Forms, i.e., by virtue of concepts' being shares of appropriate Name and Knowledge Forms. Aristotle, though denying Platonic 'separated' Forms, insists that there are Forms *in* things and that the universals in the soul (concepts) can conform to them.

3. Conceptual improvement: From 'clear to us' to 'clear in nature'

What I hope to show in this section is that Plato's procedures of collection and division and Aristotle's procedures for arriving at explanatory middles are, in slightly different ways, procedures for improving our concepts (or, if you please, for improving the clarity and precision of the concepts we have). Both procedures are in a sense to be explained, empirical; and both allow for an appropriate sort of testing of proposed definitory formulae. In section 4 I shall make brief reference to the rationalists' (especially Leibniz') use of analysis and synthesis and attempt to show the similarity of that use to the procedures ascribed to Plato and Aristotle.

Plato. According to my story so far, Plato assumes that Attic Greek is a perspicuous language (i.e., one produced by a Name-Giver who constructed it to accord with the Name-Forms) and that people who have learned it in standard perception and conversational contexts are generally capable of recognizing a variety of things around them as what they are and as being characterized in a variety of ways. They may well have that capability, however, and yet never have reflected on the usage of those terms, much less have attempted to determine or discover precise definitions. The procedure of collection and division seems clearly to be a remarkable aid to such reflection, determination, and discovery.

One starts with a given term, say, X, assumes for the purpose of inquiry that A (another term) is such that anything which is X must surely be A. But if, after a moment's thought, one can suppose that an X might not be A, one had better try again. Still let us stay with A. Then one tries to determine what 'immediate' kinds of A there are, say A_1, A_2, and A_3. Once again one must be sure that anything which is any of these must be A. And one must be sure that anything which is X must be one or the other of them, say A_1. Then one must ask what feature it is which divides A into A_1, A_2, and A_3. Call that feature f. f must be, as it were, a generic feature, species of which characterize A_1, A_2, and A_3 and which, together, exhaust the immediate species of f. Call them f_1, f_2, and f_3. The presumed appropriate *differentia* of A is thus f, species of which *differentia* characterize the immediate species of A. X, whose definition is sought, must be such that whatever is X must be A, as well as f (and thus either f_1, f_2, or f_3), and A_1. Strictly A_1 is redundant, for to be A_1 is simply to be A which is f_1.

It must be noted that there is an 'empirical' test for each of these moves. Can one imagine anything which falls under the concept of X which fails to fall under the concept of A? One may in response to the question 'discover' that one is not quite clear about A and temporarily or permanently adjust the concept. Or, in the search for what might do the job of A, one might temporarily or permanently adjust one's concept of X. Can one imagine anything which falls under X failing to fall under f_1? Again adjustments may be necessary, this time in the concept of f or of A or of X. And one may not simply imagine. If the concept of X is that of a natural kind, one may, at any stage, go out and look at or handle or listen to things which one takes to be of that kind. At any stage in the division, one may, e.g., go out and look and discover that one's concept of squirrel confuses squirrels and chipmunks. If X is a concept of social interaction, say, lying, one may discover by investigation that one's concept confuses saying what is false with deliberate deception in speech. My point is, of course, that the Platonic procedure, undertaken seriously, is very likely to involve empirical investigation and thus attention to likenesses and differences with consequent adjustment of the proposed explanatory scheme of classification.

Even in earlier dialogues, Socrates is represented as testing proposed definitions by reference to what would fall under a proposed definition. In the famous first definition of *Euthyphro*, if piety should be dearness to the gods, one and the same act could be both pious and impious. In *Meno*, if virtue could be taught, then there would be teachers of it. But, after a brief survey, it is concluded that there are no teachers of virtue. In *Phaedo*, if the soul were a harmony of bodily parts, then it could not be an initiator of changes in the body. But we observe (and attribute to soul) any number of such initiations of bodily changes. Examples abound. Never mind that some of the concepts do not appear to be 'empirical'. And never mind that some of Plato's own examples, notably those in *Statesman*, seem pretty farfetched and remote from empirical test. The back and forth procedure of collection and division, with the making of appropriate conceptual adjustments, in principle at least requires the applicability of the relevant concepts and, with increasing use of the procedure in different contexts, a virtual mapping of the sensible world.

To return to the example of the application of collection and division to X and its 'one', namely, A, it should be clear that the process of division can go on in the manner which used f (and its species) as the *differentia* of A. The choice of *differentia* for A_1 is, of course, somewhat determined by the choice of f for A. (This is what I was getting at in the reference in earlier sections to 'congruent' *differentiae*.) If one chooses, e.g., 'linearity' as the *differentia* for plane figure, the species will be straight and curved. For straight plane figure, the appropriate *differentia* will probably be sidedness, and species of it will be three-sided, four-sided, and so on (not, say color);

for one is dividing plane figure and not colored plane figure or color. And, to return to the definition of X, there will be introduced once more with the *differentia* of A_1, the back and forth checking, now with A_1, its proposed *differentia*, and X.

If one supposes the completion of the process, what is obtained is an explanation-pattern, a pattern explaining why anything which is X must be f_1 or whatever. And, with sufficient sophistication, the pattern provides means of explaining why X *may* be characterized in some way or other (or, *pari passu*, why it is impossible for X to be characterized in some way or other).

Plato, of course, is interested in our being able to discern the patterning of his changeless and timeless Forms, not simply with the improvement of our conceptual life by revising and adjusting our concepts to fit some conceptual ordering. But, as we have noted, Plato does not invite us to some sort of non-visual staring at forms. He insists that our awareness of the structure of the forms is the product of διαλεκτική. He thinks that the practice of collection and division leads to discernment of the structure of Forms. And he thinks that exposing the structure of the language and so improving our conceptual life leads to improved sharing in the structure of name-forms and thus, in reflective use, awareness of what the Name-Forms are of.

Aristotle. In *Posterior Analytics* Aristotle's announced purpose is to state, explain, and defend an account of demonstration, where demonstration is syllogism whose premisses are apodeictic (or, at any rate, likely to be necessary). And Aristotle regards demonstrated results (as demonstrated) as constituting science (ἐπιστήμη). The search for premisses in such demonstration is, he believes, a search for middles. (Thus, science is 'middled' knowledge.) As noted earlier, the desirable middles are definitory formulae which assure that C must hold for A, given that B is a definitory formula for A, and C is contained in B.

Aristotle is a bit bedeviled by the problem of what assures that a string of words constituting the definitory formula is one and not merely a string (like the *Iliad*).[10] In *De interpretatione* he finds the ὄνομα-ῥῆμα linkage to constitute one λόγος, provided that neither of the items linked overtly or covertly proves to be compound. (If one of them should be, then there is more than one λογός linked by 'and'.) But what makes a definition or definitory formula one? Without going into the detail of the *Analytics*, I

[10] See Cohen 1981, 229–240 for an account of the unity of definition in the general terms assumed here.

think that he finds the ground for a formula's being one in the form, *genus-cum-differentia*; and he takes this to be as one-making as the ὄνομα-with-ῥῆμα formula for single sentences. Thus, definitory formulae can function as single terms constituting middles in demonstrative syllogisms.

Aristotle's procedure for arriving at such middles is, obviously, very close to Plato's procedure of collection and division; and I shall ignore the details of what he takes to be differences from Plato's procedure for the present purposes. As noted concerning *An. post.* ii 19, Aristotle assumes that we acquire our first universals as a result of sense experience and memory. He assumes as well that we are capable of ἐπαγωγή (assemblage), and so able to move from original sense universals to 'higher' species and genera universals. And, as noted, this movement parallels the increasingly remote applicability of, say, the concepts, man, animal, living thing, and so on to one and the same thing which he speaks of in *Categories*.

Physics i 1 speaks of original sense universals as blurring together a number of components (even principles: ἀρχαί) which are only later separated out, where 'later' seems to mean 'after appropriate inquiry'. I think it clear enough that the sense universals are those spoken of as 'clear and lucid to us' and those resulting from serious inquiry are or may be those 'clear and lucid in nature'. It should be noted, however, that Aristotle is, in practice, fairly modest in claiming any sort of certainty for the results of inquiry. He is not so modest, however, in his claims concerning either what scientific demonstration must be or how one should proceed in arriving at definitory middles.

The actual procedure of inquiry, utilizing ἐπαγωγή in working from sense universals, taking note of differences and trying explanatory patterns on for size, differs little from that described by Plato in *Philebus*. In the schematic example used in discussing Plato, what is needed is explanation of X's being F. For the explanation, Aristotle needs a definition of X which includes F, and to get it he may have to find a 'one' which like A in the example is rather removed from X (and thus requires a good deal of division). And, in the process of inquiry, there may well be a great deal of what I called 'back and forth' in discussing Plato. Thus, there will either be concept change or careful articulation of concept contents in the process, and much of it will be due to the 'empirical' business of attending to sensible things made known by the use of the tentative or permanent concepts.[11]

---

[11] I am indebted for the general spirit and some of the details of my account of Aristotle's procedure for arriving at definitions which can be used in scientific explanations to Bolton 1976.

4. Concluding comments

I hope that the above remarks, though all too brief for the complex matters involved, make it clear enough that both Plato and Aristotle have intelligent and intelligible procedures for explanation and that those procedures allow for testing proposed definitions and correcting mistaken or omitted steps in the genera/species trees. There remain three comments which I believe will help to clarify those procedures. The last two of those comments, though clarifying, introduce complex and difficult discussions that can only be hinted at in this paper.

First, I must note that philosophers using the apparatus of contemporary logic are prone to locating necessity in combinatorics. And they are equally prone to finding scientific explanation in syllogisms consisting of a universally quantified 'if/then' proposition (major premiss), an existentially quantified (or individual) proposition (minor premiss), and an existentially quantified (or individual) proposition (conclusion). In such explanatory patterns, terms for individuals have the form of proper names, and such individuals figure in explanations only as having some characteristic or other and not as being the individuals they are.

In the reading I have been giving Plato and Aristotle, individuals are not 'bare' but, if you please, 'sortally' qualified or 'natured', as, in the example cited from Aristotle, the man, Callias. Such individuals, by virtue of being what they are (i.e., gotten at by definitory formulae), are necessarily (καθ' αὑτό) qualified in various ways. Obviously individual characterization of this sort equally supports universal quantification. If, e.g., this triangle necessarily (καθ' αὑτό) has three sides, then all triangles necessarily have three sides. But there is in this no notion that universally quantified propositions of this sort can be falsified simply by finding that some 'bare' particular fails to be appropriately qualified. Even so, as I have tried to show above, there is an empirical element in the development of classificatory schemata; and neither Aristotle nor Plato is prepared to say that such schemata are not subject to improvement.

Second, I noted earlier that there is in the procedures of Plato and Aristotle some similarity to what the so-called rationalists speak of as analysis and synthesis, in particular, Leibniz. Analysis for Leibniz is the effort to clarify an 'idea', to be assured of its possibility, and to arrive at a definition. He distinguishes several sorts of definition: nominal, merely real, real, and causal. Real and causal definitions figure into the highest level explanatory patterns. They are arrived at by something akin to what I have called above the back and forth of testing possible definitions against their fit in genera/species trees, their providing clear-cut explanations, and their fit in a pattern which leaves out no steps. Synthesis is in Leibniz' practice the process of determining the place of a definition in an explanatory pattern. In Leibniz' view, both analysis and synthesis are involved in serious scientific investigation. With the additional dictum, *praedicatum inest subjecto*, Leibniz is prepared to carry the principle of καθ' αὑτό-predication far beyond the standard uses in Aristotle (though our finite intellects are incapable of determining the content of individual concepts). The use of definitions and definitory formulae in scientific explanation was widespread in the 17th and 18th centuries, even among philosophers who expressly rejected Aristotelianism.

Third, with the mention of Leibniz and analysis and synthesis, it is difficult not to say something about axioms and theorems. In the first chapter of *Physics* (to which I attended earlier), Aristotle links together 'principles (ἀρχαί), causes (αἰτίαι), and elements (στοιχεῖα)', with the suggestion that we really get to know something when we have followed through 'right up to the elements (στοιχεῖα)'. The notion of an element here is that of an indefinable which figures into the definition (and explanation) of the genera and species which make a subject-matter. Aristotle illustrates which he has in mind in *Physics* i by arriving at the principles (ἀρχαί) needed for the general science of physics, namely, some sort of continuing matter or substratum that remains through a change and the coming to be of a having or lacking.[12] But what I wish to focus on is the notion of elements (στοιχεῖα).

Euclid's *Elements* (same term) is a paradigm of the use of primitive terms in axioms, some definitions, and some common notions in demonstrating a set of theorems and corollaries for a given subject matter. Ignoring for the present purpose the problems (if they are such) of construction-proofs, I wish to point out that the use of στοιχεῖα by both Aristotle (in *Physics*) and Euclid is not adventitious. Aristotle's use is to call attention to the ultimate terms used in demonstrations of the sort this paper discusses, and Euclid's use is to provide those terms by means of which a coherent and successive set of demonstrations can be made. Though the procedures described in this paper do not have the form of demonstrations from axioms, the στοιχεῖα of a given subject-matter provide for the genera/species patterns which make possible the sorts of demonstrations I have been calling 'explanations'. Indeed, I am prepared to argue that the procedure of explanation defended in *An. post.* ii and the axiomatization procedure are formally equivalent. The procedure of arriving at definitory formulae for

---

12 For a defense (against Barrington Jones) of taking of matter as persisting through change, see Code 1976.

use as middles in demonstrations would, in the form suggested both by Aristotle and by Plato, be excessively cumbersome; but the relation between the 'elements' and the explanations or demonstrations is essentially the same. The argument to show their formal equivalence is, however, well beyond the scope of this paper.

5

On the Notion of a Mathematical Starting Point in Plato, Aristotle, and Euclid

IAN MUELLER

In this paper I wish to discuss the question of the status of the starting points of mathematics in the philosophies of Plato and Aristotle and in Euclid's *Elements*.[1] I will be mainly concerned with Aristotle since he has a good deal more to say on the question than Plato, and for Euclid we have only his practice to interpret. It is useful to have as a model for discussion some modern conception of mathematical starting points. I here present one briefly; it is designed to accommodate discussion of the ancients. We may begin with a division of starting points into terms, assertions, and rules; and a second division into logical and material. I give a rough illustration of each of the six resulting categories:

- primitive logical term: not ($\neg$)
- primitive material term: point
- primitive logical assertion

  For any assertion P, $\neg$ both P and $\neg P$ (law of non-contradiction)
- primitive material assertion

  If a and b are distinct points, there exists a third point c between a and b

[1] I use the term 'starting point' as a general term to cover a multitude of Greek expressions. I do not use the word 'principle', which scholars often use in the way I am using 'starting point', in order to avoid giving the impression that I am discussing the Greek word ἀρχή. In the *Elements* the starting points are the propositions Euclid labels definitions (ὅροι), postulates, and common notions. In the case of Plato I will be dealing primarily with the passage in the *Republic* in which Socrates talks about what he calls the hypotheses of the mathematicians. Aristotle uses a variety of terms in this connection, as will be seen in Section 2.

- primitive logical rule

  If the assumption of P leads to a contradiction, then you may take $\neg P$ to be true (rule of *reductio*); if you have proved P and if P then Q you may say you have proved Q (*modus ponens*)

- primitive material rule

  If you have proved A and B to be congruent to C, you may take A to be congruent to B.[2]

The distinction between logical and material starting points is usually made by specifying certain terms as logical, and making it a necessary condition for something to be a logical assertion or rule that it employ only logical terms. The distinction between logical and material has been argued to be arbitrary or at least unjustified; but these arguments are of no concern here. I shall simply assume that we have made a division about which we can all agree, and that the primitive logical terms include those of a standard formulation of the predicate calculus. I shall also assume that all theories use the same primitive logical terms, assertions, and rules, since to do otherwise needlessly complicates discussion. It will, however, be necessary to return briefly to the distinction between the logical and material in discussing Aristotle.

I shall also have nothing more to say about the notion of a primitive material rule until I discuss Euclid in the next section, since one of the simplifying features of standard specifications of the logical is to make material rules unnecessary. So, for example, standard specifications of the logical will allow one to infer that A is congruent to B from 'A is congruent to C', 'B is congruent to C', and the material assertion 'If A and B are congruent to C, then A is congruent to B'; hence, the example of a material rule just given can be replaced by this assertion as a starting point. We can, therefore, assume that material rules are absent from a typical theory.

I shall also assume that our primitive logical rules and assertions and our primitive material assertions are minimal in the sense that we could not prove the same set of assertions using a proper subset of the starting points.

[2] The qualifying adjective, 'primitive', indicates that what is qualified is a starting point.

In standard accounts the only material rules will be ones permitting the transformation of given assertions into other assertions on the basis of material facts, e.g., that congruence is a transitive relation. Primitive material assertions are themselves a rule of this kind, since they permit one to make assertions on the basis of no previous ones. In the *Elements* the first three postulates are material rules of a kind quite unlike any in standard modern theories, since these postulates license the construction of new objects rather than the introduction of new assertions.

And I shall assume as well that the set of primitive material terms is minimal, though saying precisely what that means is too complicated to merit the time it would take.[3] The rough idea is that the provable assertions of the theory do not include a 'definition' of any of the primitive material terms of the theory, i.e., an assertion enabling one to translate every assertion P of the theory into an assertion P' not containing a primitive term t and such that P is provable if and only if P' is. I make an analogous assumption for primitive logical terms, but only because it enables me to say the following briefly: I call a theory the starting points of which are minimal, minimal. Hereafter, unless I use the qualifier 'logical', terms should be understood to be material terms.

There are two further points I wish to make about the starting points of a theory. The first is that a standard theory may have no logical assertions, but it cannot get by without at least one logical rule. However, it is of some significance, at least historically, that there is a logical assertion corresponding to any rule, one which, it is tempting to say, is the assumption made by the person using the rule. One might think of the law of non-contradiction as an expression of the rule of *reductio*, and of the assertion 'If both P and if P then Q, then Q' as an expression of *modus ponens*. However, the more important philosophical point is that even if it is possible to replace any particular logical rule with a logical proposition, no reasoning can proceed without some rules. In general, switching back and forth between rules and their propositional expression obfuscates issues, and it seems best to imagine that the distinction between the two is fixed for any particular theory.

The second point concerns definitions. In modern discussions definitions are not treated as starting points. They are simply abbreviations of complex expressions introduced to make complex assertions more intelligible to us, e.g., enabling us to say '28 is perfect' rather than '28 is the sum of all its factors less than it, including 1'. The only terms which are starting points are the primitive ones. However, Aristotle seems to think of definitions as starting points, and they are the most common kind of starting point in the *Elements*. Perhaps the simplest way to accommodate this discrepancy is to add to the starting points of a minimal theory, a set of defined terms and a set of definitions, where for simplicity one assumes that the definition of a defined term contains only primitives. Since I want to use the word 'definition' in my discussion of Greek authors I shall call defined terms non-primitive and definitions abbreviations. Ignoring logical terms, we can

---

[3] Throughout this introductory discussion I pass over formal complexities involved in the treatment of definitions because taking them into account would not affect the issues I treat.

say that the starting points of a minimal theory with abbreviations include the following:

the material component—
- primitive terms
- non-primitive terms
- primitive material assertions
- abbreviations,

the logical component (identical for all theories)—
- primitive logical assertions (possibly empty)
- primitive logical rules.

1. Euclid's *Elements*

Euclid's *Elements* may be divided as follows:

a. books 1–4, plane geometry
b. book 5, proportion theory
c. book 6, plane geometry, presupposing proportion theory
d. books 7–9, number theory
e. book 10, plane geometry presupposing proportion theory and number theory
f. books 11–13, solid geometry presupposing plane geometry, proportion theory, and (*via* book 10) number theory.

I have formulated this description to stress the sense in which the *Elements* builds on previously developed theories, even though it is clear that Euclid develops theories much further then his subsequent applications of them require.[4] However, despite this building it is also the case that new theories are introduced in books 5, 7, and 11, that is to say, theories with previously unused primitive material terms and assertions. Obviously I make this point from a modern perspective; and I mean that if we were to represent the *Elements* as a formal theory corresponding as closely as possible to the original, we would be forced to introduce new primitives at those

[4] The only case of possible building which I have not included is Euclid's alleged use of proportion theory in number theory. I have omitted this because I believe Euclid conceived book 5 as a geometric proportion theory, introduced a theory of proportions for numbers in book 7, and then took for granted correlations between the two theories in book 10. However, nothing I say here is altered by supposing that Euclid's number theory presupposes the proportion theory of book 5. For this and other claims about the *Elements*, see Mueller 1981.

points. However, if we look at the *Elements*, although we find at the beginning of book 1 definitions, postulates, and common notions—postulates corresponding loosely to primitive material assertions, common notions to either primitive material assertions or primitive logical assertions—at the beginning of the remaining books we find only definitions. I believe there are two related inferences we can draw from this: (1) Euclid did not believe that proportion theory, number theory, or solid geometry required its own postulates; (2) at the end of the fourth century there were no accepted presentations of these theories which included postulates, and probably no such presentations at all, presumably because no mathematician recognized the need for them. A further inference I draw is that the idea of such presentations of any mathematical theory was relatively new in Euclid's time, i.e., did not precede Plato's maturity. I believe the evidence suggests that Euclid himself is responsible for the postulates, but for the moment I will only say that, even if they are thought to predate, say, Plato's *Republic*, they should still be seen as the exception rather than the rule by Euclid's time.

The rule in the *Elements* and, I am suggesting, earlier in the history of Greek mathematics is a theory, the only explicit starting points of which are definitions. These definitions are, for the most part, either explications, which perhaps clarify the significance of a term to the reader but play no formal role in subsequent argument, or abbreviations in the modern manner. Examples of the former are 'A point is that which has no part' and 'A unit is that in virtue of which each thing is called one'; an example of the latter is 'An obtuse angle is an angle greater than a right angle.' Occasionally an assertion creeps its way into a definition as when Euclid adds to the definition of the diameter of a circle that the diameter bisects the circle; but these exceptions may, I think, be disregarded as indications of what Euclid thought he was doing; and, in any case, the assertions which do appear in the *Elements* after book 1 come nowhere close to overcoming the absence of postulates. In Euclid's practice the terms which are explicated play something like the role of primitive terms in modern theories; but, except in his practice, Euclid shows no sense of a distinction between abbreviations, which play or could play a role in argument, and explications which do not and hardly could. Moreover, the comparison with primitive terms is very limited, since in a modern presentation one expects all and only primitive terms to occur among the material terms of the primitive assertions in their unabbreviated form; whereas in book 1 Euclid does not include (even implicitly) all explicated terms in the postulates and common notions, and he uses a lot more terms than anyone with some notion of modern axiomatic method could possibly hope to characterize satisfactorily in five propositions. The impression one gets from reading the whole

Elements is that the fundamental operative notion of a material starting point is a definition. One defines the things one is going to reason about in order to make sure others understand what one is talking about; some of these definitions are formally usable abbreviations; the others serve only to help the reader grasp what is being talked about.[5]

We know that the question of the appropriate postulates and common notions for book 1 was a matter of much discussion in later antiquity, and that the manuscripts of the *Elements* were affected by that discussion. We have no way of being certain what Euclid's lists included, but the most plausible course would seem to be to follow Heiberg and Proclus, and accept the following:

Postulates

1. Let it be postulated to draw a straight line from any point to any point, and
2. to produce a limited straight line in a straight line,
3. to describe a circle with any center and distance,
4. that all right angles are equal to each other,
5. that, if one straight line falling on two straight lines makes the interior angles in the same direction less than two right angles, the two straight lines, if produced *ad infinitum*, meet one another in that direction in which the angles less than two right angles are.

Common Notions

1. Things equal to the same thing are also equal to one another.
2. If equals are added to equals the wholes are equal.
3. If equals are subtracted from equals the remainders are equal.
4. Things which coincide with one another are equal to one another.
5. The whole is greater than the part.

The first thing I wish to point out is that the postulates include both assertions and rules. There corresponds to this division not only the distinction between theorems and problems, the latter being what we would call constructions, but also the distinction between the reasoning part of a proof (ἀπόδειξις in Proclus' terminology) and the construction (κατασκευή) which precedes it. Euclid's geometric reasoning is highly constructional in this way, and I see no reason to doubt that Greek geometry always was. However, even when this aspect of geometric reasoning is recognized,

[5] For detailed discussion of some of the material in this paragraph, see von Fritz 1971, 393–414.

there is a tendency to focus on assertions and proofs rather than rules and constructions and, in particular, to speak of geometry as a matter of proving assertions from assumed assertions. This tendency may represent a philosophical bias, but, at least since Aristotle, accounts of reasoning have standardly focused on procedures by which assertions are transformed into other assertions and not on procedures by which constructions are built out of other constructions. The fact that Aristotle does not recognize primitive constructions as starting points of geometry suggests, although it hardly proves, that they did not occur among the presentations of geometry accessible to him. Since it also seems likely that the assertional postulates 4 and 5 are no earlier than the other three, there is some reason to think that nothing like Euclid's postulates was known to Aristotle [cf. Heath 1956, i 202].[6] This is a point to which I return at the end of my discussion of Aristotle.

The common notions appear to be assertions relating to quantitative reasoning. Each could be transformed into a rule for such reasoning, common notion [5], for example, allowing one to go from 'a is part of b' to 'b is greater than a'. Euclid's formulation of them as assertions is perhaps another reflection of the tendency to think of rules as founded on assertions. In any case, Euclid's list is quite inadequate to the quantitative reasoning he actually applies; and it is sufficiently inadequate to make me believe that Euclid had no desire to formulate a complete list, but settled for the most prominent principles he employs. A perhaps more interesting question is whether Euclid thought of the common notions as logical or material. From the point of view of standard predicate logic there is no question that the common notions are material; but I know of no fully satisfactory reason for denying argument about equality, addition, subtraction, coincidence, parts, and wholes, the status of logical reasoning. Here, as in the case of set theory, the division between logical and non-logical may be arbitrary. However, in the case of Euclid the issue may be refined by asking whether Euclid has a notion of rules of reasoning corresponding to our predicate calculus (or Aristotle's syllogistic). To be more specific: we know that Euclid follows such rules, and we know that he did not try to formulate the rules. Should we say that the absence of such an attempt reflects a lack of self-consciousness about these rules and, hence, an at least tacit belief that quantitative principles are the closest one comes to logical principles? Or should we suppose that Euclid acknowledged the use of logical principles,

[6] The fourth postulate is much the most difficult to explain. Heath [1956, i 201] argues for an association with the fifth, but see Mueller 1981, 29–30.

but did not consider them to be his concern? The text does not allow us to decide between these alternatives, and to that extent supports the first.[7]

I conclude this section with a summary. The explicit starting points in the *Elements* are definitions, postulates, and common notions. Of these the definitions either correspond to abbreviations or they are what I have called explications. In thinking about starting points in the *Elements* and, hence, in thinking about them in Greek mathematics, we ought to think primarily about these definitions, even though explications play no official role in modern theories and abbreviations are starting points only by courtesy. In book 1 Euclid adds to the first definitions for plane geometry the postulates and common notions. The postulates correspond to primitive material rules and assertions. The common notions are general truths about quantities almost certainly intended to apply to numbers as well as geometricals. These assertions could be turned into rules without altering the character of the *Elements.* The questions whether they are logical or material starting points and whether they are the most general reasoning principles recognized by Euclid does not admit a clear answer. Certainly Euclid uses general logical principles, just as he uses primitive material and quantitative rules and assertions he has not made explicit. But using such principles does not constitute recognizing them. The following, then, is my list of acknowledged starting points in the *Elements*: explications, abbreviations, material rules, material assertions, quantitative assertions.

2. Aristotle

Aristotle's notion of mathematical starting points has been much discussed by historians of mathematics and historians of philosophy. In general the main passages which have to be looked at are well known, but no consensus on an overall reading of them seems to have emerged. In this section of my paper I will go through the passages in an order which facilitates what I think is their correct interpretation. For I believe that it is possible to find a relatively coherent and uniform view of mathematical starting points in Aristotle, and that standard accounts of the relationship between this view and Greek mathematical practice are not justified. Unfortunately, the content of the relevant passages overlaps and diverges in ways which necessitate discussing a variety of topics partially until, if all goes well, a total picture emerges.

[7] That is to say, the absence of a distinction in an author is *prima facie* (but only *prima facie*) evidence that the author did not make it.

When Aristotle sets out in *An. post.* i 7 to establish the impossibility of showing something by applying a proof in one genus to another, he announces that there are three things involved in proofs:

> One is what is proved, the conclusion (this is a matter of belonging to some genus *per se*), one is the axioms (axioms are from which), and third is the subject genus, the properties and *per se* attributes of which are made clear by the proof.[8] [*An. post.* 75a39–b2]

I will refer to this triad as the elements of a deductive science, and I will try to render plausible the view that these elements also represent Aristotle's basic conception of the starting points of a science. Aristotle offers other versions of the triad in other places.[9] For example, in *An. post.* i 10 he writes,

> Every demonstrative science concerns three things: the things it hypothesizes to be (these things constitute the genus of which it studies the *per se* properties), the so-called common axioms from which first things it proves, and third the attributes of which it assumes what each signifies. [*An. post.* 76b11–16]

---

[8] My translations are not always literal. They are designed to facilitate my argument, but only by taking for granted what I think are relatively non-controversial interpretations.

[9] In addition to the passages quoted in the text, the following one is generally thought to express the same doctrine.

> Every demonstrative science investigates concerning some subject the *per se* attributes from the common opinions. Therefore, it belongs to the same science to investigate concerning the same genus the *per se* attributes from the same opinions. For that concerning which belongs to one science, that from which to one, whether it is the same or a different one, so that also either they investigate the attributes or one science composed of them does. [*Meta.* 997a19–25]

It seems clear that the common opinions in this last passage are the same as the common axioms: cf. Lee 1935, 113–114. But Aristotle's considered view seems to be that although there is a single genus belonging to each science, all sciences share the common axioms, which are themselves the domain of no single demonstrative science. Hence, Aristotle's view on the point raised in the last sentence quoted would be something like that one science investigates one genus from the common axioms, so that this science must necessarily investigate the attributes of that genus. Here he wants to leave open the possibility that there is a science of the common axioms (an issue raised in the preceding ἀπορία); but it is surprising that he goes so far in the direction of openness as to omit his own view, leaving only the possibility that two sciences or a composite science investigates the attributes.

After a brief excursus in which he points out that sometimes one or the other of these things is not explicitly hypothesized, he insists that by nature there are three things—the thing about which one shows, the things one shows, and the things from which one shows (περὶ ὅ τε δείκνυσι καὶ ἃ δείκνυσι καὶ ἐξ ὧν).[10] In *Meta.* B 2 Aristotle writes,

> If there is a demonstrative science of them, there will have to be some subject genus, and some of the principles will have to be properties, some axioms... ; for it is necessary for proof to be from some things, about some thing, and of some things. [*Meta.* 997a5–9]

On the basis of these and other passages it seems to me reasonable to say that for Aristotle the elements of a demonstrative science are the common axioms, the subject genus, and the properties associated with the genus. But there are several things to notice about Aristotle's characterization. First, the axioms are thought of as the premisses of scientific proof, the things from which one proves;[11] but the genus and the properties are apparently not thought of in this way. Secondly, Aristotle can speak of the genus in the singular or the plural, but presumably when he speaks of hypothesizing the existence of the genus he means hypothesizing the existence of things in the genus. However, it is important to see that even if Aristotle has in mind the hypothesis that, say, number or numbers exist, he does not seem to think of the hypothesis as a premiss of mathematical argument. In this sense, it does not matter much whether one speaks of hypothesizing the genus or hypothesizing its existence. Similarly, it seems to make no difference to Aristotle whether one speaks of the third element in demonstrative science as the conclusions or the properties shown in the conclusions to hold of subjects in the genus, but the latter formulation is somewhat more typical. Finally Aristotle speaks of assuming what the properties signify, and although this almost certainly relates to definitions, again the definitions are apparently not seen as premisses of argument. The picture one gets of a science then is that it proves properties of subjects in a genus from the common axioms. To do so it must take for granted the existence of the subjects and the significations of the properties. There

[10] See also *An. post.* 77a27–29: 'I call common the things which all sciences use in the sense of proving from them, but not that about which they show things or the thing they show.'

[11] I emphasize that for Aristotle the common principles are assertions (προτάσεις), things from which one proves, premisses, and not rules. Ross [1949, 531] places great stress on two passages, *An. post.* 76b9–11 and 88b1–3, in which Aristotle speaks of proving through (διά) common things; but I agree with Barnes [1975a, 135] that no great significance should be attached to this preposition.

are many problems in this picture, including questions of its consistency with other things Aristotle says. I shall attempt to address these problems only after I have attempted to clarify the picture.

a. *The common axioms*

> Since the mathematician also uses the common things but restricted to his own science, it also belongs to first philosophy to investigate the principles of mathematics. For that when equals are taken from equals the results are equal, is common to all quantities, but mathematics studies a certain part of the domain of the axiom in isolation, e.g., it studies lines or angles or numbers or some of the other quantities... [*Meta.* 1061b17–24]

Here Aristotle explicitly mentions as an axiom or common principle of mathematics what we know as common notion 3 of the *Elements.* However, in the parallel passage at the beginning of *Meta.* Γ 3 he only refers to 'what are called axioms in mathematics', without giving any examples, and he stresses the idea that these axioms are true of all things and are used by all reasoning. This characterization is obviously more appropriate to the context in which Aristotle is concerned with versions of the fundamental logical laws which we call non-contradiction and excluded middle, a point which is brought out clearly in the two statements of the ἀπορία which is being addressed in both passages:

> Whether it belongs to the science [first philosophy] to consider only the first principles of substance or whether it also deals with the principles from which everyone proves, e.g., whether or not it is possible simultaneously to assert and deny one and the same thing, and the other things of this kind [*Meta.* 995b6–10]

and

> It is an open question whether it belongs to one science or several to deal with the principles of proof. By principles of proof I mean the common opinions from which everyone shows things, e.g., that it is necessary to affirm or deny each thing, and that it is impossible for something simultaneously to be and not be, and all other such assertions. [*Meta.* 996b26–31]

One sees from these passages that Aristotle includes among the common principles of the special sciences clear instances of what we would call logical assertions and the common notions of Euclid's *Elements*, which I have called quantitative assertions to avoid having to settle the issue of whether they are logical or material. It is possible that Aristotle acknowledges some distinction between these two kinds of common assertion, since he rarely mentions both kinds together.[12] However, he talks about the two kinds in essentially the same way, so that the one passage [*An. post.* 77a26–31] in which he does mention the two together as common things can be taken as quite decisive evidence that Aristotle does not distinguish them.[13]

b. Problematic passages in the Posterior Analytics. In his translation of and commentary on the *Posterior Analytics*, Barnes [1975a, 136] provides a list of 'various classifications of the elements of demonstrative science' given by Aristotle, and says that 'Aristotle himself makes no attempt to coordinate them.' I believe that one can make reasonably good sense of all of these classifications in terms of the triad genus, properties, common axioms, and in this section I attempt to do so.[14]

At the start of *An. post.* i 10 Aristotle says,

> I call the principles in each genus the things which cannot be shown to exist. Thus, what the first things [i.e., principles] and the things composed of them are is assumed; on the other hand it is necessary to assume that the principles exist, but to show that other things

[12] In *An. post.* i 10 76a41, 76b20–21 Aristotle mentions 'equals from equals' as a common thing, but never gives a logical example of an axiom in that chapter. At 88a36–b1, he cites the law of the excluded middle as a common principle.

[13] Theophrastus apparently did distinguish them. For Themistius [*In an. post.* 7.3–5] tells us that he defined axioms as certain opinions, some concerning things of the same category, e.g., 'equals from equals', some concerning absolutely everything, e.g., the law of the excluded middle.

[14] I do not discuss the last two paragraphs of *An. post.* i 10 (76b23–77a4). In the first of these Aristotle distinguishes propositions which must be believed, hypotheses, provable propositions which the teacher assumes without proof and the student accepts, and postulates which the teacher assumes and the student does not accept. This categorization does seem to me quite independent of all the others. (See on this passage von Fritz 1971, 365–366. I note that there are no other passages listed under αἴτημα in Bonitz' index which help to clarify the possible logical or scientific sense Aristotle attaches to the word 'postulate', although chapter 20 of the *Rhetorica ad Alexandrum* discusses rhetorical postulates.)

In the other paragraph Aristotle distinguishes ὅροι from hypotheses. I take hypotheses to be premisses in general (ὅσων ὄντων τῷ εκεῖνα εἶναι γίγνεται τὸ συμπέρασμα), but I am not certain whether a ὅρος is a definition or a term. (I am not inclined to think it is a *definiens*, although this cannot be ruled out. See, e.g., Mignucci 1975, *ad loc.* or Landor 1981.) Since the point Aristotle is making is that a ὅρος is not an assertion, taking ὅροι to be definitions would lend support to my overall interpretation; but I do not see any way to rule out the possibility that ὅροι here are terms and Aristotle's point a relatively trivial one.

> [i.e., the things composed of the first things] exist. For example we assume what monad or straight and triangle signify; and we assume that monad and magnitude exist, but we show that the others exist. [*An. post.* 76a31–36][15]

Aristotle here contrasts the principles in a genus with the things composed of them, giving monad and magnitude as examples of principles, straight and triangle as examples of composites. The contrast between monad and magnitude represents the contrast between arithmetic and geometry, and it seems reasonable to assume that they represent the genera of the two sciences. Other passages suggest that magnitude is a stand-in for point, line, surface, and solid, although Aristotle usually uses as illustrations only the first one or two of these rather than all four. Similarly he sometimes speaks of number rather than monad as the genus of arithmetic, as he does in the passage from the same chapter quoted above. The fact that Aristotle here speaks of hypothesizing both that these principles exist and what they signify should be seen as an amplification of passages in which Aristotle only mentions the first kind of hypothesis in connection with the genus. For it seems obvious that one must know what the genus signifies as well as that it exists if one is to prove things about it; however, for Aristotle the hypothesis of existence is associated uniquely with the genus and, hence, is the most interesting one to mention in connection with the genus.

Triangle and straight are both geometric items, and I believe they should be placed in the class of what I have been calling properties. It does not seem to me to count heavily against this assumption either that Aristotle calls these things composites or that he speaks of proving their existence. For in the next passage in chapter 10 he speaks of proving the existence of the properties. In the passage Aristotle makes explicit another distinction he sometimes invokes in discussions of mathematical starting points, the distinction between two kinds of things used in demonstrative sciences, common ones, such as 'equals from equals', and special ones, the examples of which include both definitions ('A line is such and such, and the straight is such and such' [*An. post.* 76a40]) and

> the things which are assumed to exist and concerning which the science investigates the *per se* features, e.g., monads in arithmetic, points and lines in geometry. For it is assumed that these things

[15] This passage should be compared with the more obscure *An. post.* 87a38–40 for which Barnes' notes [1975a] are very helpful:

> The science of one genus, i.e., things which are composed from the primary things and are parts or *per se pathē* of these things, is one science.

> exist and that they are such and such. But what each of the *per se* properties of these things signifies is assumed, e.g., in the case of arithmetic, what odd and even and square and cube signify, and, in the case of geometry, what irrational and being broken and verging signify; but that these things exist is proved through the common things and from what has already been proved. [*An. post.* 76b3–11]

It is difficult to see that any sense can be attached to the notion of proving that these properties exist other than proving that they apply to their subjects [cf. Ferejohn 1982–1983, 394–395].[16] Thus, if the composites mentioned at the beginning of chapter 10 are to be identified with these properties, proving their existence will also be proving that they apply to their subjects. Aristotle speaks of these properties as composites because he is thinking of them as defined in terms of the simple objects of the subject genus. In this sense the properties are not principles or starting points since they depend for their definition on other things, just as they exist only as belonging to other things.[17] But they or their definitions are starting points in another way since they are taken for granted by the mathematician in his argumentation. What Aristotle has in mind by assuming existence is brought out in the next passage in chapter 10, the beginning of which was quoted early in this paper. In the part not quoted Aristotle mentions cases in which a science does not assume explicitly one of the three elements he has identified. He contrasts the necessary assumption of the existence of number, the genus of arithmetic, and assuming the existence of hot and cold, which it is not necessary to do because existence in this case is obvious. It seems to me relatively clear that the notion of existence involved here

[16] Such a proof might be a construction; for example, Euclid's construction of an equilateral triangle in book 1 prop. 1 might constitute a proof of existence for Aristotle; but there is no textual reason for denying that Aristotle would think of the proof of the incommensurability of the side and diagonal of a square as a proof of the existence of incommensurability. I know of no good evidence for the frequently repeated suggestion that for Aristotle existence in mathematics was somehow connected with constructibility. Cf. Barnes 1975a, 92.

[17] Cf. *Meta.* 1077b3–4 (where Aristotle says that a is prior in definition to b if the definition of b is composed out of the definition of a), 1035b4–14. Elsewhere Aristotle illustrates this priority in terms of point and line, and of line and triangle:

> A belongs to B *per se* if A is in the definition of B; for example, line belongs *per se* to triangle and point to line since the substance of triangle and line are composed from line and point, which are present in the formula which says what they are. [*An. post.* 73a34–37]

In *Top.* 108b26–31 Aristotle mentions definers who treat the point as principle of the line and the monad as principle of number.

is neither technical nor profound. The 'assumption' that hot and cold exist is something we all make in our everyday conversation about the weather. When the arithmetician hypothesizes that there are numbers or units, he is only insisting that he is talking about something and asking that philosophical or ontological questions be put aside, just as, for the most part, such questions can be put aside when one talks about heat or cold. Only a philosophically trained (or mistrained) person would ask why he should believe there are such things as hot and cold. In the *Republic* Socrates indicates the kind of question which might be asked of the arithmetician:

> What kind of number are you talking about in which the one is such as you demand, each equal to every other and not differing in the least and having in itself no part? [Plato, *Resp.* 526a2–4]

Socrates goes on to suggest that the arithmetician's response will bring out the intelligible non-sensible character of his objects, but Aristotle's position seems to be that the arithmetician will simply insist that he be granted that there are such things so he can proceed. My suggestion, then, is that when Aristotle speaks of hypothesizing the existence of the subject genus of a science, he has in mind a broad sense of existence precision about which serves no scientific purpose. The idea that a single science deals with a single genus is very important in Aristotle's doctrine that there cannot be a single universal science. But where did Aristotle get this idea? He does not offer any real argument for it. And it seems quite independent of syllogistic, which is a purely formal theory, although the idea of proving properties of a subject is undoubtedly related to the subject-predicate conception of an assertion (πρότασις) underlying Aristotle's syllogistic. Nor, I think, should the idea be connected with the doctrine of categories or highest genera of being. For although that doctrine might be taken to exclude the possibility of a transcategorial science, Aristotle's favorite example to illustrate the 'one science/one genus' doctrine is arithmetic and geometry, both of which presumably deal with species of the category quantity. In fact it seems likely that Aristotle takes restriction to a single genus as an observed fact about actual sciences, and that in this connection he uses the word 'genus' rather informally. All he is saying is that every demonstrative science deals with only one kind of thing.

In *An. post.* i 1 Aristotle describes the kinds of prior knowledge presupposed by learning:

> In the case of some things it is necessary to assume in advance that they exist, and in the case of others it is necessary to apprehend what the thing said is; and in still others both are required. For example, one must assume that the law of the excluded middle exists, what

> triangle signifies, and for the monad, both what it signifies and that it exists. For each of these is not equally obvious to us. [*An. post.* 71a12–17]

It is clear that we have to interpret the assertion that 'the law of the excluded middle exists' as the assertion that the law is true, but it should also be clear that the need for this interpretation in this passage does not by itself warrant interpreting 'exist' (εἶναι) as 'is true' in cases where the text does not require it. For in this passage Aristotle is trying to illustrate the trichotomy he applies in chapter 10 to the special starting points only (ὅτι ἔστι· τί σημαίνει· τί σημαίνει καὶ ὅτι ἔστι) in terms of the trichotomy of common axioms, properties, and subject genus. The result is perfectly defensible, but misleading in so far as it blurs distinctions made clearly elsewhere.

In *An. post.* 72a7 Aristotle defines a principle as an immediate πρότασις, and goes on to describe προτάσεις as assertions and denials.[18] He then says,

> I call an immediate syllogistic principle which cannot be shown and which it is not necessary for a person to have to learn something a θέσις. But an axiom is something which a person must have if he or she is to learn anything whatsoever; for there are some things of this kind, and it is our custom to apply the term 'axiom' to them especially. One kind of θέσις is a hypothesis; it assumes one half of a contradiction, e.g., I mean, that something exists or does not exist; another kind, without this, is a definition. For a definition is a θέσις, since the arithmetician lays down that the monad is indivisible in quantity. But a definition is not a hypothesis; for what a monad is and that a monad exists is not the same. [*An. post.* 72a14–24]

All commentators point out that Aristotle rarely, if ever, uses the words θέσις and 'hypothesis' in the way explained here; and normally he does not refer to the learning situation in explaining axioms. But there seems to me no reason to doubt the text: the things which a person must acknowledge to be able to learn anything are the axioms, principles presupposed in all scientific argument,[19] and θέσεις are the special principles for individual sciences. 'Hypothesis' appears to have the more general sense of assumption

[18] The fact that Aristotle's formulation very early in the *Posterior Analytics* is thoroughly propositional has had great influence on accounts of his doctrine of the starting points. For one attempt to minimize this passage, see Ferejohn 1982–1983, 382–383, n16.

[19] Cf. *Meta.* 1005b5–23, where Aristotle describes the law of non-contradiction as something which one must know if one is to know anything.

at the beginning of this passage, but the contrast between hypotheses and definitions depends on treating them as existential assumptions. The transition from the general to the specific sense proceeds by the opaque phrase 'e.g., I mean, that something exists or does not exist', which is sometimes interpreted as 'i.e., I mean, that something is or is not the case'. This interpretation has the advantage of giving a clearer sense to the negative alternative, but makes Aristotle's opposition of definitions and hypotheses an apparent equivocation. It seems to me preferable to say that the negative alternative is included because of the general sense of 'hypothesis' introduced here, but that the concrete examples Aristotle has in mind are affirmations of the existence of the genus of a science.

I conclude that Aristotle's doctrine of the starting points of demonstrative science involves the division of starting points into common and special ones. The common ones or axioms include both quantitative and logical assertions, but Aristotle probably does not distinguish the two. The special ones are the subject genus and the properties of the genus. Aristotle frequently speaks of hypothesizing the existence of the genus or its members, and refers to definitions of the properties as well as of the genus and its members.[20] In this sense the special starting points can be thought of as propositional, but it is important to bear in mind that Aristotle thinks of the axioms as the only premisses used in demonstration. The hypothesis of the existence of the genus is the assumption that one is talking about something real in a science, and the definitions are simply determinations of the genus and the properties one is going to discuss. However, to say that the axioms are the only premisses of a science is not to say, at least for Aristotle, that all the theorems could be derived from them. For the axioms are too general to permit the derivation of specific truths. One needs to particularize them by bringing in a genus and its properties.[21]

[20] There is a close correlation between the elements of Aristotle's subject genus and the things whose definitions in the *Elements* I have called explications, and also between his properties and those whose definitions I have called abbreviations. However, I am not sure that Aristotle noticed this difference. And I certainly agree with Barnes [1975a, 134] that he did not distinguish primitive and defined terms.

[21] See *An. post.* 88a36–b3, where Aristotle, in arguing that 'it is impossible for all syllogisms to have the same principles', considers the possibility that some of the common principles (of which he gives excluded middle as an example) might play the role of universal principles. Aristotle does not say that these principles are insufficient but only that the genera are different and that one proves with these genera through the common things. Shortly thereafter, in a very difficult passage, Aristotle considers the possibility that the primary immediate propositions are the principles and makes the curious remark that there is one for each genus, by which he perhaps means the definition of the genus [so Ross 1949, *ad loc.*].

That is to say, even if one could prove all the premisses of a science (the axioms) in a higher science, one would not thereby be able to prove all the theorems in the higher science.

c. The mathematics known to Aristotle. Aristotle sometimes mentions the common axioms in ways which would seem to insure that he is talking about a feature of mathematics known to his audience,[22] and it seems safe to assume that mathematical texts known to him included 'equals from equals' and presumably at least the first three of Euclid's common notions. On the other hand, the absence of fundamental logical laws from Euclid's list of common notions suggests to me that Aristotle's inclusion of them among the axioms is a reflection of philosophical discussion in the Academy concerning the general principles of reasoning, discussion having no direct impact on Euclid. Philosophical discussion in the Academy and the alleged Platonic 'reform of mathematics' may also underlie the inclusion of the common notions in mathematical texts, but I can think of no considerations which weigh particularly heavily for or against this suggestion.

From a modern point of view the idea that the common axioms might be the only premisses of, say, geometric proof is incredible. There are at least two factors which may help explain why Aristotle adopted it.[23] The

---

Finally, Aristotle takes up the suggestion that although different principles are used in different proofs, they are all of a piece (συγγένειος). Aristotle responds by reasserting his doctrine that 'the principles of things differing in genus are different in genus'. 'For', he says, 'the principles are twofold, those from which and those concerning which; the former are common, the latter, e.g., number and magnitude, special.' Thus, even here, in the context of a discussion which relies heavily on the doctrine of the categorical syllogism, Aristotle on the whole treats the common principles as the only premisses and makes their non-universality turn on the fact that they are specialized through restriction to a genus.

22 See especially *Meta.* 1005a20, where Aristotle mentions 'the things called axioms in mathematics'. von Fritz [1971, 421–422] argues with considerable plausibility that Aristotle's analysis of the axioms as common as opposed to special starting points is Aristotle's own contribution and not a reflection of the mathematicians' understanding of their own practice.

23 The fact that the common notions could not be employed reasonably in an Aristotelian syllogism does not strike me as particularly problematic in this connection, since Aristotle does not attend to issues of formalization in a rigorous way: see Mueller 1974, 48–55. The same relaxed standards are evident in his discussion of the way in which the laws of non-contradiction and excluded middle are used in demonstrations: cf. *An. post.* 77a10–25. He says that the latter is assumed in every *reductio* argument, but that the former is only used when the conclusion is in the form P & $\neg\neg P$. This second claim is bizarre not only because it is hard to envision a scientist trying to prove such a proposition, but also because non-contradiction is used in any *reductio*. To argue for the claim

first is drawn from the *Elements*. As I have already mentioned, what we would call a Euclidean geometric proof is customarily divided into two parts called by Proclus the κατασκευή (construction) and the ἀπόδειξις (proof). Roughly, the κατασκευή depends on the postulates and previously established constructions, whereas the ἀπόδειξις depends on the common notions and previously proved theorems. Thus, there is at least the possibility of thinking that the only ultimate assumptions used in geometric proofs (i.e., the 'real' proofs, the ἀποδείξεις) are the common notions. The supposition that Aristotle did think of proof this way would be strengthened if it could be rendered plausible that the geometry texts known to Aristotle included no postulates among their starting points. For if they contained only definitions and common notions, then Aristotle would have at least empirical grounds for thinking of the common axioms as the only substantive assumptions made by the mathematician. The argument from Aristotle's silence with regard to the Euclidean postulates seems to me quite strong in this case [cf. Heath 1921, i 336],[24] but a number of scholars[25] have argued that at least Euclid's first three postulates correspond to Aristotelian existence assumptions. I wish to argue briefly that the correspondences are at best very tenuous and probably non-existent.

I have argued elsewhere [Mueller 1981] that the first three postulates are not existence assertions at all, but licenses to carry out certain constructions. This position is, of course, quite compatible with the fact that they play a role analogous to the existence assumptions in modern formulations of geometry, as well as with the possibility that Aristotle thought of the postulates as existence assertions. However, Aristotle's description of scientific existence hypotheses corresponds neither to Euclid's postulates nor their modern analogues, but to the modern logical notion of a theory

Aristotle invokes features of the categorical syllogism, and it is true that any provable categorical proposition is provable without using non-contradiction.

24 However, Heath [1949, 56] maintains that Euclid's first three postulates 'are equivalent to existence assumptions and therefore correspond to Aristotle's "hypotheses"'. I discuss briefly what seems to me the strongest evidence for a pre-Euclidean formulation of the postulates in two appendices.

25 Most notably Lee 1935, 115–117. It is difficult to characterize von Fritz' position on the question of the correlation between Aristotelian existence assumptions and Euclid's construction postulates. He seems to concede all the difficulties but nevertheless insist on the correlation.

presupposing a domain or having an intended interpretation.[26] Euclid's postulates allow one to move from given objects of a certain kind (two points, a straight line, a point and a 'distance') to others (a straight line, a longer straight line, a circle). None of the objects constructed using Euclid's postulates is mentioned by Aristotle as an element of the genus; indeed, straight is mentioned as something whose signification we assume but whose existence we prove, and it seems to me reasonable to suppose that circle would fall into the same category, as triangle does. Moreover, Aristotle thinks there are existence hypotheses in arithmetic, but there is no trace of postulates of any kind ever being used in ancient number theory [cf. Kullmann 1981, 248–249].

For these reasons the attempt to correlate Aristotle's existence assumptions with Euclid's constructional postulates seems to me quite implausible. Moreover, the intepretation I have offered of these assumptions seems to me to correlate well with what Aristotle says about them and to cohere with Aristotle's general conception of reasoning and scientific knowledge. We cannot know whether or not the geometry of Aristotle's time included postulates in Euclid's manner; but Aristotle, who provides us with our best evidence for fourth-century mathematics, gives us no grounds for thinking it did.

Prima facie it would seem highly likely that the texts in mathematics known to Aristotle included definitions of the kind familiar to us from Euclid. Moreover, some passages in Aristotle suggest that definitions are the only starting points of a science. For example, at *An. post.* 90b24 (an aporematic passage) he calls definitions the principles of proofs, and at 99a22–23 he says that 'all sciences come about through definitions'.[27] Aristotle's recognition that these definitions do not or should not involve any existential implications, and that the source of these implications must lie elsewhere is a tribute to his powers of analysis. On the other hand, it is somewhat curious that he downplays the use of definitions as premisses. For in Euclid and in mathematical reasoning generally they do have this role; and Aristotle himself treats definitions as premisses in book 2 of the *Posterior Analytics*. However, the discussion of definitions in book 2 is notoriously problematic in itself and in relation to the account of starting

[26] I am not sure what von Fritz means when he says [1971, 393] that Greek mathematicians did not find it necessary to formulate the Aristotelian existential starting points explicitly. If he means that they were aware of these assumptions but did not formulate them, he is indulging in pure speculation. And if he means that their mathematics commits them to these assumptions, he is making a philosophical rather than a historical claim.

[27] Barnes [1975a, 109] lists other passages which he thinks express the same view.

points in book 1. Here I wish only to make a few suggestions which may help to clarify Aristotle's conception of starting points.

Aristotle's specific words for definition are ὁρισμός and ὅρος, but he frequently refers to definitions by using the expressions 'what it is' (τί ἐστι) and 'what it signifies' (τί σημαίνει). In book 2 Aristotle consistently uses the former expression until chapter 6 in which he raises objections to the view that one might be able to prove what something is. His second objection goes as follows:

> How is it possible to show what something is? For it is necessary that a person who knows what a human or anything else is also know that humans exists; for no one knows what something is if it does not exist. But when I say 'unicorn' I may know what the expression or name signifies, but it is impossible to know what a unicorn is. [*An. post.* 92b4–8]

The text and interpretation of the next objection is disputed, but I need only a small and relatively clear part of it:

> Therefore, there will be a proof that something exists, which is what sciences now provide. The geometer assumes what triangle signifies and shows that it exists. [*An. post.* 92b14–16]

The distinction Aristotle makes in the first of these passages is normally expressed as the difference between a real and a nominal definition. Apprehension of a real definition involves an apprehension of the existence of its subject, whereas a nominal definition only relates to words and, hence, bears no existential import. The second passage suggests that, at least in the case of properties, the mathematician uses nominal definitions. I believe that this is Aristotle's conception of all mathematical definitions, although this claim cannot be proved by arguing that Aristotle always uses the expresssion 'what something signifies' in connection with mathematical definitions. He does not, but it is striking how frequently he does. For example in *An. post.* i 1–10, there are seven occurrences of expressions related to 'what something signifies' in the vicinity of references to mathematics [71a14–16, 76a32–36, 76b6–12, 76b15–21][28] and only two related to 'what something is' [72a23; i 10]; moreover, both of these occur in conjunction with expressions related to 'that something exists', so that speaking of what something is produces a more elegant coupling.

---

[28] The phrase 'what the thing said is' (τί τὸ λεγόμενόν ἐστι) of *An. post.* 71a13 seems to me more likely to fall on the side of 'what something signifies' than on that of 'what something is'.

The suggestion I wish to make is that in the early chapters of *An. post.* i and in other discussions which clearly focus on mathematics, Aristotle thinks of definitions as nominal, and that the distinctive doctrines concerning definition in book 2 relate to real definition [cf. Gómez-Lobo 1981, Leszl 1980]. This assumption would explain why Aristotle speaks in book 1 of assuming or proving the existence of things the signification of which has been determined, but in book 2 he holds that knowing what something is entails knowing that it exists. It might also help to explain why Aristotle does not treat definitions as premisses in his descriptions of the three elements of deductive science, but does treat them as premisses in book 2. The idea would be that real definitions can play this role, but nominal ones cannot. Moreover, there seems to be some plausibility in the idea that nominal definitions are not assertions at all, e.g., that they are not really capable of truth and falsehood, and so could not play the role of genuine premisses in a science. In a sense the only truths assumed in mathematics are the common axioms since mathematical objects do not really exist and the definitions are purely nominal.

If my interpretation of Aristotle is correct, then the only real common ground between Aristotle's theory and Euclid's practice is the common notions. There is also a kind of commonality in the case of definitions, but we have no way of knowing whether Euclid understood definitions in the way Aristotle did. He may have, and he may also have believed that the different sciences in the *Elements* treat different genera assumed to exist. Neither belief is reflected in the way Euclid presents his starting points; he never asserts the existence of a genus, and he presents his definitions as if they were premisses of his arguments, or, at least, he uses them in that way.

To conclude my discussion of Aristotle I want to address what I take to be the most problematic aspect of my interpretation, my attempt to deny that for him the special starting points function as ultimate premisses of scientific proof. Clearly much of Aristotle's discussion of science is built around the idea of chains of deductive argument starting from unproved or immediate premisses frequently called principles (ἀρχαί). And, as I have mentioned, his discussion of definitions in *An. post.* ii does treat them as premisses in scientific arguments. The difficulties involved in harmonizing all major points made or apparently made by Aristotle in the *Posterior Analytics* are well known, but I doubt that it is worthwhile to try to defend my account by arguing that it is no worse off than other available ones. Instead I shall attempt to offer a way of explaining this inconsistency.

The logical picture of deductive science provides Aristotle with a strong argument that the observed situation in which mathematicians do not try to prove their starting points is what must always be the case: deductive proof presupposes premisses which are not proved. These premisses are principles (ἀρχαί) for Aristotle, but obviously this fact does not entail that for him all principles are such premisses. Nor does it even follow that whenever Aristotle is discussing the question whether the principles can be proved, he is thinking of the principles as these premisses. That is to say, it is possible for Aristotle to ask whether it can be proved that number exists or that a number is a system of monads without his thinking of these statements as ultimate premisses. My proposal, then, is that we separate the purely logical notion of a principle as an ultimate premiss of proof from what might be called the analytic notion of a starting point, analytic because it is based on an analysis of mathematical practice. The two notions coincide in so far as one kind of starting point is an ultimate premiss and in so far as neither starting points nor ultimate premisses are provable, but they do not coincide completely since not all starting points are ultimate premisses. This lack of coincidence explains, I believe, the difficulty of mapping the analytic notion of a starting point onto the logical conception of an ultimate premiss.[29]

3. Plato

Toward the end of book 6 of the *Republic* Socrates introduces Glaucon to what turns out to be a distinction between two kinds of reasoning, one exemplified in mathematics, the other in dialectic. He explains one feature of mathematical method in the following way:

> I think you are aware that those who concern themselves with geometrical matters and calculations and such things hypothesize the even and the odd and figures and three kinds of angles and other things related to these in the case of each subject; they make these things hypotheses, as if they were known; they do not see fit to give any account of them either to themselves or others, as if they were evident to everyone; they begin from these things and proceed through the others until they reach by agreement that which they started out to investigate.
>
> I know that perfectly well, he said. [Plato, *Resp.* 510c2–d4]

[29] The most interesting attempt at a mapping known to me is found in Hintikka 1972. However, the criticisms of it voiced by Ferejohn [1982–1983] and Frede [1974] seem to me very weighty. I note, however, that in his response to the latter, Hintikka [1974] virtually abandons the hope of understanding Aristotle's conception of starting points independently of his logic.

In terms of Aristotle's categorization of starting points in *An. post.* the examples of hypotheses mentioned by Socrates would most plausibly be interpreted as properties. But it is not clear what Socrates means by giving no account of these things. The most direct interpretation would be that the mathematician uses terms like 'odd', 'even', 'square', 'hexagon', 'right', 'obtuse', and 'acute' without defining them. I do not wish to rule out this interpretation, but, as I have indicated, it seems to me quite unlikely that these terms were used without definition in the mathematics known to Plato.[30] Hence, if Socrates' description is as obviously correct as Glaucon's answer suggests, Socrates may simply mean that the mathematician does not give justifications for his definitions. He or she says that an even number is one which is divisible into two equal parts and expects everyone to agree. In either case the important point is that Socrates focuses on what Aristotle calls properties, although there is no indication that Plato would draw any distinction between the underlying genus and these properties. That is to say, Socrates' list might have included 'point' or 'line' without affecting anything Plato says.

It is also important that Plato does not show any awareness of anything corresponding to either the axioms or the underlying genus mentioned by Aristotle or to Euclid's postulates, although Socrates does mention the active character of geometry [Plato, *Resp.* 527a6–b1]. In the *Meno* Socrates gives a relatively clear description of a mathematical hypothesis which is propositional but not a definition; however, the hypothesis is not intended to be a starting point in the sense I have been discussing, but a provisional assumption to which an unanswered question can be reduced [cf. Solmsen 1929, 104n1]. Of course the hypotheses of the *Republic* are also provisional in a way, but there is no indication that they are conceived propositionally except possibly in the sense that definitions are propositions. This point, of course, bears on the interpretation of Socrates' claim that dialectic can somehow do away with the hypothetical character of mathematics. The vocabulary Socrates uses in this passage makes it possible to interpret what he says as a matter of deducing hypothetical assertions from a single unhypothetical one. But this interpretation is certainly not necessary, and the fact that the mathematical hypotheses are properties or definitions and the unhypothetical starting point is the Good, makes it rather implausible. To be sure, we do not know what, if any, more precise picture underlies Socrates' account of dialectic in the *Republic*; but it seems to me that, from a more or less logical perspective, it is best to imagine the task of dialectic

[30] This is Solmsen's view [1929, 96–97]. He imagines that Socrates' description is a Platonic transformation of a mathematics based entirely on drawn figures. The interpretation I offer here is parallel to that of Sidgwick 1869.

with respect to mathematics as the rendering perspicuous of definitions through a systematic ordering of the concepts involved with some kind of non-deductive *Ableitung* of central concepts from highest ones [cf. Solmsen 1929, 101–103]. We do not know how such an *Ableitung* would work or even what good it would do, but we do no service to Plato and we read him inaccurately if we suppose that he believed in the existence of some transparent proposition from which all propositions could be deduced.

I suggest then that if we take what Plato says about mathematical hypotheses in the *Republic* at face value, then the mathematics, or at least the geometry,[31] with which Plato was familiar contained as starting points at most, and probably at least, definitions. His view was that mathematicians proved things from these definitions or from undefined terms. I have already remarked that the conception of mathematics as resting on definitions alone is the dominant one in Euclid's *Elements* and that there are certain passages in Aristotle which suggest a similar conception. In any case Plato saw the hypothetical structure of mathematics as a shortcoming, which he thought could be overcome by justifying correct definitions of certain terms. Perfect justification would involve incorporation in a conceptual structure covering the whole of reality. This structure is sometimes called a universal science by modern scholars, but it is probably wrong to think of it as a universal deductive science. No doubt deduction and argument would be a part of it, but its upper level, that closer to the absolute starting point would involve derivation and justification in a much looser sense. We might then view the Platonic universal science as a two-tiered system with the following structure:

the ἀρχή of all
('dialectical *Ableitung*')
the 'hypotheses' of the special sciences
the special sciences

Of course, one point of the notion of a universal science is precisely to deny the special sciences their special or isolated position. But to say that a discipline is part of a whole is not necessarily to deny that the part can be pursued on its own.

The *Timaeus* gives us some picture of how Plato conceived one special science, physics, which for him is, at least in part, subordinate to geometry. That dialogue also gives some hints that Plato might have espoused a development of geometry quite different from what we find in the *Elements*,

---

31 Plato's notion of arithmetic or logistic, as he usually calls it, is not as clear as it is frequently taken to be. But this is not an issue into which I can enter here.

but it seems to me reasonable to suppose that Plato also looked with favor on more straightforward deductive reasoning of the Euclidean kind. If we take this part of Platonic geometry as the relevant material for this paper, we may say that Plato views the science of geometry as using as starting points, or hypotheses as he calls them, only primitive terms or, more probably, definitions, where the definitions would presumably include both explications and abbreviations.

4. Justifying the starting points

For us a proof is primarily a means of justification. To prove P is to show that P is true in a way which justifies belief in P. However, one can also prove P as a way of teaching somebody that P is true. I will distinguish these two ways of using proof by speaking of proof as justification and proof as instruction. We, I think, tend to play down the notion of proof as instruction, particularly if the conception of proof is formal. We are willing to say a person has been taught and hence knows that the continuum hypothesis is independent of standard axioms of set theory if all he or she knows is classical set theory and that Paul Cohen was given a prize for the proof of independence. And we would certainly say that a person who knew the rudiments of Cohen's proof but not the details of forcing techniques knew Cohen's result. But Aristotle holds that we do not know anything provable unless we know its proof. Hence, if teaching is making known, teaching provable things has to be teaching their proofs; and teaching proofs is quite naturally identified with presenting them. Thus, for Aristotle, proof is both a means of justification and of instruction; proof serves to make known and to justify the theorems of a science.

However, for Aristotle, proof can do neither in the case of the starting points of a science, since he believes those starting points are unprovable. Sometimes he appears to defend this belief by turning it into the apparently less controversial doctrine that a science cannot prove its own starting points. But the more important form of the doctrine is that for certain sciences, including geometry and arithmetic, there is no higher science from which their starting points can be derived. In the case of the common axioms Aristotle appears to believe that they are not only unprovable, but that there is no way of making them known since he says [*An. post.* 72a16–17] that the axioms are a presupposition of learning anything. However, there are for him ways of making known the other kinds of starting points. Toward the end of his discussion of definition in *An. post.* ii, Aristotle concludes that definitions of the derived concepts are made known through their use in proofs, even though they are not themselves proved:

> Some things have a cause different from themselves, and some do not. So it is clear that some definitions are immediate and principles, namely, the definitions of those things for which it is necessary to hypothesize or make evident in some other way both that they exist and what they are. The arithmetician does this, since he or she hypothesizes what a monad is and that it exists. Of things which have a middle and of which there is a different cause of the οὐσία it is possible, as we have said, to make what something is clear through proof without actually proving it. [*An. post.* 93b21–28]

Aristotle's best known discussion of making known the starting points is the last chapter of the *Posterior Analytics*, where he describes a process of induction and speaks of apprehension of the principles by νοῦς. The description suggests that induction and νοῦς relate first and foremost to the primary concepts or subject genus of a science.[32] It is sometimes supposed that the topic of ii 19 is both the learning of starting points and the justification of our claim to have knowledge of them. The basis for this supposition is Aristotle's comparison of νοῦς and ἐπιστήμη as conditions of knowledge (ἀληθῆ ἀεί, *An. post.* 100b7–8) and contrast of them by saying that νοῦς is the more accurate of the two [cf. *Eth. Nic.* vi 6]. Thus, the impression arises that, although induction by itself does not justify our apprehension of the starting points, there supervenes as a result of it a self-justifying intuition of them, νοῦς.[33] I do not believe that it is possible to dismiss this interpretation entirely, but it does not seem to me to represent adequately all of Aristotle's thoughts on the justification of the starting points of the sciences [cf. Barnes 1975a *ad* ii 19; Burnyeat 1981, 130-133]. For there are clear indications in other treatises, notably the *Topics* and the *Metaphysics*, that he thinks it possible to provide justifications of a kind for them. I shall deal briefly with the three kinds of starting points in turn.

The only possible candidate for justification in the case of properties would seem to be justification of their definition and, in so far as a genus or its elements is also defined in the special sciences, the same notion of

---

32 Kahn [1981] stresses the fact that this chapter is most simply read as a description of concept formation. Traditionally it has been assumed that Aristotle must be giving an account of how primary propositions become known: see, e.g., Ross 1953, 58. Barnes [1975a, *ad loc.*] shows that one *could* read the text in terms of the apprehension of propositions.

33 Cf. Ross 1953, 217: '[Induction is] the process whereby after experience of a certain number of particular instances the mind grasps a universal truth which then and afterwards is seen to be self-evident. Induction in this sense is the activity of "intuitive reason".'

justification would be relevant to the genus. Evans [1977, 50] argues that 'Plato conceived dialectic as essentially involving a search for definitions', but that Aristotle abandons this conception. Evans does not seem to me to do justice to a variety of passages in the Aristotelian corpus, including *Topics* vi–vii, and, in particular, to Aristotle's assertion that there can be a syllogism of the definition and essence [*Top.* 153a14–15]. This assertion is, to be sure, compatible with the position that there cannot be a deductive proof of a definition; but it equally does not mean just that there can be a valid deductive argument with a statement of a definition as conclusion. Aristotle seems to be saying that there is a kind of reasoning, usually called dialectical, which can be used to establish definitions. It goes without saying that this reasoning is not scientific because scientific reasoning is characterized by proceeding from starting points, including definitions. Equally, because the reasoning is dialectical, it can only be *ad hominem*, not absolute.

I have not found evidence that Aristotle thinks the existence of a genus can be justified by means of dialectical arguments. And I think it is reasonably clear that he does not think this. Two passages are particularly useful in this respect. The first occurs in *Phys.* ii 1 where Aristotle, after indicating what nature is, says,

> To try to show that nature exists is laughable. For it is evident that many such things exist. But only a person unable to distinguish what is known through itself and what is not would show evident things through unevident ones. [*Phys.* 193a3–6]

In this case Aristotle is dealing with a starting point the existence of which he thinks is so obvious that anyone who asked to be convinced of its existence could be dismissed as stupid or merely contentious. The case of fundamental mathematical objects is not at all the same, but neither is Aristotle's attitude to the question of their existence at all as clear. At the end of *Meta.* M 1, before turning to this question, Aristotle says,

> It is necessary that if mathematical objects exist, they either exist in sensibles, as some people say, or separate from sensibles—and some people do say they exist in this way—or, if they exist in neither way, either they do not exist or they exist in some other way. So that the issue for us will not concern their existence, but the manner of their existence. [*Meta.* 1076a32–37]

Here Aristotle seems to entertain the possibility that one might deny the existence of mathematical objects and then leave it out of consideration. Annas [1976, 136] calls the denial of existence absurd, and I suppose that if one attenuates the notion of existence enough it is absurd: there must be some sense in which mathematical objects exist, e.g., as figments of the imagination, so the only task is to figure out the sense. In *Meta.* M 2 Aristotle rejects the two alternatives he mentions, and in M 3 he gives his account of the way in which mathematical objects do exist. This account, I suggest, provides the justification of the geometer's or arithmetician's postulating of a genus. It embeds those genera into an ontology, but neither the ontology in general nor the embedding of mathematical objects in it is a subject of scientific proof. We can, I think, be certain that Plato tried to embed mathematical objects into a general ontology, and, if we can believe Aristotle, the method of embedding was some kind of derivation from first principles. Aristotle has many detailed objections to the derivation, but his main procedural difference from Plato seems to me to be his insistence on the difference between scientific proof (in the sense of the *Posterior Analytics*) and other kinds of argument.

Aristotle's complex and obscure treatment of the common axioms in *Meta.* Γ has been the subject of much discussion which I cannot go into here. Instead I content myself with some general points. Aristotle deals only with the logical principles and not the quantitative ones; but I am inclined to think that he would suppose the quantitative principles could be dealt with in much the same way, although he might imagine a sequence in which the logical laws were established first and then the quantitative ones. The method is dialectical or, as Aristotle calls it at 1006a12, refutational (ἐλεγτικῶς; at 1062a3 it is called πρὸς τόνδε or *ad hominem*). The passage in which Aristotle discusses the method is very important for my purposes. Aristotle first asserts that proof has to start from something unproven to avoid an infinite regress and that anything one might start from in trying to prove the law of non-contradiction would be more in need of proof than it. He continues,

> But one can prove that the denial of the law of non-contradiction is impossible by refutation, if only the person who denies it says something. But if he or she says nothing, it is absurd to try to say something against a person who has nothing to say in so far as he or she has nothing to say. For such a person, in so far as he or she has nothing to say, is like a vegetable. I say that proof by refutation differs from proof because a person who proves might be thought to be taking as a starting point what is to be proved, but if another person provides the starting point, there will be refutation and not proof. The starting point in all such cases is not the demand that the person assert or deny some proposition, since one might take this to be a begging of the question, but that the person signify

> something both to himself or herself and to someone else. For this is necessary if the person is to say anything at all. But for someone who will not signify anything there will be no such thing as speaking either to himself or herself or to another person. But if someone will give this much, there will be a proof. [*Meta.* 1006a11–24]

As I understand Aristotle's position, it is that the law of non-contradiction is an assertion P which cannot be proved in the strict sense because the premisses needed for such a proof would be more doubtful than P. But P can be derived from any premiss at all, so that all we need to refute anyone who denies P is the person's willingness to say something and mean it. Now from a modern point of view if P can be derived from any premiss at all it is provable, since, e.g., it can be derived from its own denial. The fact that the would-be prover has to provide this premiss is irrelevant, since as the argument proceeds the premiss is eliminated, and P is proved without assumption.[34]

It seems to me that Aristotle has been misled here by a certain asymmetry in the way he treats dialectic and demonstrative science. At its center dialectic is for Aristotle a procedure of argumentation involving two people, a questioner (Q) and an answerer (A). Q's questions are designed to elicit assertions from A from which inferences are drawn until a proposition (possibly the denial of one of A's original assertions) is reached. Aristotle frequently abstracts from the human situation of dialectic to the extent of ignoring the possibility that inferences are incorrectly drawn, but not to the extent of thinking that the results of dialectic could be severed from their connection with the opinions of A. For Aristotle dialectic can serve to defend one person's opinion against another's objections or refute a person's opinions, but its success is strictly relevant to individuals. Plato, on the other hand, seems to have felt that prolonged and strenuous dialectical exercise could yield a profound insight, an insight transcending what we would call the strictly logical implications of dialectical exchange.

In this respect Aristotle's treatment of dialectic is quite unlike his treatment of scientific reasoning, which he seems to sever more or less completely from its human practitioners. In doing so he gives the appearance of a total and naive faith in the science of his day; but we may, if we like, suppose him to be adopting the position that, at least with respect to the mathematical

---

[34] Further evidence that Aristotle misses this point is provided by *An. post.* 77a31–35. There Aristotle argues that dialectic cannot prove any proposition because the dialectician argues on the basis of answers to questions and so could prove a proposition only if he could establish it from opposite assumptions. Aristotle assumes this cannot be done, but, of course, it can if the proposition is logically true.

sciences, his role is descriptive rather than prescriptive. However, it seems fairly clear that Aristotle thinks some of the features of science he takes for granted must be the way they are. In any case, for Aristotle science consists first and foremost, if not entirely, in the correct derivation of truths on the basis of the starting points. Moreover, he thinks of derivation as direct, since he believes that *reductio* arguments play no essential role in science [see *An. prior.* 62b38–40; *An. post.* i 26]. If we allow indirect proof based on the refutation of assumptions introduced by the prover, then there is no reason why the refutation Aristotle thinks possible could not be counted as an assumptionless proof.[35]

It is frequently pointed out that the refutations Aristotle provides in *Meta.* Γ presuppose the law of non-contradiction. I do not think Aristotle would find this presupposition an objection to his procedure, for, as we have seen, a person who does not already know the law of non-contradiction is incapable of learning anything and, hence, in particular, of learning something by having it proved. Another way of putting this point is to say that Aristotle thinks the non-vegetable already 'knows' the law of non-contradiction and merely has to be shown that his pretence not to know it is indefensible.[36] From the formal point of view it seems to me best to say that Aristotle uses a rule corresponding to the law in order to prove its formulation as an assertion. This way of putting the matter makes clear how little Aristotle's refutations actually accomplish, while also making clear that there is a sense in which he accomplishes what he sets out to do, i.e., to justify one of the common axioms of the sciences. The real shortcoming in Aristotle's approach to the common principles is his failure to recognize explicitly that these principles also include rules, and that reasoning cannot justify rules of reasoning. But for purposes of my historical analysis the important point is that Aristotle thought the common principles were assertions which could be justified dialectically, and that, from our point of view, the justification, if it were possible, would constitute a proof. By insisting that the justification is not a proof Aristotle separates himself from Plato, but once again the separation would seem to be much more a matter of distinguishing what Plato does not than of refusing to

[35] In my discussion I have not made use of the interesting suggestion by Irwin [1977–1978] that the treatment of the law of non-contradiction in book Γ is a model for a broadened Aristotelian conception of science which treats a restricted class of dialectical arguments as scientific. As far as I can see, what I have said would not be much affected if my contrast between scientific and dialectical argument was transformed into one between a narrower and a broader kind of scientific argument.

[36] Cf. *Meta.* 1005b23–26 where Aristotle suggests that anyone who denies the law of non-contradiction does not believe what he says.

engage in a kind of reasoning about mathematical starting points which Plato enthusiastically espoused.

6. Summary and conclusion

I conclude by presenting the sketch I have offered in a more chronological sequence, leaving out certain alternative possibilities I have considered, and adding some small details.

1. The mathematics known to Plato at the time of the writing of the *Republic* probably acknowledged only definitions as starting points. Plato called these definitions or the things defined hypotheses and believed that the definitions and hence mathematical theories could be encompassed in a universal body of knowledge which justified the definitions and performed some kind of ontological derivation of all entities. The evidence suggests that Plato did not distinguish, at least clearly, between these justifications and derivations, on the one hand, and strict deduction of the kind associated with, say, Euclid, on the other.

2. The mathematics known to Aristotle included as starting points in addition to definitions at least the first three common notions of the *Elements*, perhaps called axioms or common axioms. We cannot know why or how these principles were added, but they may be associated with Academic reflection on reasoning and argument. Aristotle includes fundamental logical laws among the axioms, and tends to think of them as the only premisses used in mathematical demonstration. The other starting points of a science for Aristotle are unique to each science. He sometimes thinks of these special starting points as things: the underlying genus consisting of fundamental objects and the properties which are proved of these objects. But frequently he treats them as assertions, namely, the assertion of the existence of the fundamental things and the definitions of them and of the properties. But for Aristotle these starting points, even construed propositionally, function as presuppositions of argument rather than as premisses. This conception of the genus and its properties as starting points of the science is Aristotle's philosophical interpretation and not a pure description of the science of his day.

Like Plato, Aristotle thinks of the practice of science as the derivation of conclusions based on starting points not to be discussed in the science. But whereas Plato sees this practice as inadequate and to be superceded by a universal science, Aristotle sees it as inherent in the nature of science. Hence, he argues that no science can justify its own starting points because the only justification it can offer is a proof based on its starting points. Aristotle also argues that no higher science can prove the starting points of sciences like geometry and arithmetic, but here he ultimately has to rely on his theory that every science must presuppose a genus. Aristotle usually makes this point by denying a universal science. Here he is arguing against Plato, but ultimately his disagreement comes to an insistence on the distinction between deduction and looser forms of reasoning. For Aristotle allows metaphysical or dialectical justifications of the starting points; and in the case of the common axioms the justification he envisages would, if it worked, amount to a proof.

3. In Euclid's *Elements* we find definitions, postulates, and common notions as starting points. The definitions predominate, and confirm one's sense that the introduction of postulates and common notions into Greek mathematics was relatively late. Indeed, it seems to me reasonable to think that the postulates are due to Euclid himself, and result from an analysis of the propositions and constructions needed to reach the major results of the end of book 1. The common notions are more problematic, but we can be virtually certain that their explicit formulation in mathematical texts predates Euclid. In any case, I see nothing in Euclid's starting points which would suggest to an unbiased reader influence from the work of Plato or Aristotle. If I had to choose between Plato and Aristotle in this regard, certainly I would choose Aristotle. But the greater plausibility of this choice is surely satisfactorily explained in terms of Aristotle's concern to describe the sciences as they are rather than in terms of his alleged influence on the way sciences turned out to be.

Appendix 1: On Speusippus and Menaechmus in Proclus

In his commentary on Euclid's *Elements* Proclus says some things about the fourth-century figures Menaechmus and Speusippus. What he says about the latter has been taken by some[37] as evidence that Euclid's constructional postulates were already known in the fourth century. I have nothing to add to the arguments which have already been given against this reading of the evidence,[38] but it is worthwhile to look at the relevant passages since they provide a good example of how cautiously the reports of Proclus on early mathematics and philosophy have to be treated. At Friedlein 1873, 77.7 Proclus introduces the now commonplace division of propositions (προτάσεις) into problems (προβλήματα) or constructions and

[37] Notably von Fritz [1971, 392]. von Fritz 1969, 94–95 offers a brief response to critics, which perhaps shows that Speusippus might have formulated the constructional postulates but does not make the possibility any more likely.

[38] Notably by Tarán [1981, 427–428].

theorems (θεωρήματα). In practice the distinction is quite clear, although formulating it in general terms is not entirely easy. Proclus says,

> Problems include the generations of figures, the divisions of them into sections, subtractions from and additions to them, and in general the characters that result from such procedures [i.e., the objects constructed?], and theorems are concerned with showing the essential attributes of each [of the things constructed]. [Friedlein 1873, 77.8–11]

He then tells us that certain fourth-century thinkers, notably Plato's nephew and successor Speusippus, thought it right to call all these things theorems rather than problems on the grounds that theoretical sciences deal with eternal things in which there is no generation. Hence, it is better to say that constructed objects exist and that 'we look on our construction of them not as making but as understanding them, taking eternal things as if they were in a process of coming to be' [Friedlein 1873, 78.4–6].

This passage clearly suggests that Speusippus collapsed an already existing distinction of mathematical propositions into theorems and problems by insisting that all problems are really theorems. But there are a number of reasons for initial skepticism about this. One is that we have no independent evidence for the existence in the fourth century of the later distinction between theorems and problems. In the fourth century theorems are things contemplated, problems are things proposed for investigation. Second, the attempt to collapse the distinction seems misguided: to say that constructions are ways of apprehending eternal things is not to deny that there is a difference between constructing a square and proving the Pythagorean theorem, a distinction which is marked grammatically by Euclid, who formulates theorems as assertions, problems using the infinitive ('to construct a square on a given straight line', and so on). I suggest that if Speusippus wanted to substitute the word 'theorem' for the word 'problem', he simply wanted to get away from the conception of science as answering questions raised or carrying out tasks assigned (whether of constructing or proving) and over to the conception of it as apprehending eternal truths. He might have referred to constructions[39] to underline the inappropriateness of geometrical language (as Plato does in the *Republic*), but his doing so need not imply that the process of apprehending truths through proofs is any less misleading about the character of the world of theoretical science.

---

[39] Proclus gives three examples of constructions (corresponding to *Elem.* i props. 1, 2, and 46) in his presentation of Speusippus' view. I see no more reason to suppose that these particular examples derive from Speusippus than any of Proclus' other examples which I discuss in this appendix.

This interpretation is confirmed by what Proclus says about Speusippus' alleged adversary Menaechmus. Menaechmus, he tells us, wanted to call all inquiries problems, but he distinguished two kinds of problems which might be proposed: one to provide what is sought, the other to see whether a thing has a certain property.[40] To suppose that Menaechmus abolished the distinction made by Proclus necessitates saying that he restored it as a dichotomy in the class of problems. The word 'inquiries', which Proclus has no motivation to supply, is a good indication that Menaechmus was speaking not about propositions, but about kinds of things into which one might inquire, i.e., problems in the standard dialectical sense.[41] Menaechmus' division of problems may, indeed, be the origin of Proclus' (or even Euclid's) division of propositions into theorems and problems; but it is important to see that Menaechmus' relates to kinds of inquiries, not to mathematical texts like Euclid's *Elements*. Proclus' report on Menaechmus confirms what one would already expect, namely, that fourth-century geometers both proved theorems and carried out constructions; but it does not provide any evidence that the geometry textbooks of the fourth century marked the distinction in anything like the way Euclid does.

We are entitled to infer from this passage only that Speusippus called all geometrical knowledge theorems, i.e., matters of contemplation, for platonist reasons, and that Menaechmus made a distinction between two kinds of things into which a mathematician might inquire, i.e., between two kinds of problems. Sometime later Menaechmus' distinction was turned into one between two kinds of results (propositions) a mathematician might achieve, a construction (problem) and a theorem.

At Friedlein 1873, 178.1 Proclus turns to Euclid's postulates and axioms, which he considers to be kinds of principles (ἀρχαί). He suggests that the distinction between postulates and axioms parallels that between problems and theorems; but that principles must always be superior to the things after them in simplicity, unprovability, and self-evidence. He then cites Speusippus:

---

[40] Friedlein 1873, 78.10–13: ὅτε μὲν πορίσασθαι τὸ ζητούμενον, ὅτε δὲ περιωρισμένον λαβόντας ἰδεῖν ἢ τί[ς] ἐστίν, ἢ ποῖόν τι ἢ τί πέπονθεν, ἢ τίνας ἔχει πρὸς ἄλλο σχέσεις. I see no reason to accept Becker's suggestion [1959, 213] that περιωρισμένον should be πεπορισμένον and, therefore, no reason to accept his claim that the distinction made by Menaechmus was between solving a problem constructively and then investigating what has been constructed (for which reading one might expect τὸ πεπορισμένον).

[41] Bowen [1983, 27n36], whose account of the Proclus passages discussed in this appendix differs considerably from mine, suggests that προβλήμα has another sense in the fourth century, namely geometrical demonstration or scientific deduction. The passages he lists do not seem to me to support the claim.

> In general, says Speusippus, in the hunt for knowledge in which our mind is engaged, we put forward some things and prepare them for use in later inquiry without having made any elaborate excursion and our mind has a clearer contact with them than sight has with visible objects; but others it is unable to grasp immediately and therefore advances on them step by step and endeavors to capture them by their consequences. [Friedlein 1873, 179.14–22: trans. in Morrow 1970, *ad loc.*]

Proclus goes on to give examples to illustrate the difference between principle and subsequent result, and returns to the comparisons of postulates with problems and of axioms with theorems. He then says,

> However, some people think it right to call all principles postulates, just as they call all things sought problems. Thus, Archimedes at the beginning of book 1 of *On Equilibria* says, 'We postulate that equal weights at equal distances are equally balanced.' But one might rather call this an axiom. Others call them all axioms, just as they call all things which need proof theorems. It would seem that these people have transferred words from special uses to common ones in accordance with the same analogy. [Friedlein 1873, 181.16–24]

It seems clear that Proclus is talking about Menaechmus and Speusippus; but it is striking that the example for calling all things postulates is drawn not from the fourth century but from Archimedes, who died at the end of the third and certainly did not call all principles postulates, but rather more or less completely disregarded the terminological distinctions Proclus thinks are important. We cannot exclude the possibility that Menaechmus used the word 'postulate' in something like the way suggested by Proclus, but the passage on problems suggests that at most he called anything taken for granted (or conceded) in a mathematical inquiry a 'postulate'.[42] We have no reason to suppose that he distinguished kinds of postulates as he distinguished kinds of problems, nor that in calling them postulates he was reacting against a distinction between constructional and propositional 'principles'.

Proclus gives two pairs of examples to illustrate Speusippus' distinction between principles and things subsequent to them: Euclid's first postulate and his first proposition (the construction of an equilateral triangle); and his third postulate and the generation of a spiral by the motion of a point along the revolving radius of a circle. Tarán [1981, 427–428] has argued that

[42] Possibly relevant to Menaechmus' discussion is the distinction Aristotle makes between a postulate and a hypothesis. See n14, above.

the examples are not Speusippus'. For my purposes it is sufficient to say that we cannot assume they are Speusippus' and, hence, cannot infer from this passage that Euclid's first and third postulates were already formulated as 'principles' in the mid-fourth century. We may, I suppose, accept that Speusippus called all principles axioms, but we have no very clear notion of why he would choose that word over other possibilities.[43] The last sentence of the last quotation suggests that Proclus had no information about the reason, but only assumes that the choice of 'axiom' as a name for principles is related to the choice of 'theorem' as a name for the things after the principles.

My conclusion is that the passages from Proclus which I have discussed tell us very little about fourth-century mathematics and philosophy of mathematics that we might not have guessed already. The most interesting information we get is perhaps that Menaechmus made a distinction between assertions to be proved and constructions to be carried out; for we have no explicit recognition of that distinction in Plato or Aristotle, although I think it must have been applicable to the mathematics they knew. We get the philological information about Speusippus' use of the word 'theorem' and Menaechmus' of 'problem' and perhaps about the former's use of 'axiom' and the latter's of 'postulate'. But none of this information seems to me to relate in any specific way to the content of fourth-century mathematics.

Appendix 2: Oenopides and Zenodotus

Proclus mentions [Friedlein 1873, 65.21–66.4] Oenopides of Chios in the so-called Eudemian summary of the history of geometry. It is natural to infer from this mention that Oenopides was active *ca.* 450 BC. Proclus also tells us [Friedlein 1873, 283.7–8] that Oenopides was the first to investigate the problem of dropping a perpendicular from a point to a straight line, a problem he thought useful for astronomy, and that [Friedlein 1873, 333.5–6] Oenopides was the first to discover [the solution to] the problem of copying an angle. In the second of these passages Proclus mentions Eudemus as source of information, so it is likely that Eudemus is Proclus' ultimate source for the first as well; there is also no reason to doubt that we are still dealing with the fifth-century Oenopides of Chios. Since the time of Heath [cf. 1921, i 175] it has been customary to say that Oenopides' innovation was to carry out the constructions in question with a ruler and

[43] It is interesting to recall Aristotle's conceptions of axioms as the common premisses of all sciences.

compass, since the dropping of a perpendicular could easily be solved using a draftsman's right angle. To this conjecture Szabó [1978, 275] has added another: Oenopides made conscious use of Euclid's first three postulates and is perhaps their originator. There is, however, a big difference between reducing certain constructions to others and laying down postulates as starting points. As for Heath's conjecture itself, it is very probable that Eudemus attributed to older geometers the solution of problems and proofs of theorems which he thought were presupposed by other knowledge ascribed to them [see, e.g., Dicks 1959, 302–303; Gigon 1945, 55; Wehrli 1969, 116]. Discussion of this point has largely focussed on Eudemus' ascription of certain results to Thales, but there is every reason to think he did the same sort of thing in the case of Oenopides.

Proclus mentions an Oenopides one other time in connection with a more philosophical matter:

> Those around Zenodotus, who belonged to the succession of Oenopides and was a pupil of Andron, distinguished theorems from problems in the following way: a theorem inquires what property is predicated of its subject matter, a problem what is the case given that such and such is the case. [Friedlein 1873, 80.15–20]

This passage tells us all we know about Zenodotus, Andron, and the succession of Oenopides, so there is no real ground for von Fritz' assertion [1937, col. 2267] that Zenodotus was an '*Enkelschüler*' of Oenopides of Chios.[44] The terms in which the distinction between theorem and problem is made are through and through Peripatetic,[45] suggesting a *floruit*

[44] And even if Zenodotus were the pupil of a pupil of Oenopides, there would be no more basis for inferring Oenopides' concerns from Zenodotus' than for inferring Socrates' from Aristotle's.

[45] As formulated by Proclus the distinction in question is almost certainly that between a categorical and a hypothetical assertion. There is no doubt that the description of a theorem is a description of a Peripatetic categorical assertion, the predication of a property of a subject. For evidence that a problem is being characterized as a hypothetical, consider Galen's description:

> Another kind of proposition is that in which we do not maintain something about the way things are but about what is the case given that such and such is or what is the case given that such and such is not the case; we call such propositions hypothetical. [Galen, *Inst. log.* iii 1]

Cf. Aristotle's use of τίνος ὄντος τὸ προκείμενόν ἐστι at *Top.* 111b17–18 with Alexander's comment *ad loc.*

If Proclus gives an accurate representation of Zenodotus' meaning, then Zenodotus presumably compared the givens of a problem with the antecedent of a conditional, the object constructed with the consequent. Thus, he might have

no earlier than the late fourth century for Zenodotus. Immediately after Zenodotus Proclus mentions the way Posidonius made the distinction (problems ask whether or not something exists, theorems what or what kind of thing something is) as if it were somehow derived from Zenodotus' (ὅθεν). This indication provides some support for a *terminus ad quem* of the 1st century BC, but in the absence of information about Andron or what is meant by the succession of Oenopides, the question of dating must be left open. Considerations of simplicity suggest that we identify this Oenopides with the fifth-century one mentioned by Proclus elsewhere, but this identification does not help to clarify the character of fifth- or fourth-century mathematics.

Acknowledgement. The first version of this paper was written while the author held a research fellowship from the National Endowment for the Humanities. Parts of it were read to groups in Los Angeles and Davis, California before the presentation of its main ideas at the Pittsburgh conference. The discussions which followed those meetings affected this paper in more ways than I can now recall in detail, but I would like to thank Alan Bowen, Jim Lennox, Geoffrey Lloyd, Tom Upton, and especially the late Joan Kung and Henry Mendell.

read Euclid's first proposition as 'If there is a straight line, then an equilateral triangle can be constructed on it.'

6

Ratio and Proportion in Early Greek Mathematics[1]

D. H. FOWLER

Let Aristotle introduce the subject. His *Topics* is a manual of syllogistic dialectic, a kind of formal debate between two people called here the 'questioner' and the 'answerer'. At *Top.* 158a31–159a2 Aristotle writes:

> There are certain hypotheses upon which it is at once difficult to bring, and easy to stand up to, an argument. Such (e.g.) are those things which stand first and those which stand last in the order of nature. For the former require definition, while the latter have to be arrived at though many steps if one wishes to secure a continuous proof from first principles, or else all discussion about them wears the air of mere sophistry: for to prove anything is impossible unless one begins with the appropriate principles, and connects inference with inference till the last are reached. Now to define first principles is just what answerers do not care to do, nor do they pay any attention if the questioner makes a definition: and yet until it is clear what it is that is proposed, it is not easy to discuss it. This sort of thing happens particularly in the case of the first principles: for while the other propositions are shown through these, these cannot be shown through anything else: we are obliged to understand every item of that sort by a definition. The inferences, too, that lie too close to the first principle are hard to treat in argument.... The hardest, however, of all definitions to treat in argument are those that employ terms about which, the first place, it is uncertain whether they are used in one sense or several, and, further, whether

[1] I use the phrase 'early Greek mathematics' to denote the period up to and including the time of Archimedes.

> they are used literally or metaphorically by the definer. For because of their obscurity, it is impossible to argue upon such terms; and because of the impossibility of saying whether this obscurity is due to their being used metaphorically, it is impossible to refute them.... It often happens that a difficulty is found in discussing or arguing a given position because the definition has not been correctly rendered.... In mathematics, too, some things would seem to be not easily proved for want of a definition, e.g. that the straight line, parallel to the side, which cuts a plane figure divides similarly (ὁμοίως) both the line and the area. But, once the definition is stated, the said property is immediately manifest: for the areas and the lines have the same ἀνταναίρεσις and this is the definition of the same ratio.... But if the definitions of the principles are not laid down, it is difficult, and may be quite impossible, to apply them. There is a close resemblance between dialectical and geometrical processes.[2]

In brief: Define your terms! The mathematical proposition that Aristotle is describing, in his typically vague fashion, must be the following:

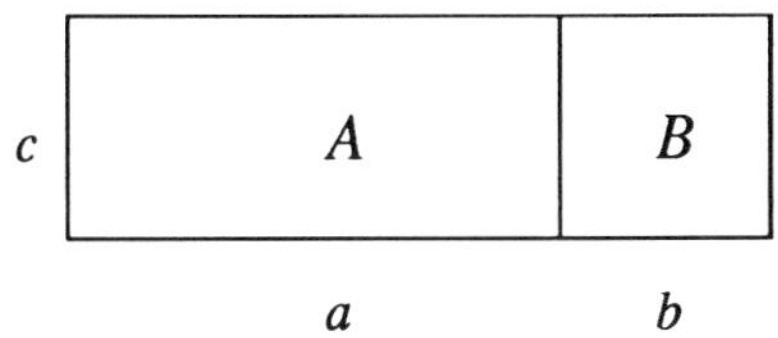

Figure 1

> If a rectangle or parallelogram is divided by a line parallel to a pair of sides, as in Figure 1, then the ratio of the bases, $a{:}b$ is equal to the ratio of the areas, $A{:}B$.

I shall refer to this proposition hereafter as 'The *Topics* proposition': a similar result is proved by Euclid at *Elem.* vi prop. 1, where it forms the link between the study of the equal[3] figures of books 1–4 and the similar figures of book 6 and later. It is no exaggeration to say that the *Elements*

[2] Most of this translation is taken from Ross 1908–1952 i; the mathematical example is adapted from Heath 1949, 80.

[3] That is, equal in magnitude, according to the Euclidean conception of equality and inequality, set out in the Common Notions of book 1. This is sometimes described by mathematicians as 'cut and paste' equality.

hinges on this proposition,[4] and so it is worth considering its proof in detail. This will be the subject of Section 1.

We start from Aristotle's 'first principles in need of precise definition':

What are 'ratio' (λόγος) and/or 'proportion' (ἀνάλογον)?

One difficulty with discussions about these words is that it is often uncertain, as Aristotle says, whether they are being used 'in one sense or in several... and literally or metaphorically'.[5] They are often treated by ancient and modern writers as synonymous; the underlying concepts are expressed using a number of other apparently equivalent descriptions, and the words themselves have a very wide variety of non-mathematical connotations. Rather than introduce yet further words to identify the distinction I want to make and maintain, I shall hereafter use them with the following precisely differentiated meanings.

Ratio. Euclid states at *Elem.* v def. 3, that

Λόγος ἐστὶ δύο μεγεθῶν ὁμογενῶν ἡ κατὰ πηλικότητα ποιὰ σχέσις.

A ratio is a sort of relation in respect of size between two magnitudes of the same kind.

So, given these two homogeneous objects,[6] a and then the ratio of a to b (abbreviated $a{:}b$) will be no more and no less than a description of just how this relation is conceived and expressed.

---

[4] The role of the *Topics* proposition in the formal theories of proportion and the classification of incommensurables in the *Elements* is analysed in detail in Knorr 1975, ch. 8. However, his accompanying thesis depends heavily on the surprising difficulty of his anthyphairectic proof of *Elem.* v prop. 9 (that if $a{:}c :: b{:}c$; then $a = b$): see Knorr 1975, 338–340. But surely a practicing mathematician, faced with the difficulty that Knorr has uncovered, would proceed indirectly *via* the *alternando* property that he had just proved. For then $a{:}c :: b{:}c$ is equivalent to $a{:}b :: c{:}c$, whence the result follows immediately.

[5] See, for example, the useful description in Mueller 1981, 138 (quoted in this very context in Berggren 1984, 400) of '[Euclid's] conception of definition as characterisations of independently understood notions'. Here we must draw out what these 'independently understood notions' might be and examine them against the historical background, such as we know it.

[6] I shall use this word 'object' as synonymous with 'magnitude' or μέγεθος. Euclid does not give a definition of the ratio of two μεγέθη, though he does introduce the idea, as will be noted below.

Proportion. We then read in *Elem.* v defs. 5 and 6 that

> Ἐν τῷ αὐτῷ λόγῳ [= ἀνάλογον] μεγέθη λέγεται....
>
> [Four] magnitudes are said to be in the same ratio (that is, proportional) if....

This means that proportionality is a condition that may or may not hold between four objects. Given a, b, c and d, we then answer with either 'They satisfy the condition and so are proportional' (abbreviated $a{:}b :: c{:}d$); or 'They do not satisfy the condition so they are not proportional'; or 'They fail to satisfy some homogeneity condition so the question of proportionality does not apply.' For example, it would be meaningless to ask, in the context of the *Topics*, whether $a{:}b :: A{:}5$.[7]

The *Topics* proposition has now two different formulations which I shall distinguish. Either a, b, A, and B are proportional, i.e., $a{:}b :: A{:}B$, or the ratio of a to b is equal to the ratio of A to B, i.e., $a{:}b = A{:}B$ (where we must now also describe the conditions under which two ratios are equal).

Having taken care of the definitions of the first principles, I shall go on to the 'inferences that lie too close to the first principle and which are therefore hard to treat in argument', and discuss how the proof of the *Topics* proposition depends on the underlying definition of ratio or proportion. Only when we are aware of the range of possible ways of giving sense to the concepts involved should we consider what might constitute a deductive proof of this result in a given historical context. These different proofs will then illustrate two other contrasts which I wish to emphasise and discuss: that of arithmetised *versus* non-arithmetised mathematics, and that of early *versus* later Greek mathematics.

---

[7] There are several possible homogeneity conditions: we may have—(i) a, b, c, and d all homogeneous magnitudes; (ii) a and b homogeneous magnitudes, c and d homogeneous magnitudes, but a and c not homogeneous; (iii) a, b, c, and d all ἀριθμοί; (iv) a and b homogeneous magnitudes, c and d ἀριθμοί; or (v) a and b ἀριθμοί, c and d homogeneous magnitudes. Euclid is ambivalent about (i) and (ii) in *Elem.* v [see Mueller 1970]; (iii) is the topic of *Elem.* vii; and the absence of any link between the theories of *Elem.* v and vii—hence, the absence of any discussion of (iv) and (v)—leads to a notorious lacuna in the proof of *Elem.* x prop. 5.

1. Seven proofs of the *Topics* proposition

1.1 *The naive 'proof'*

Find a common measure d of a and b. Then the rectangle D [see Figure 2] will be a common measure of A and B, and as many times (say n) that d goes into a, D will go into A; and.... Therefore,....

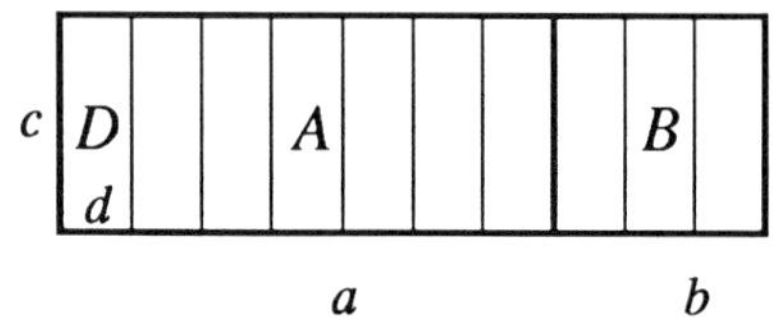

Figure 2

Comments. What does this tell us about ratio or proportion? What is the underlying definition? Neither of these ideas is mentioned explicitly and a generous interpretation of the 'proof' may suggest no more than an underlying definition that, for four ἀριθμοί, $m{:}n :: m'{:}n'$ if $m = m'$ and $n = n'$; and a similar definition for four magnitudes $a{:}b :: A{:}B$ where $a = nd$, $A = nD$, $b = md$, and $B = mD$, for some magnitudes d and D.

What explicit evidence do we have for this proof, or some variation of it, as an argument of Greek mathematics of the fifth or fourth centuries BC? I know of none.

Notwithstanding these reservation, this kind of proof, often only implied, seems to dominate discussions of pre-Eudoxan ratio- and proportion-theory.

1.2 *A geometrical definition and proof*

Let a, b, c,... denote lines, and $A, B, C,\ldots$ plane regions of some suitably restricted kind; here, for example, rectangular regions will suffice. We suppose that these geometrical objects can be manipulated in the style of *Elements* i–iv. Define an operation, written here and later as a multiplication—$\otimes$—(but any other word or abbreviation would serve equally well) as follows:

> The product $a \otimes b$ of two lines is the rectangle with adjacent sides a and b, and the product $a \otimes B$ of a line and a plane region is the rectangular prism with base B and edge a.

And define

> The objects x, y, z, and w are proportional if $x \otimes w$ and $y \otimes z$ make sense and are equal [see n3].

The proof of the *Topics* proposition then follows immediately [see Figure 3] since $a{:}b :: A{:}B$ means $a \otimes B = b \otimes A$,which is true since both are rectangular parallelopeds with sides a, b, and c.

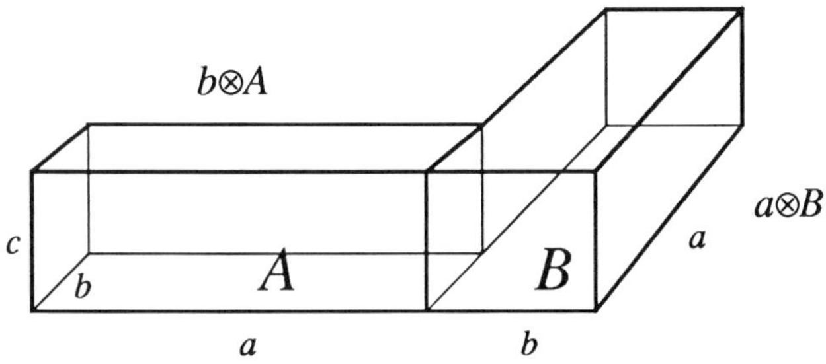

Figure 3

Comments. This proof is constructed from impeccably Euclidean ingredients [see *Elem.* ii def. 1 and vi prop. 16, where Euclid refers to 'the rectangle contained (περιεχόμενον) by two lines'; and vii prop. 19, which gives a similar manipulation for four ἀριθμοί], and it is generally believed that this material dates from well before the proportion-theory of *Elements* v. The other basic results of Euclidean proportion-theory can be handled by an extension of this procedure. In other words, there is no difficulty in constructing a theory of proportionality from the basic techniques available, say, to Hippocrates and Theodorus. But I am not advancing this here as a proposal for a reconstruction of an early definition of proportionality; of this we have little or no evidence one way or the other, and all that we can say is that this kind of manipulation had become, by the time of Euclid's *Elements*, a standard part of formal proportion-theory in geometry and ἀριθμητική. It is presented here in this form only to refute common assertions that fifth-century mathematicians did not have available the means to develop a theory of proportionality that would handle incommensurable magnitudes.

This is a proportion-theoretic proof: the ratio $x{:}y$ is not defined.

1.3 *An arithmetised interpretation*

For a long time, and explicitly since Descartes, geometry has been 'arithmetised'. In this type of interpretation

> The letters $a,\ b, \ldots;\ A,\ B \ldots$ denote, ambiguously, either geometrical objects that are manipulated geometrically, or 'numbers' (or 'numerical quantites') that are manipulated arithmetically.
>
> The area of a rectangle, a number, is the product of the lengths of its base and height, e.g., $A = a \times c$ in Figure 1.
>
> The ratio of two magnitudes is the quotient of the corresponding numbers, $x{:}y = {}^{x}/_{y}$.

These definitions, and the permitted manipulations of arithmetic, then yield the following proof:

$$a{:}b = \frac{a}{b} = \frac{a \times c}{b \times c} = \frac{A}{B} = A{:}B$$

Comments. This has nothing to do with early Greek mathematics. The first time that anything like this is found in Greek geometry is in the metrical geometry of Heron, in the first century AD. I shall discuss this issue further in Section 2.

What is a number? Answers to this question are, in fact, easy to supply and any of the different definitions of ratio to be given below may be considered as 'numbers' in some sense. But a really difficult question is, How can we describe, correctly and completely, arithmetic with these numbers? I believe that this question posed a profound and perplexing problem to arithmetised mathematics, though the deceptive ease with which post-Renaissance mathematicians were apparently able to manipulate decimal numbers, newly introduced in the West at the end of the sixteenth century, enabled them to set the problem to one side for two centuries. But no satisfactory answer to the question was known before Wednesday, November 24th, 1858, the day when Dedekind says he conceived his construction of the real numbers. In his *Stetigkeit und die irrationale Zahlen* [1872, see Dedekind 1901], Dedekind defines addition in detail and then goes on to say:

> Just as addition is defined, so can the other operations of the so-called elementary arithmetic be defined, viz., the formation of differences, products, quotients, powers, roots, logarithms, and in this way we arrive at real proofs of theorems as, e.g., $\sqrt{2} \cdot \sqrt{3} = \sqrt{6}$, which to the best of my knowledge, have never been established before.[8]

This is a ratio-theoretic proof: ratio is first defined and then proportion is defined to be equality of ratio. This last step is often assumed to be a mere formality, but in fact it can be very subtle. For decimal numbers, for example, the statement '$0.999\ldots = 1$' evokes the mathematical ingredients of Zeno's paradox of Achilles.

Interlude: the historical context

In *Elem.* v we find a theory of proportion, based on v def. 5, which is believed either to be due to Eudoxus or to be a development of Eudoxus' ideas, and which is dated, in conception at least, to around 350 BC, just before Plato's death.

The *Elements* is dominated by the sheer bulk of book 10 and the subtlety of its application in book 13: book 10 sets up a classification of certain kinds of mutually incommensurable lines, and book 13 applies this classification to the lines that arise in the construction of regular polygons and polyhedra. Although it contains few explicit references to the idea of ratio, this material is clearly connected with the idea of the ratio (*not* proportion) of these lines. See, for example, the terminology of the definitions of *Elem.* x, where new lines are described by their relation to an assigned line and distinguished according as this relation is either expressible (ῥητός) or without ratio (ἄλογος). See also the description in the culminating *Elem.* xiii prop. 18:

> The said sides, therefore, of the three figures, I mean the pyramid, the octahedron, and the cube, are to one another in expressible ratios (λόγοι ῥητοί). But the remaining two, I mean the side of the icosahedron and the side of the dodecahedron, are not in expressible ratios either to one another or to the aforesaid sides; for they are ἄλογοι, the one being minor, the other apotome.

[8] Translation from Dedekind 1901, 22. This book also contains the translation of *Was sind under was sollen die Zahlen* [1888], in which Dedekind returns to, repeats, and emphasises this view. He also has some very apposite remarks about the relation between his definition of the real numbers and *Elem.* v def. 5. See Section 1.7 for further comments on arithmetic.

This programme of *Elem.* x and xiii is attributed, on reasonably good authority, to Theaetetus. Plato's eponymous dialogue is an encomium to the dying Theaetetus who has just been carried from what is believed to be the siege of Corinth in 369 BC.

Hence, we have evidence of massive activity in the study of *ratios* of incommensurable lines, in which the expressible ratios such as the diagonal to side of a square play a prominent role, well before the development of the proportion-theory of *Elem.* v.

It does not take much space to describe our positive and negative evidence concerning the early Greek mathematical idea of ratio (*not* proportion). Note first that *Elem.* v and vii describe theories of proportion for magnitudes and ἀριθμοί respectively, though some definitions in book 5—namely, definitions 3, 4, 9, 12, 13, 14, 16, and 17—refer to ratio. If we exclude material which, though it is expressed in terms of ratio is then immediately reformulated and used in terms of proportion, we have in the *Elem.* v def. 3 (quoted above), x and xiii (described above), vi def. 5, vi prop. 23, and vii prop. 5 (which refer to an operation of compounding ratios which then plays no further part in the *Elements*),[9] and some definitions and propositions on the extreme and mean ratio, reciprocal ratios, and duplicate and triplicate ratios mainly to be found in *Elem.* vi. To this material in the *Elements*, we can add *Data* def. 2 ('A ratio is said to be given when we can make another equal to it') and the passage in Aristotle, *Top.* viii 3 with which we started and which may refer either to ratio or proportion. There are allusions to ratio in technical passages in Plato and Aristotle. That exhausts the positive surviving evidence. Our negative evidence is that we have no explicit sign whatsoever that early Greek mathematicians worked with any arithmetised conception of ratio: see Section 2 for further elaboration of this remark.

Let us now return to the *Topics* proposition.

1.4 *Aristotle's proof*

Aristotle summarises his proof thus:

> τὴν γὰρ αὐτὴν ἀνταναίρεσιν ἔχει τὰ χωρία καὶ αἱ γραμμαί· ἔστι δ' ὁρισμὸς τοῦ αὐτοῦ λόγου οὗτος.
>
> For the areas and the lines have the same *antanairesis* and this is the definition of the same ratio.

[9] This is very clearly and thoroughly analysed in Mueller 1981, 88, 92–93, 135–136, 154, 162, 221, 225–226, and 229.

'*Antanairesis*' is an ordinary Greek word used to describe the subtraction of one thing from another; for example, we find it used interchangeably with ἀνθυφαίρεσις in commercial accounts on papyrus.[10] The corresponding verb, with an adverb indicating a repetitive action, is used by Euclid in *Elem.* vii props. 1 and 2, x props. 2 and 3. Consider, for example, *Elem.* x prop. 2:

> If, when the less of two unequal magnitudes is continually subtracted in turn (ἀνθυφαίρειν ἀεί) form the greater, that which is left never measures the one before it, the magnitudes will be incommensurable.

We can perform this operation of anthyphairesis on any pair of homogeneous objects. For example, given the ἀριθμοί (51,15), we subtract the smaller from the larger to get (36,15), then (21,15), then (6,15). The originally larger term is now smaller, so we now subtract it: (6,9), then (6,3). At this stage we see that the less or term 3 measures the term before it 6, and so [see *Elem.* vii prop. 2] the greatest common measure of 51 and 15 is 3. But also note, with Aristotle, that the relation in respect of size between 51 and 15, the 'anthyphairetic ratio', is characterised by this pattern: three subtractions, two subtractions, two subtractions, and no more.[11] If performed on two ἀριθμοί, the subtraction process will always terminate [see *Elem.* vii props. 1 and 2]; for two magnitudes, it may or may not terminate [see *Elem.* x props. 2 and 3].

Now consider the *Topics* proposition. We can characterise the relation of size, both between the two lines a and b and between the two areas A and B, by this subtraction process. But each subtraction of the line can be made to correspond to each subtraction of the rectangle standing on that line, and vice-versa [see Figure 4]. Hence, the pattern of the two subtraction processes will be the same. Moreover, since Aristotle says that this is the definition of the same ratio, the proposition is proved.

[10] For examples taken from the same set of documents, the Zenon archive, see P. Lond. vii 1994.164, 176, 223 and 321 and vii 1995.333 (both dated 251 BC; here ἀνθυφαίρειν); and P. Cair. Zen. iii 59355.95 and 150 (243 BC; ἀνταναίρειν). Another equivalent, ἀνταφαίρειν, is found in Nicomachus, *Intro. arith.* i 13.11.

[11] One standard modern notation for this, used in my book and elsewhere, is to write 3:2 = [3, 2, 2]. But the mathematical explorations can be carried through in natural language, without any symbolism, or using notations like this only as a convenient shorthand.

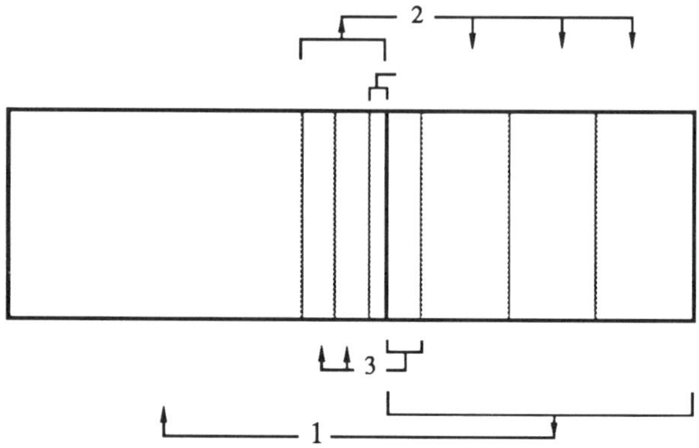

Figure 4

Note that this proof can be interpreted either in terms of ratio or of proportion: it is not entirely clear to which Aristotle refers.

1.5 *A proof using astronomical ratios*

We consider an idealised model of astronomy, in which all motions are uniform and go on forever without change or the least deviation. So, for example, we tally each sunset, the beginning of each day, starting the record at some arbitrary point:

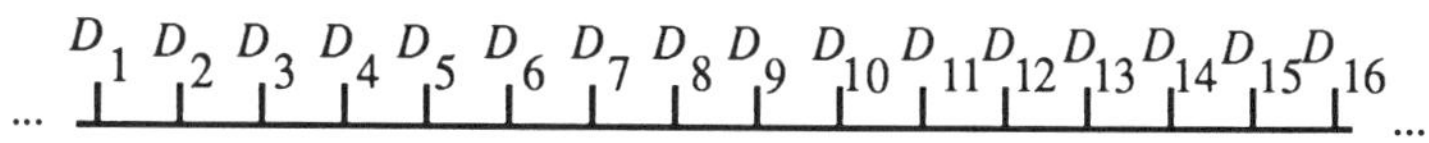

Then on the same tally, we can mark off some other uniform astronomical phenomenon like the conjunction of Sun and Moon which marks the succession of months. Suppose the period of this second motion is between one and two days long, else our tally will have to go on for a very long time before we see anything happening. We then will get a pattern like this:

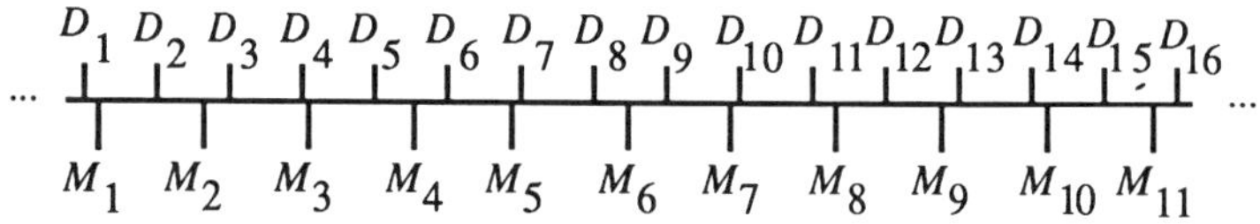

Note that this pattern only describes the order of successive events, and that we have no precise idea of the distance between them; indeed, for temporal events, the measurements of these time-intervals would pose serious practical and theoretical problems. So we could just as well represent this pattern as

$$D,\ M,\ D,\ M,\ D,\ D,\ M,\ D,\ M,\ D,\ D,\ M,\ D,\ M,\ D,\ D,\ M\,D,\ldots$$

or as

$$\ldots\ 1 \text{ or } 2,\ 1,\ 2,\ 1,\ 2,\ 1,\ 2,\ 1,\ 1,\ 2,\ 1,\ 1 \text{ or } 2,\ldots$$

or in any other such equivalent way. These patterns may or may not contain coincidences, and may or may not contain repeating blocks. And, in this theoretical astronomy, we also suppose that there is no problem in detecting which of two events occurs first, or whether there is a coincidence.

Our fundamental insight is that these patterns also characterise the relation of size between the period of the two events: they define what may be called the astronomical ratio. Moreover, we can apply the same procedure to two geometrical objects. Take, for example, two lines a and b. We can now start the tally with a coincidence

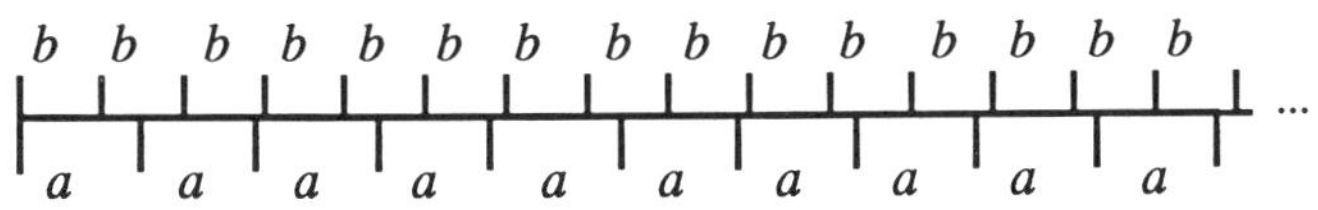

and so get a pattern

$$a \text{ and } b,\ b,\ a,\ b,\ a,\ b,\ b,\ a,\ b,\ldots.$$

Or we can do the same process with two rectangles [see Figure 5].

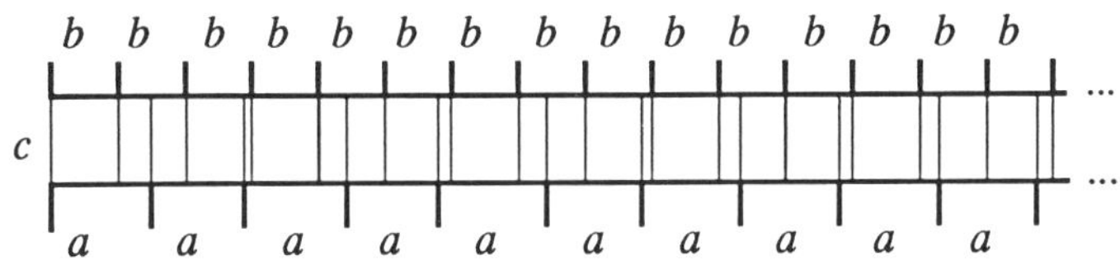

Figure 5

In the configuration of the *Topics* proposition, the tallies of the bases, a and b, and the rectangles, A and B, will clearly give rise to the same pattern. So, again, with this underlying definition of astronomical ratios, the proposition is proved.

Again note that this is a ratio-theoretic formulation of the *Topics* proposition; indeed it is the definition of ratio that underlies the theory of proportion developed in *Elements* v. Also, in the geometrical context, when we can arrange for a coincidence with which to begin our tallying, there is no difficulty in identifying when two ratios are equal, and so converting this ratio-theoretic formulation and the proof into proportion-theory; we then get Euclid's proof of *Elem.* vi prop. 1. But we have no such liberty in the astronomical context, and different phase shifts between the two events will give rise to different patterns. How to recognise when ratios are equal in respect of size, and only differ in respect of phase, is then very far from obvious;[12] so one sees again that it is not always easy to pass from ratio-theory to the corresponding proportion-theory.

1.6 *Variations on a theme*

If we contemplate the last two proofs, we see that the pattern of many different addition or subtraction processes, performed on two homogeneous objects a and b, may be used to characterise the relation of size between a and b. For example, instead of using a process of alternating subtraction, we can always subtract from the first object, or from the second; instead of always undershooting, so subtracting with remainder, we can overshoot, and continue with the excess; at each step we can perform some specific predetermined scaling operation; and so on. The only general properties we use of the underlying objects is that we can compare any two to determine if they are equal or which is the greater, and that we can add any two or subtract the less from the greater.

Any such process, consistently applied, will generate a pattern that will then characterise the relation of size between the two original objects; and hence from each idea of ratio there will be a corresponding proof of the *Topics* proposition. Here are two examples of what I shall call decimal ratios and accountant's ratios.

Decimal ratios. Suppose that $a \geq b$. Compare a with $b,\ 10b,\ 10^2b, \ldots$ to locate that index k for which $10^k \leq a < 10^{k-1}b$. Then define

$$\begin{array}{llllll} a & = n_k & \times 10^k b & + & a_{k-1} & \text{where} \quad 0 \leq a_{k-1} < 10^k b \\ a_{k-1} & = n_{k-1} & \times 10^{k-1} b & + & a_{k-2} & \text{where} \quad 0 \leq a_{k-2} < 10^{k-1} b \end{array}$$

and so on.

[12] Such a procedure, found by E. C. Zeeman, is described in Zeeman 1986 and Series 1985.

The process will continue indefinitely; each n_j lies between 0 and 9, and if any remainder a_j is zero, then all subsequent terms $n_j, n_{j+1}, n_{j+2}, \ldots$ will be zero. Also the procedure has been arranged so that this sequence is uniquely defined and cannot finish with an unending sequence of nines.

The pattern described by this sequence $n_k,\ n_{k-1}, \ldots,\ n_0,\ n_{-1},\ n_{-2} \ldots$ will then characterise the relationship of size between the pair a and b. For the first time in these illustrations, the second term b plays a privileged role, here as a 'unit' in a scale of measurement. Conventionally we write this particular ratio as a decimal number $a{:}b = n_k n_{k-1} \ldots n_1 n_0 \cdot n_{-1} n_{-2} \ldots$. (If $a \leq b$, we compare a with b, ${}^b/_{10}$, ${}^b/_{10^2}, \ldots$ and get $0 \cdot 0 \ldots 0 n_k n_{k-1} \ldots$.) Here again we have a proof of the *Topics* proposition, since the pattern of the subtraction process generated by a and b will again be the same as the pattern generated by A and B.

The importance of this algorithm for decimal subtraction is that there is an almost universal delusion, even among mathematicians, that we can easily perform arithmetical operations on these decimal ratios—that we can add, subtract, multiply, or divide decimal 'numbers'—and that it is obvious that this arithmetic satisfies the usual manipulations of arithmetic like

$$x \div y = (x \times z) \div (y \times z) = (x \div z) \div (y \div z)$$

for any three 'numbers' x, y and z. Also ':' and '$\div$' are now treated as being virtually synonymous. This leads to the following version of the earlier arithmetised proof of the *Topics* proposition in Section 1.3.

> Let a, b, c, A, and B be as in Figure 1, and fix some standard line 1, the unit of length. This unit determines a standard square 1^2, the unit of area. Denote the decimal ratios (or 'numbers') $a{:}1$, $b{:}1$, $c{:}1$, $A{:}1^2$, and $B{:}1^2$, a', b', c', A', and B', respectively. Then, by the supposed basic properties of arithmetic:
>
> $$a{:}b = (a{:}1){:}(b{:}1) = a'{:}b' \text{ and } A{:}B = (A{:}1^2){:}(B{:}1^2) = A'{:}B'.$$
>
> Use multiplication to define a numerical area, where
>
> $$\text{area} = \text{base} \times \text{height}.$$
>
> This turns out to be equivalent to the earlier numerical definition of area:
>
> $$A' = a' \times c', \qquad B' = b' \times c'.$$
>
> Hence,
>
> $$a{:}b = a'{:}b' = (a' \times c'){:}(b' \times c') = A'{:}B' = A{:}B$$

This 'proof' is ridiculous, though it would take a lengthy disgression to analyse fully its mathematical and historical solecisms. As pointed out earlier, the flaw is that the proof depends on the underlying arithmetic, but this will not be properly set up before the development of the idea of the real numbers. Fortunately this digression is not necessary here; my concern is early Greek mathematics, and problems with and delusions about the role of decimal numbers in the arithmetisation of mathematics are not part of this. However, there is a very similar process that differs only by a change of base—sexagesimal ratios—that it is necessary to consider briefly, since sexagesimal arithmetic is found in Babylonia some 1500 years before the development of early Greek mathematics. Concerning this, I here make only the following two observations, and defer further comment to Section 2, below.

First, the problems with decimal and sexagesimal arithmetic arise with 'non-terminating' decimal numbers, those ratios in which an infinite number of the n_k are non-zero, and the consequent difficulties that arise from the possibility of a 'carry' through an arbitrarily long sequence of digits.[13] Babylonian arithmetic shows a proper caution about this, since many (thought not all) of the manipulations that are found are restricted to the terminating or 'regular' sexagesimal numbers.

Second, our earliest trace of sexagesimal numbers in Greek mathematics are found around the second century BC, in the work of Hypsicles and Hipparchus. We have as of yet no explicit evidence of any influence of Babylonian arithmetical procedures on early Greek mathematics.[14]

Accountant's ratios. Let me illustrate this final definition by an example. The ratio 65:24 is more than twice, less than three-times; that is,

$$65 = 2 \times 24 + 17 \quad \text{or} \quad a = n_0 b + a_1 \quad \text{with} \quad a_1 < b.$$

We now describe 17:24. Since 17 goes once, not twice into 24, this ratio is more than half:

$$2 \times 17 = 24 + 10 \quad \text{or} \quad n_1 a_1 = b + a_2 \quad \text{with} \quad a_2 < a_1.$$

[13] See Fowler 1985a and 1985b for illustrations of the difficulties with decimal arithmetic.

[14] On this topic, see Berggren 1984, 397–398: 'If the event [of Babylónian influence on pre-Euclidean mathematics] cannot be located historically one must recognise the possibility that it may not have occurred.'

We continue by comparing the remainder 10 with 24; it is more than the third:

$$3 \times 10 = 24 + 6 \quad \text{or} \quad n_2 a_2 = b + a_3 \quad \text{with} \quad a_3 < a_2.$$

Finally

$$4 \times 6 = 24 \quad \text{or} \quad n_3 a_3 = b.$$

Hence, the process is described by the pattern 2, 2, 3, 4, and no more, a kind of pattern which again gives an immediate proof of the *Topics* proposition. Again, this is not a reciprocal subtraction process: after the first step, the current object is always subtracted from the second term up to the first overshoot, which then becomes the object for the next step.

The special interest of this process is that we again have an arithmetised interpretation. In our modern notation,

$$\frac{65}{24} = 2 + \frac{1}{2} + \frac{1}{2 \times 3} + \frac{1}{2 \times 3 \times 4}$$

or

$$\frac{a}{b} = n_0 + \frac{1}{n_1} + \frac{1}{n_1 \times n_2} + \frac{1}{n_1 \times n_2 \times n_3} + \cdots.$$

(This is, in fact, not unlike the way fractional quantities were expressed by Greek accountants and mathematicians, though our evidence also makes it quite clear that this particular algorithm was not used to generate the expressions we find them using, since their expressions do not exhibit the characteristic pattern n_1, $n_1 \times n_2$, $n_1 \times n_2 \times n_3 \ldots$.) Again, although this suggests that there may again be an underlying arithmetised proof of the *Topics* proposition, the details of such a proof are far from obvious.

2. Arithmetic and ἀριθμητική

The previous section illustrated a series of distinctions between ratio and proportion, between complete proofs (seven, on my count) founded on explicit definitions and optimistic pseudo-proofs in which the crucial and difficult details were omitted (three, one in Section 1.1, two in Section 1.6), and between arithmetised and non-arithmetised mathematics. It is that last distinction that I now wish to consider.

In brief, my proposal is that early Greek mathematics and astronomy show no trace of influence of arithmetisation. The emphasis on 'early' in 'early Greek mathematics' is essential: after the amalgamation of Babylonian and Greek techniques, which seems to take place from the second century BC onwards, we do find examples of a Greek arithmetised mathematics in Heron and thereafter, and a Greek arithmetised astronomy in

Ptolemy and thereafter. Thus, there is a profound split between the aims and conceptions of early and later Greek[15] mathematics, such as I conceive them: note, for example, how neither Theaetetus' programme of *Elements* x and its application in *Elements* xiii nor the sophisticated Eudoxan astronomy seem to have any place in these later arithmetised traditions. What is more, today's mathematics is profoundly arithmetised, founded as it is on the use and intuition of what has come to be known, comically, as 'the real numbers'.[16] This again interferes with our understanding of early Greek mathematics; indeed one of the difficulties in the way of understanding my reconstruction is the problem of purging one's mind of this arithmetised way of thinking.

Early Greek mathematics does, of course, draw on the use and intuition of some kinds of numbers, as I shall describe briefly. Most fundamental are the ἀριθμοί, conceived in a very concrete sense which is best conveyed in English by the series

solo, duet, trio, quartet, ...

Moreover, the words usually occur with the definite article, which enhances further their concreteness. In formal mathematics the unit has a different status from the rest [see *Elem.* vii defs. 1–2], which means that sometimes this case has to be distinguished as, e.g., in *Elem.* vii prop. 2. The ἀριθμοί are also found in different grammatical forms, such as the repetition-numbers

once, twice, three-times, four-times, ...

The grammar of natural language describes the manipulations of the ἀριθμοί. For example, contrast 'four-times the duet gives the octet' with the abstract manipulations of abstract symbols, '$4 \times 2 = 8$', that we tend to learn and use today. So there would not be the same temptation among early Greek mathematicians to extend the scope of these abstract manipulations and objects, for example, to extend $8 \div 4 = 2$ to the case of

[15] 'Greek' means 'written in Greek'. Early Greek mathematics is geographically Greek: our evidence points to Ionia, Magna Graecia (Southern Italy and Sicily), mainland Greece, and then the Greek colony of Egypt. This *portmanteau* designation, 'Greek', later comes to encompass a vast collection of different traditions and influences.

[16] See, for example, Mueller 1981, ch. 7 for recognition of the difficulty of incorporating *Elem.* x into arithmetised mathematics (e.g., 1981, 271: 'One would, of course, prefer an explanation that involved a clear mathematical goal intelligible to us in terms of our own notions of mathematics . . . . Unfortunately book 10 has never been explicated successfully in this way, nor does it appear amenable to explication of this sort').

$8 \div 3 = ?$, or to move to higher degrees of abstraction such as $a \times b = c$. The Greek formulation of division in formal mathematics tends rather to the use of manipulations such as 'the trio goes into the octet twice leaving a duet as remainder' [see *Elem.* vii props. 1–4], or the general verbal descriptions to be found in *Elem.* vii defs. 5–10. I shall refer to these Greek investigations of the ἀριθμοί by their Greek name, as ἀριθμητική, and thus distinguish between that which we find in early Greek mathematics and the later arithmetic which concerns more general and abstract kinds of numerical quantities.

Everyday Greek accounting does employ a system of describing fractional quantities, traces of which are also found in formal mathematics. This is based on the system of the μέρη,[17] which are best conceived as the series:

the half, the third, the quarter, the fifth, ...

and represented by a transcription of their Greek notation: $\angle$, $\acute{\gamma}$, $\acute{\delta}$, $\acute{\epsilon}$, ..., for example, as $\acute{2}$, $\acute{3}$, $\acute{4}$, $\acute{5}$, ... Here the definite article is an essential aid to understanding: neither the words nor the notation contain those features that lead easily to our conception of our common fractions, where we can pass almost imperceptibly from 'one fifth' and '$^1/_5$' to 'two fifths' and '$^2/_5$', and so on. Moreover, Greek fractional quantities are always expressed and always seem to be conceived as sums of different μέρη, in what is often called the 'Egyptian' system. In fact, I do not believe we have any convincing evidence for anything corresponding to our common fractions $^m/_n$ in Greek scientific or everyday life. What is usually taken as the notation for common fractions[18] seems rather to be an abbreviation employed extensively by Byzantine scribes but found in very few documents before then, in which the phrase τῶν m τὸ $\acute{n}$ ('of m the nth') is abbreviated as $\overset{\acute{n}}{m}$. But this phrase 'of m the nth', the standard phrase used to describe division, always seems to be conceived and is almost always immediately expressed as a sum of μέρη, for example

τῶν	$\iota\beta$	[τὸ $\iota\acute{\zeta}$]	$\angle\,\iota\acute{\beta}\,\iota\acute{\zeta}\,\lambda\acute{\delta}\,\nu\acute{\alpha}\,\xi\acute{\eta}$
of the	12	[the 17th is]	$\acute{2}\,\acute{12}\,\acute{17}\,\acute{34}\,\acute{51}\,\acute{68}$

for what we now write as

$$\frac{12}{17} = \frac{1}{2} + \frac{1}{12} + \frac{1}{17} + \frac{1}{34} + \frac{1}{51} + \frac{1}{68}.$$

[17] Greek has two words, μέρος and μόριον (plural: μέρη and μόρια), which appear to be perfectly synonymous.

[18] For an influential description, see Heath 1956, i 42–45.

These expressions would have been looked up in division tables, of which many examples have now been published. The only manipulations of these expressions that are found are very restricted: 'of m the nth' is seen to be the same as 'of km the kn^{th}', and 'of m the nth' and 'of p the nth' can be added or subtracted to give 'of $m \pm p$ the nth', where all of the expressions are, I repeat, still conceived as sums of μέρη. Nowhere, to my knowledge, do we get an example where two general expressions 'of m the nth' and 'of p the qth' are directly combined without going through some sequence of these basic manipulations.

Most of our evidence comes from school or commercial texts, far removed from the kind of mathematics in which we are primarily interested here. Unfortunately the most pertinent mathematical text, Archimedes' *Measurement of a Circle*, survives only in a very late and corrupt version which shows clear signs of interference by scribes and commentators. Aristarchus' *On the Sizes and Distances of the Sun and Moon* has survived in a less corrupt state though, here again, as with almost all of our evidence, our only text is a Byzantine copy made in the ninth century AD. Nevertheless, my description above also fits the evidence that we find in both of these calculations.[19]

The arithmetic of these sums of μέρη is very clumsy and does not show any promise of an interesting or useful mathematical theory. This might be interpreted as an explanation why early Greek mathematicians do not seem to bring to bear on their mathematics any intuitions about arithmetical manipulations with fractional quantities, if such an explanation is needed for us, today, to come to terms with what seems to be an uncomfortable feature of our evidence. My own preference is to state boldly and accept completely that such evidence we have of early Greek mathematics shows no influence of arithmetisation; and not, at this stage, to attempt to fabricate any further explanation.

3. *Envoi*

It may be difficult for someone brought up within the now universal and highly successful tradition of arithmetised mathematics to conceive that there are many ways of handling ratios other than as something that is, or is approximated by, some suitably formulated kind of numerical quantity, such as common fractions $^m/_n$ for commensurable ratios, or some systematically organised collection of common fractions, like decimal or sexagesimal

[19] A much more complete description of our evidence concerning Greek calculations appears in Fowler 1987.

fractions, for incommensurable ratios. I, at least, once found this difficult, so that the substantial explorations of non-arithmetised ratio theories set out in my book [Fowler 1987] were a liberating experience. Moreover, the exploration of these ideas revealed many remarkable and unexpected mathematical and historical insights. I do not wish to attempt to summarise this material here, so I will finish with one illustration and refer the reader to the book for more details.

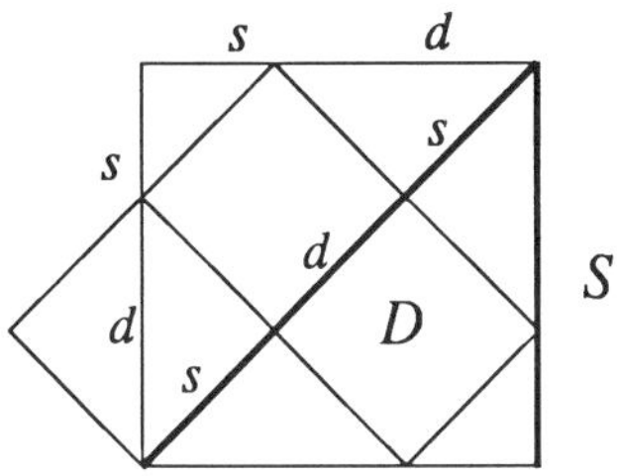

Figure 6

Let us work out the anthyphairetic ratio of the diagonal to side of some regular polygons. We start with the square. Let s and d denote the diagonal and side of some given square. The beginning of Socrates' encounter with the slaveboy at Plato, *Meno* 82a–85c brings out that $s < d < 2s$; hence, the first step of the anthyphaireses of $d{:}s$ will be one subtraction and not two, and the remaining steps will then be described by the ratio $s{:}(d-s)$. In the notation of n11 we can write this as $d{:}s = [1,\ s{:}(d-s)]$. The idea is general:

$$a_0{:}a_1 = [n_0,\ a_1{:}a_2] = [n_0,\ n_1,\ a_2{:}a_3] = \ldots.$$

We now need to evaluate $s{:}(d-s)$. Perhaps, like Meno's slaveboy, we also need the help of a diagram such as is given in Figure 6. Here we construct a new larger square whose side S is the side plus diagonal of the smaller oblique square in the left-hand corner:

$$S = s + d;$$

then, by filling in some lines in the figure, we see that the larger diagonal is equal to two small sides plus the small diagonal:

$$D = 2s + d.$$

Since the size, location, and orientation of our square are immaterial,

$$s{:}(d-s) = S{:}(D-S) = (s+d){:}s$$

which we can now evaluate as two subtractions, followed by the ratio $s{:}(d-s)$. We now go round and round: $s{:}(d-s)$ is therefore, twice, twice, followed by itself, so it is twice, twice, twice, twice, followed by itself, and so on. Hence, the anthyphairetic ratio of the diagonal to side of a square is once, twice, twice, twice, ...

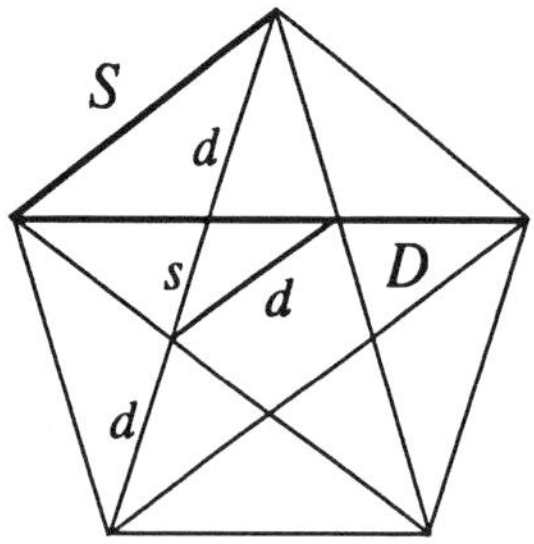

Figure 7

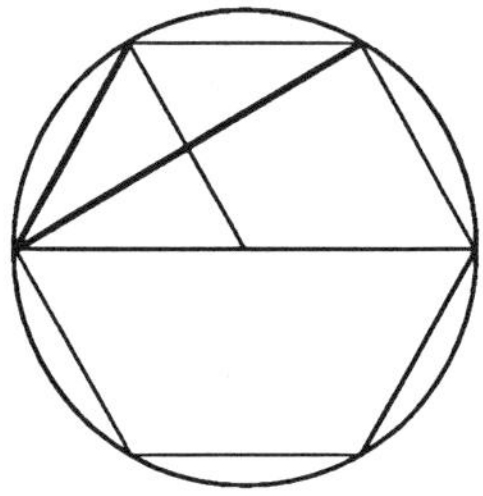

Figure 8

The reader is recommended to use a similar argument, applied to Figure 7, to evaluate the ratio of the diagonal to side of a pentagon. Now consider the ratio of a diagonal to the side of a hexagon [see Figure 8]. The longer diagonal is twice the side—that is a description of the anthyphairetic ratio—while the square on the shorter diagonal is three-times the square on side [for details, see Euclid, *Elem.* xiii prop. 12]. This ratio can be evaluated in the context of the following more general programme: Given a line and two ἀριθμοί n and m, we can use *Elem.* ii prop. 14 to construct squares equal to n–times and m–times the square on the given line. What can we say about the anthyphairetic ratio of their sides? The answer to this question plays a central role in my reconstruction: it involves heuristic explorations followed by a range of different proofs based on the figures of *Elements* ii; and it leads to a motivation for *Elements* x, and to a new description of the problems and motivations of early Greek mathematics.[20]

[20] An outline of some of these interpretations may also be found in Fowler 1979 and 1980–1982.

An earlier version of this paper was presented at a colloquium, 'Logos et théorie des catastrophes: A partier de travail de R. Thom' (Centre culturel Cerisy la Salle, September 1982), and has subsequently circulated in duplicated form. I wish to thank the many people who have offered comments. Some of the topics discussed here are treated more fully in Fowler 1987.

7

What Euclid Meant: On the Use of Evidence in Studying Ancient Mathematics

WILBUR R. KNORR

For most historians of mathematics the principal data are documents—records of past thoughts preserved in writing. It follows that the interpretation of documents is central to the methodology of historians and, hence, that discussions of the principles of interpretation can be brought to bear on efforts in this field.

As a specialist in mathematical history, I have found that my colleagues in the areas of literary studies tend to register surprise at the thought that mathematical texts are subject to *interpretation*, even as they take for granted that all literary texts require interpretation. Moreover, I would anticipate that associates in the disciplines of mathematics and the physical sciences would be surprised—perhaps appalled—at the suggestion that the understanding of technical documents could be illuminated through the insights of theorists of literary criticism. Somehow, the patent universality of mathematical discourse might be construed as precluding the relevance of critical principles whose objective is to offer guidance in the study of individuals in their special historical circumstances.[1]

My project in the present essay is to explore this meeting ground between historical study and literary theory. My focus will be on the particular issue of the role of authorial meaning in the work of the critic. After a brief synopsis of some ideas from recent debates, I will discuss their bearing on three problems in the interpretation of Euclid's mathematics: his conceptions of ratio and proportion, his notion of fraction, and the aim over all of the *Elements*. From the outset I must emphatically disclaim any special expertise

---

[1] This ahistorical, Platonizing tendency of mathematicians and mathematical historians is noted and criticized in Unguru 1979.

in the wide-ranging field of hermeneutics. My approach here is entirely pragmatic: to select from the diversity of views those which I perceive can assist the historian in the effort to understand why disagreements arise in the examination of such problems and what is implied within the different options that one might espouse in their interpretation.

1. Authorial meaning in literary criticism

We all continually subscribe to the view that we can formulate our ideas in writing and successfully communicate them to others. After all, did the ancients not invent writing precisely for this end? But contemporary critics have come to recognize the difficulties in applying this common sense notion toward the interpretation of literature.[2] In the old *régime* one approached a text with the assumption that there was a *datum*, designated as the author's meaning or intention, which was the object of critical exegesis. Within the 'new criticism' of this century, however, profound doubts were expressed: one could multiply examples of how one and the same case had received diverse, incompatible accounts of its author's meaning; one could note the drastic consequences that the assumption of irony has for the interpretation of a text, yet the frequent difficulty of establishing an author's ironic intentions; and so on.[3] By way of reaction, comes scepticism which emphasizes the problems of *access* and *relevance*: How can we presume to enter into the mind of an author? and Why should we even want to do this as part of our critical efforts?

A particularly trenchant statement of the sceptical position was put forward by W. K. Wimsatt, Jr. and M. C. Beardsley under the rubric of the 'intentional fallacy'.[4] In the course of time, their essay has been invoked in support of positions far more extreme than theirs, so that the 'intentional fallacy' has come to signify for some the impossibility of any critical use of the concept of authorial meaning.[5] In such exaggerated formulations, one maintains that the special, private circumstances of an author are beyond

---

[2] H. Parker [1984, 213–243] sketches some main currents in modern criticism, with particular emphasis on the teaching of American literature. He is decidedly more antithetical to the new criticism, even than Hirsch [see below].

[3] See the synopsis of the arguments in Hirsch 1967, ch. 1.

[4] Cf. Wimsatt and Beardsley 1946: their sceptical position on authorial meaning extends the views expressed in Richards 1929.

[5] Hirsch [1967, 11–12] notes that the 'popular version' has severely exaggerated the claims that Wimsatt and Beardsley actually maintain in their essay. Cf. also H. Parker 1984, 214–215.

our ken and irrelevant to the critical task anyway; that one can consider only the public meanings attributable to the text. A text has no fixed meaning; the process of interpretation is dynamic, as readers respond to it in their individual ways. As critics, we are bound to our own historical circumstances. Instead of aiming to render a historically correct account of the text's original meaning, then, we should seek to articulate our own responses.[6]

Even those unsympathetic to this position admit to the positive effects its adoption has had on the critical disciplines and the teaching of literature in recent decades.[7] Whereas it had been common earlier to glean literature as a source of historical, social or political information, for instance, one now could analyze literary products for themselves: a poem is a poem, and only incidentally, say, a record for the reconstruction of the author's biography.[8] Nevertheless, the sceptical position effectively abandons the historical project: if no interpretation of a text is privileged, all are equivalent and it becomes meaningless to examine historical texts for their historical content.

While the sceptical vein represented by the intentional-fallacy argument has been intensified in some circles, others have proposed counter-arguments in defense of a more traditional literary methodology. A particularly thorough venture of the latter type is the hermeneutical study by E. D. Hirsch, Jr.[9] A brief account can presume neither to do justice to the richness and subtlety of his discussion, nor to give due coverage to the rejoinders from advocates of other positions. I hope merely to provide here a synopsis of

[6] I venture to note a certain parallel with contemporary developments in the philosophy of science: the older objectivist–positivist views now seem naive, as the subjective elements implicit in scientific theory and research have come to be recognized even by those who would still favor some form of scientific realism.

[7] Cf. Hough 1966, 62 which maintains that for students of literature, Richards' theory was 'extremely fruitful' for providing 'a means of compelling close attention to the work itself and the processes involved in reading it, as a prophylactic against conventional and secondhand judgments.... But it is not the normal kind of reading.'

[8] A pertinent example appears in the essay by Charles Kahn [see ch. 1, above]: that the older view of Hesiod's work had inclined toward an anthropological analysis, while recent efforts have attempted instead to grasp the impact and purposes of his poetry *as poetry*.

[9] Hirsch's account is widely known among a diverse range of scholars in the humanities, and is highly respected even among those who do not adopt his relatively traditional position. Needless to say, the field of criticism over the past two decades has grown enormously, and one might incline to view such efforts as Hirsch's as outdated. My objective here, however, is not to show that one or the other critical theory is *correct* (whatever that could mean), but rather that Hirsch's views in particular can be *useful* for the practicing historian.

some key notions that will be of service in the subsequent discussion of ancient texts.

In Hirsch's view [1967, ch. 4], the author's meaning is not only a legitimate aim of criticism, it is the only possible aim: for it alone is shared by all interpreters of the given text.[10] The critic's work, he maintains, is of two basic sorts: to give an account of the *meaning* of the text and of its *significance*. The latter embraces the major portion of criticism as such (indeed, the whole of it, in the sceptical view): the connections between the text and whatever else the critic chooses, the critic's personal response to the text, and so on. But a precondition of any discussion of a text's significance, Hirsch continues, must be an accurate grasp of its meaning. This entails two projects: to *understand* the text and to *explain* (or interpret) it. In explaining a text, the interpreter seeks to communicate its meaning to others. To this end, one typically resorts to paraphrases, recasting the text in terms calculated to be familiar to the audience.[11] One may well introduce elements entirely extraneous to the text itself, and it is a subtle demand on the interpreter to make sure the meaning of the text is not violated in the process.

To explain the text, the interpreter must already understand the meaning of the text, that is, the meaning intended by the author. The sceptics maintain, however, that this sort of meaning is inaccessible to us. But Hirsch [1967, ch. 5 and app. 1, sect. c] here introduces another distinction: if we insisted on *certainty* in our understanding of the author's meaning, the sceptics would be sustained. But the critic seeks not certainty, but *validity* of interpretation. Validity is a probabilistic notion; the interpreter engages in a heuristic process, refining and modifying tentative conceptions of the text's meaning, and so achieving interpretations of progressively increasing probability of being correct. In effect, the author's meaning is the limit of this heuristic process; without it, the process would have no object or criterion of accuracy. But how does one gauge the validity of one's account—that is, as being highly probable, or plausible, or merely possible? Hirsch cites four criteria: *legitimacy* (e.g., the account must

[10] A useful synopsis of Hirsch's position on the verification of meaning appears in 1967, app. 1, esp. sect. c.

[11] Hirsch's distinction of the interpretive and critical functions of textual commentary [1967, ch. 4, sect. b] might be used to suggest a position on the issue of geometrical algebra, currently debated among historians of ancient mathematics [see Unguru 1979]. To explain certain aspects of ancient geometry, it may become advisable, even necessary, to import notions from more recent fields, like algebra. This raises the possibility of anachronism, as is present in all analogical forms of exegesis. That risk becomes acceptable if the alternative is the learner's incomprehension.

assign to words of the text only those meanings which are possible for the author and his contemporaries); *correspondence* (each linguistic component of the text must be accounted for); *genre appropriateness* (where 'genre' embraces those conventions and expectations pertinent to the text which the author and his audience will share); and *coherence* (the interpretation must be plausible in the context of the whole of which it is part).

Hirsch [1967, 76–77, 237–238] observes that the criteria of genre and coherence sometimes lead into a hermeneutic circle. The broad notions of genre with which we initiate the examination of a text, for instance, have limited explanatory value; indeed, they function only as heuristic guides, as one refines one's conception of the text's meaning. Ultimately, knowing the intrinsic genre of the text is tantamount to understanding its meaning.[12] Similarly, in assessing the coherence of our interpretation, the whole against which we set our text will depend on our interpretation. Initially, when our view of its meaning is still open, the correlative ensemble of texts will be large; but as we sharpen our conception, the context will narrow. Testing an interpretation will involve showing that the author means precisely this in texts just like our text. As before, Hirsch obviates the problem of circularity by consideration of the heuristic element in interpretation. If we aspired to certainty, he argues, the reservations of the sceptics would be sustained, rendering the quest for author's meaning futile. But we do not demand certainty in most contexts of thought and action, and need not do so in hermeneutics either. Our aim ought to be to hit upon accounts of high probability; the process of *validation* of interpretations is precisely that of gauging the relative probabilities of competing interpretations. Thus, the uncertainties, the possibility of alternative views, the role of subjective factors—altogether familiar within a spectrum of human pursuits—are natural adjuncts of the interpretive enterprise. To propose these as fundamental objections against the viability of the search for author's meaning merely misconstrues what one's goals ought to be.

Thus, interpretation is not mechanical: it is a dynamic heuristic process, where one seeks to measure the validity of interpretations. This is enough, in Hirsch's view, to dispel the greatest difficulties raised by the sceptics. One can engage in an orderly quest for a valid interpretation of author's

[12] Hirsch [1967, 86] defines 'intrinsic genre' as 'that sense of the whole by means of which an interpreter can correctly understand any part in its determinacy'. By this he of course specializes the notion of genre, which in common usage denotes much broader categories of literary effort. Hirsch hereby captures the extremely close connection between grasping the meaning of a text and specifying its genre, but avoids the tautology whereby every text would constitute its own separate intrinsic genre.

meaning; author's meaning is the objective toward which this process is and must be directed. Hirsch observes that the sceptics admit this *de facto* by virtue of their participation in criticism: for inquiry would be senseless if all conceivable claims about a text had equivalent validity. Finally, understanding the author's meaning is the precondition for all the other inquiries in which critics engage, since it is the only feature of a text common to all potential critics.

This critical scheme provides a basis for examining the interpretation of texts from ancient mathematics. It also provides a cautionary note, by alerting the interpreter to the subtle difficulties that this activity poses, in particular, the hazards entailed in the interplay between the objective content of the texts and the subjective elements present in the experience of the interpreter. Mathematical texts are especially susceptible to being read in the context of the philosophical and mathematical predispositions of readers trained in the modern disciplines. Avoiding the misconstructions of authorial meaning that can result becomes, as we shall see in the following examples, the particular concern of the historian of mathematics.

2. The Euclidean concepts of ratio and proportion

As one would expect, the interpretation of ancient mathematical texts is strongly influenced by considerations grounded in modern mathematical theory, and these may introduce anachronizing tendencies. The discussion of the ancient convergence principles, specifically as they relate to the definitions given by Euclid at the beginning of his proportion theory [*Elem.* v], provides an interesting example. It is widely maintained that Euclid's definitions (in particular, def. 4) have the aim of excluding non-Archimedean magnitudes from the domain of geometry, a claim that is supported through consideration of subtle requirements of the Euclidean proofs. Similar observations are made for Archimedes' convergence theorems and his application of the so-called Archimedean axiom, with which the Euclidean definition is typically associated. But if one moves from the mathematics to the text, a different story emerges.[13]

Among the definitions prefacing the theory of proportions in *Elem.* v are the following:[14]

---

[13] My account will be in substantial agreement with that presented in Mueller 1981, 138–145.

[14] My translation from the text of Heiberg [1883, ii 2].

> 'Ratio' is of two homogeneous magnitudes the manner of relation [they have to each other] with respect to size.[15] [def. 3]

> 'Having a ratio to each other' is predicated of magnitudes which when multiplied can exceed each other. [def. 4]

With reference to the latter definition, T. L. Heath offers this commentary:

> De Morgan says that it amounts to saying that the magnitudes are of the same species. But this can hardly be all; the definition seems rather to be meant, on the one hand, to exclude the relation of a finite magnitude to a magnitude of the same kind which is either infinitely great or infinitely small, and, even more, to emphasize the fact that the term *ratio*, as defined in the preceding definition,... includes the relation between any two *incommensurable* as well as between any two commensurable finite magnitudes of the same kind. [Heath 1956, ii 120: his emphasis]

By the phrases 'to be meant' and 'to emphasize', Heath clearly indicates his own intent to articulate the meaning Euclid himself had in mind. But it must seem remarkable that three such different meanings—homogeneity, the exclusion of non-finite (i.e., non-Archimedean) magnitudes, and the inclusion of incommensurables—could be covered in a single expression, and further, that Euclid could emphasize a claim about incommensurables without actually using the term.[16] We thus confront a situation where, to use Hirsch's terminology, the effort to *understand* (or interpret) Euclid's text is separate from its *criticism*, that is, the elaboration of its mathematical implications.

Read in isolation, Definition 4 may indeed be construed as a condition intended to exclude non-Archimedean magnitudes. It would then be transcribed in the form that magnitudes A and B (for $A < B$) have a ratio

---

[15] Contrast Heath 1956, ii 114 (emphasis his): 'A *ratio* is a sort of relation in respect of size between two magnitudes of the same kind.' Doubtless, Heath's rendition now would be considered standard. But in employing the indefinite article ('a sort of relation'), he appears to have lost a nuance of the Greek definite article (cf. my 'the manner of relation' for ἡ... ποιὰ σχέσις). More important, Heath's version is vacuous, since in his rendering nothing is actually being defined (Mueller [1981, 126] calls this sense of the definition 'mathematically useless', but sets the onus of the difficulty on Euclid.) In my version, Euclid is specifying 'ratio' as a relation of *quantitative measures* of homogeneous figures. That is an essential and non-trivial condition, and would qualify as a definition on the supposition that the reader already grasps the notion of quantity or size (πηλικότης).

[16] The terms for 'commensurable' and 'incommensurable' first appear in the first definition of book 10.

if and only if there exists a finite integer m such that $mA > B$. Thus, for instance, for the indivisible element A of a finite figure B, there could be no ratio between A and B, since A taken any finite number of times could not be made to exceed B.[17] The proofs of *Elem.* v prop. 8 and x prop. 1 both depend on such an assumption: that the smaller of two given magnitudes, when multiplied, will eventually become greater than the other.[18] Since, furthermore, indivisibles were debated within early Greek natural philosophy and played a role in the heuristic analysis of figures by some precursors of Archimedes,[19] one might accept that Euclid (or Eudoxus, the author of the source version of the theory) chose to exclude such cases from the formal theory of proportions of magnitudes. Indeed, already among ancient writers, the definition was read as a condition for convergence by the exclusion of non-finite magnitudes.[20]

Nevertheless, this view of Euclid's principle runs into several difficulties. Most notably, Euclid's definition sets up a symmetrical relation between two given magnitudes, whereas the cited applications require only a property relating the smaller to the greater. These aspects of the question are well summarized by Mueller, so that I can omit their discussion here and turn at once to the presentation of an alternative view.[21]

Let us first take note of the definition of proportion on which Euclid's theory depends; it is announced immediately after the principle we have just considered:[22]

---

[17] The use of indivisible elements of figures, most familiar in the context of the work of B. Cavalieri (1598–1647), is characteristic of Archimedes' heuristic measurements in the *Method*: see Dijksterhuis 1956, ch. x, esp. 318–322. I am preparing a study of the Archimedean method of indivisibles and the evidence for precursors in the older Greek geometric tradition.

[18] A statement of their critical assumption appears in n24 below. For an account of these propositions, see Mueller 1981, 139–142; van der Waerden 1954, 185–186, 188.

[19] For a discussion of the evidence of pre-Archimedean uses of indivisibles, see Knorr 1982a, 135–142.

[20] In Hero, *Def.* no. 123, the Euclidean definition of ratio is observed not to apply to the class of points; that is, the comparison property of multiples is essential to finite homogeneous magnitudes: cf. Schöne and Heiberg 1903–1914, iv 78. By contrast, the scholiasts on Euclid's *Elements* call attention to the condition of homogeneity [cf. Heiberg and Stamatis 1969–1977, i 215–216: nos. 13, 15–17] or to the inclusion of irrationals [no. 14]. While these Heronian and Euclidean writers may draw on authentic critical traditions, their dating is obscure and their authority questionable in determining Euclid's actual purposes.

[21] For additional details, see Mueller 1981, 138–145.

[22] My translation is based on Heiberg's text [Heiberg and Stamatis 1969–1977, ii 1].

> 'Being in the same ratio'—a first [magnitude] to a second and a third to a fourth—is predicated whenever, in regard to the equimultiples of the first and third relative to the equimultiples of the second and fourth, according to any multiplication whatever, the former [equimultiples] alike exceed the latter, or alike equal [them], or alike fall short, taken in the same order. [def. 5]

That is, for magnitudes A, B, C, D, the ratio $A{:}B$ is the same as the ratio $C{:}D$ if, for arbitrary integers m, n, $mA > nB$ and $mC > nD$ obtain together, $mA = nB$ and $mC = nD$ obtain together, and $mA < nB$ and $mC < nD$ obtain together. Euclid's scheme for proportions thus turns on the formation and comparison of equimultiples of given magnitudes. Every application of the definition depends on the hypotheses, that $mA > nB$, that $mA = nB$, or that $mA < nB$.[23]

The fourth definition, which has just preceded, may be read in the context of this statement of proportionality. It thus stipulates that magnitudes A, B will be said to have a ratio if the inequalities between their multiples can be satisfied; i.e., that there exist m, n such that $mA > nB$ and also m', n' such that $m'A < n'B$. This is the reading favored by Mueller, but does not appear to have been recognized by other commentators.[24] The condition, as now formulated, simply refers to the comparability of arbitrary multiples, just as the fifth definition and all the proofs dependent on it thereafter require.

The fourth definition does of course exclude non-Archimedean magnitudes. For it supplies a gap which the fifth definition by itself would suffer: if B and D were indivisible (that is, zero) magnitudes, for instance, then only the inequalities $mA > nB$ and $mC > nD$ would be possible; it might then be technically possible to prove the proportionality $A{:}B = C{:}D$, since the condition for the opposite inequalities would be vacuously true. But in view of the terms of its formulation, the definition is not easily seen to bear this latter consequence as its principal aim. Indeed, since it has been phrased in precise conformity to the demands of the definition of proportion in def. 5—whence one would locate the intention of def. 4 as the explicit

---

[23] One may note a degree of redundancy. The condition $mA = nB$ can be satisfied only if A, B are commensurable; in this case, the conditions on inequalities are unnecessary. On the other hand, if the inequalities alone are satisfied (this, of course, is the only possible situation for incommensurables), then this would suffice for establishing proportionality, even without reference to the case of equality.

[24] The usual transcription is that 'for some m, $mA > B$, where $B > A$'. Cf. Heath 1956, ii 120; Dijksterhuis 1929–1930, ii 58; van der Waerden 1954, 186n; Frajese and Maccioni 1970, 298. This assumption, which indeed is made in *Elem.* v prop. 8 and x prop. 1, is taken by Mueller to be different from *Elem.* v def. 4 [see below].

precondition for def. 5—one may doubt that Euclid even recognized its implication for the elimination of non-finites. As all applications of the proportionality theorems (in books 6, 11–13) are in fact for cases of finite magnitudes, the difficulty does not there arise. On the other hand, the general theorems of book 5 do require a conditional restricting the domain to finites. One is free to judge as one likes how serious is Euclid's omission of an appropriate qualifying statement. It is clear, at any rate, that def. 4, as proposed in the *Elements*, does not expressly fill that role.

Evidence from Archimedes and the later commentators provides materials for sketching out the origins of these Euclidean definitions, by revealing the form of a technique of proportions alternative to what we now have in *Elem.* v. I have argued elsewhere [Knorr 1978b] that a certain technique evident in these sources can be assigned to the pre-Euclidean period, indeed, to Eudoxus himself. In the sketch I give here, I will refer to it as 'Eudoxan', with quotation marks to indicate the circumstantial nature of the attribution.

To prove a given theorem according to the 'Eudoxan' technique, one first takes up the commensurable case, usually a straightforward consequence of the assumption of a common measuring magnitude. To establish the incommensurable case, one adopts an indirect reasoning: if the ratios are unequal (say, $A{:}B$ is greater), one can construct a suitable magnitude B' commensurable with A, such that $A{:}B > A{:}B' > C{:}D$. (A constructing procedure is reported in an extant fragment and is comparable to the construction in *Elem.* xii prop. 16 [see Knorr 1978b, 187–188].) This condition is then shown to contradict geometric properties already shown for the commensurable case. In similar fashion, one shows that the contrary supposition (that $A{:}B$ is the lesser ratio) also leads to contradiction.

If we replace the ratio of commensurable magnitudes $A{:}B'$ with a ratio of integers equal to it, say $m{:}n$, the inequality just mentioned can be expressed in an equivalent form: that there exist integers m, n such that $mA > nB$ at the same time that $mC < nD$. This happens to be precisely the condition by which Euclid defines 'greater ratio' in *Elem.* v def. 7. Inverting the inequalities, we obtain the conditions for $A{:}B$ to be the lesser ratio (not, however, defined separately by Euclid). Equality of ratio would follow if there exist no integers for which either set of inequalities obtains; that is, if for all multiples, $mA > nB$ entails that $mC > nD$, while $mA < nB$ entails that $mC < nD$. In this way, the definition of 'same ratio' adopted in def. 5 can be deduced as the logical inverse of the definitions of 'greater ratio' and 'lesser ratio'.

Three considerations indicate that Euclid's definition of 'greater ratio' is a vestige of an earlier form of the theory, as my proposal for the genesis of the Euclidean definitions suggests. First, the notion of a greater ratio is not developed as such; indeed, its definition is directly invoked only twice (namely, in *Elem.* v props. 8 and 13), in lemmas auxiliary to effecting proofs of proportionality in accordance with def. 5.[25] For instance, *Elem.* v prop. 8 establishes that $A > B$ implies that $A{:}C > B{:}C$, and similarly for the reverse inequalities; it is applied in v prop. 10, but the manner of appeal here is flawed, and an alternative proof founded directly on def. 7 would have been preferable.[26] Both theorems, in conjunction with v prop. 13, lead to the establishment of inequalities critical for the proofs of v prop. 16 and the following. For such uses, the concept of greater ratio is effectively superfluous, however, since it serves merely as an abbreviation for certain inequalities among multiples of given magnitudes. By contrast, in the 'Eudoxan' technique, manipulations of greater and lesser ratios are characteristic. This manner is imitated within the indirect arguments of book 12 (specifically, the theorems on the measurement of the circle, pyramid, cone, and sphere), where also the inequalities of v prop. 8 are conspicuous. Moreover, it is surprising that these same propositions do not exploit the definition of 'greater ratio', for that would have been natural and convenient. By here invoking the alternative assumptions of the sort characteristic of the 'Eudoxan' technique, Euclid appears to preserve marks of the older base of the theory of book 5.[27] Second, Euclid provides no construction for the integers whose existence is postulated in def. 7. This would doubtless cause uneasiness, were Euclid to have proposed theorems on inequalities of ratios corresponding to the proportion theorems of book 5. A construction lemma for the equivalent assumption does, however, appear among the texts associated with the alternative 'Eudoxan' technique [see Knorr 1978b, 187–188]. Third, as noted above, the convergence assumption alleged for the meaning of def. 4 is invoked only twice (namely, in *Elem.* v prop. 8 and x prop. 1), in the form that, given magnitudes A, B, where $A < B$, there exists a multiple m of B such that $mA > B$. But even here, one may question whether Euclid intends this

---

[25] For a criticism of the proofs, see Heath 1956, ii 152–153, 161–162; Mueller 1981, 130–131, 139.

[26] On the proof of *Elem.* v prop. 10, see Heath 1956, ii 156–157; Mueller 1981, 130.

[27] The basic similarity of technique between the proofs in book 12 and those in the alternative proportion theory is used to argue the Eudoxan provenance of the latter: cf. Knorr 1978b. For an account of Euclid's proof of the circle theorem [*Elem.* xii prop. 2] and the expected alternative method in the manner of the theory of book 5, see Knorr 1982a, 124–127; 1986a, 78–80.

assumption to be covered by his def. 4. For in general, Euclid manages cross-referencing *via* the literary device of verbal reminiscence; if he requires a previous theorem or postulate to justify the step in a proof, he will typically restate the terms of its enunciation in a paraphrase tailored to the present context.[28] It is remarkable, then, that in v prop. 8 and x prop. 1 the assumption of convergence is stated in terms quite different, indeed gratuitously different, from those in v def. 4.[29] I would thus infer that Euclid himself has formulated this assumption in the proofs, without perceiving—or, at the least, without wishing to mark—the connection with def. 4.

It thus seems far less clear than the usual view supposes, that Euclid's def. 4 was intended as a convergence assumption to exclude the case of non-finite magnitudes. The present account has shown how this definition is bound into the logical structure of the whole Euclidean theory of proportion. I think it possible that Euclid, in reworking the materials on proportion and convergence, as in v prop. 8 and x prop. 1, felt that the assumption he there had to make on the comparison of magnitudes was sufficiently obvious as not to require a special postulate. Certainly, there is no attempt to derive his assumption from def. 4, even though one could do that. Whatever textual affinities one can detect between the convergence assumption and the definition—and these are surprisingly few—can be accounted for through parallel developments from older sources.

Thus, far from being the definition's primary role, these applications in the convergence theorems are at best derivable consequences from it which Euclid appears not to have perceived. Significantly, when Archimedes formulates his own condition on convergence (Archimedean axiom), his terms

---

[28] Neuenschwander [cf. 1972–1973, 339–352] has discerned a pattern of close, often literal, recapitulation as Euclid's manner of cross-referencing in the planimetric books, especially prominent in book 2 and reasonably so in books 3–4. Comparable examples can be found in books 10, 12–13. van der Waerden [1979, 352–353] cites these insights as confirmation of a Pythagorean origin of books 2 and 4, a view with which Neuenschwander [1972–1973, 369–78] himself is in substantial, if not complete, agreement. It seems to me that the device of verbal cross-referencing is established; whether that owes to Euclid's editorial hand or is a feature of his sources, however, is a separate matter, by no means as clearly decided.

[29] In both of these propositions, the following formula is used: 'X when multiplied shall sometime be greater than Y; let it be multiplied, and let Z, a multiple of X, be greater than Y.'

indicate no linguistic affinity with Euclid's definition. For Archimedes asserts,

> of unequal areas the excess by which the greater exceeds the lesser can, added itself to itself, exceed any preassigned finite area. [Heiberg 1910–1915, ii 264] [30]

Any effort to view the intent of Archimedes' postulate to be an explicit extension of Euclid's definition falters through the absence of verbal resonances between the texts. Indeed, the template for Archimedes' wording is readily detected in the applications of implicit convergence assumptions in the Eudoxan limiting theorems, e.g., statements of this sort as in *Elem.* xii prop. 2:

> cutting the arcs in half... and doing this continually, we shall leave certain segments of the circle which shall be less than the excess by which circle EZHΘ exceeds the area *S*. [Heiberg and Stamatis 1969–1977, iv 80–83]

The origins and meanings of Archimedes' postulate raise questions that go beyond the present context. But I would insist that the effort to analyze it ought to adhere to the same textual procedure that I have proposed for Euclid's definition. In both cases, the usual procedure, founded on recourse to considerations drawn from the modern mathematical field, attempts to conflate the meaning of the ancient texts with certain implications derivable from them. In general, this is a dubious interpretive procedure, and in the particular cases at issue here results in confusion instead of insight.

3. The ancient concept of fraction

Reading David Fowler's remarks on the ancient technique of ratios [see ch. 6, above], I was struck by the following remark:

> [N]either the words nor the notation [employed for the expression of unit-fractional terms, or proper parts] contain those features that lead easily to our conception of our common fractions.... In fact, I do not believe we have any convincing evidence for anything corre-

[30] The principle is restated in essentially the same terms in the preface to Archimedes, *De lin. spir.* [Heiberg 1910–1915, ii 12], but quite differently in the fifth postulate of *De sphaer.* i [Heiberg 1910–1915, i 8]. For analyses, see Dijksterhuis 1956, 146–149; Knorr 1978b, 205–213.

sponding to our common fractions $^m/_n$ in Greek scientific or everyday life.[31]

The thesis is provocative, challenging the basic intuition that anyone with the most elementary training in mathematics would today have, namely, that our own concept of fraction is essentially obvious and, hence an inevitable feature of any viable computational tradition. As before, we meet an interpretive issue centering on intention; for the most part, our texts present only calculations without conceptual elaborations, so that our attempts to formulate the ancient authors' *concept* amounts to our own view of the intention underlying their technical operations. Fowler is right to approach the question open to the possibility that the ancient and modern views *could* be different, especially in the light of certain special procedures of unit-fractions that mark the ancient practice in distinction from the modern. He is also right to insist that any conclusions be persuasively documented. Agreeing on these basic principles, we may proceed to scrutinize more closely his claim, that the Greek arithmetic tradition never evolved the conception of fraction we now take for granted. My chief interest here will be to locate the evidence bearing on the question and to grasp what the range of convincing interpretations could be.

First, a *caveat*: the discussion of conceptions is a tricky matter and perhaps better assigned to the philosopher than to the historian. Technical works, whether ancient or modern, devote little or no space to conceptual discussions; and it is too easy for us to state our own preferences in such specific terms that we can discern their presence or absence in older works, as we choose. What, after all, is the modern conception of fraction? In any relatively sophisticated modern account [see e.g., Waisman 1951] one will find an analysis which reduces all the properties of fractional numbers to relations (specifically, ordered pairs) of integers. Is it impossible, then, that Plato [*Resp.* 525e: cf. Lee 1955, 293] has some comparable objective in view when he insists that 'the unit is indivisible... ' and that the experts will 'make you look absurd by multiplying it if you try to divide it... .' Similarly, Euclid might be engaging in such a sophisticated project in the arithmetic books (7–9) of the *Elements*, where he may be viewed as handling fractional numbers under the guise of ratios of integers. In such cases, the restriction of the term ἀριθμός to whole numbers need not betoken a failure to grasp the concept of fraction.

---

[31] A similar claim is made in D. H. Fowler 1983, 557. It is developed in D. H. Fowler 1987, ch. 7 (see 193, 226), where its validation depends critically on making a sharp break within the ancient arithmetical tradition around the 1st century AD (or earlier). In this way evidence from Hero and Diophantus is taken to attest only to the later phase of arithmetic technique.

In what follows, then, I recommend adopting a naive view of the concept of fraction. Our ancient writers may handle fractions as the quotients of division of integers, or alternatively, as ratios of integers. It will be enough, I propose, if the resultant entities in either case are treated as numerical terms, combinable with each other and with integers according to the rules of arithmetic. Notations or dictions need not disqualify an ancient effort *prima facie* as an acceptable manifestation of the general notion of fraction. Of particular interest for us is the ancients' remarkably persistent adherence to unit-fraction representations. Do these actually prevent the formulation of the general technique of fractions? As our evidence will derive from the full span of antiquity, we will also have to consider in what way late evidence bears on our views of the early (pre-Euclidean) period.

A rich store of fraction computations survive in the Akhmîm papyrus (in Greek, from 6th-century AD Egypt).[32] It opens with an enormous spreadsheet, giving a systematic listing of the results of divisions of series of integers, first by 3, then 4, and so on up to 20, expressed as sums of unit-fractions. For instance, the entry for 12 among the 17ths is $\acute{2}$ $\acute{12}$ $\acute{17}$ $\acute{34}$ $\acute{51}$ $\acute{68}$.[33] This table is used repeatedly in the arithmetic problems which make up the rest of the papyrus. Typical of one kind of fraction computation, occurring in about 30 of the problems, is this (no. 8):

> From $^{2}/_{3}$ subtract $\acute{3}$ $\acute{9}$ $\acute{99}$. What computation gives $\acute{3}$ $\acute{9}$ $\acute{99}$? the 11th of 5. $^{2}/_{3}$ of 11 is 7 $\acute{3}$. From 7 $\acute{3}$ subtract 5: remainder 2 $\acute{3}$. Of 2 $\acute{3}$ the 11th is $\acute{6}$ $\acute{33}$ $\acute{66}$. [Baillet 1892, 67]

The unit-fractions strike an alien note for the modern reader. But is it implausible to describe the writer's procedure as first raising terms to a common denominator, then subtracting, and finally simplifying the remainder? The subtrahend $\acute{3}$ $\acute{9}$ $\acute{99}$ is known from the table to be $^{5}/_{11}$; taking 11 as denominator, the terms become 7 $\acute{3}$ and 5, or a difference of 2 $\acute{3}$. The difference of the given fractions is thus the quotient of 2 $\acute{3}$ by 11.

We would prefer to simplify this to $^{7}/_{33}$ (raising terms by a factor of 3). But the ancient scribe is committed to answers in unit-fractional form, and so performs the division to get $\acute{6}$ $\acute{33}$ $\acute{66}$. Nevertheless, the unit-fractions play no computational role: the scribe first eliminates them by a suitable raising of terms before actually performing the required arithmetic operation, and only at the end is the result cast back into unit-fractional form.

---

32 The Greek text has been edited with French translation and commentary by Baillet [1892].

33 For a survey of such tables, see Knorr 1982b. For other instances, see D. H. Fowler and Turner 1983; D. H. Fowler 1987, ch. 7. Our particular example of the division of 12 by17 is cited also by D. H. Fowler on pages 115–116 of this volume.

The mathematical papyri relate to the humblest stratum of ancient mathematical instruction, and P. Akhmîm happens to be late in the tradition. From several centuries earlier a set of demotic (Egyptian) papyri contain problems of a similar kind. I cite one example to illustrate:

> Subtract $\acute{3}\ \acute{15}$ from $^{2}/_{3}\ \acute{21}$: Take 5, 7 times: 35; its $^{2}/_{3}\ \acute{21}$ is 25 and its $\acute{3}\ \acute{15}$ is 14. Subtract 14 from 25: 11. Find 11 35ths: $\acute{4}\ \acute{28}\ \acute{35}$. Answer: $\acute{4}\ \acute{28}\ \acute{35}$.[34]

Like the Greek scribe, the Egyptian works within the format of unit-fractions, while performing the arithmetic operations only after first clearing fractions, here by multiplying by 35. Presumably, that choice of multiplier (i.e., 35 or $5 \cdot 7$) has been hit upon through consultation of a table of 5ths and 7ths (since $^{2}/_{3}\ \acute{21}$ is $^{5}/_{7}$ and $\acute{3}\ \acute{15}$ is $^{2}/_{5}$).

For us, of course, the problem is complete with the value $^{11}/_{35}$. The demotic scribe, however, goes on to convert this result into unit form, by working out the division as $\acute{4}\ \acute{28}\ \acute{35}$. In this he offers a clear precedent for the technique followed by the later Greek scribe of P. Akhmîm. Their procedural agreement, despite their chronological, cultural, and linguistic separation, indicates the stability of the elementary computational tradition.

From a higher stratum of Greek geometrical writing, the following examples may be noted: in *Dimen. circ.* prop. 3, Archimedes (3rd cent. BC) adjusts both terms of the ratio ($5924\ \acute{2}\ \acute{4}$):780 by $^{4}/_{13}$ to obtain 1823:240; he next approximates an irrational square root as 1838 $^{9}/_{11}$. His ultimate estimate for the ratio of the circumference and diameter of the circle (our π) is given as 'the triple and greater than 10 $\acute{71}$'. [Heiberg 1910–1915, i 242.17–18].[35] Admittedly, the manuscripts are late copies (10th century at the

[34] This is no. 60 of the demotic mathematical problems edited by R. Parker [1970]. Similar manipulations appear in nos. 56–59, 61.

[35] The result is also phrased as 'triple and greater by more than 10 $\acute{71}$' [Heiberg, 1910–1915, i 242.19–21]. Yet another phrasing is adopted in the enunciation of prop. 3: 'the perimeter is triple of the diameter and moreover exceeds... by greater than ten seventy-firsts' [Heiberg 1910–1915, i 236.8–11]. In Knorr 1989a, pt. 3, ch. 4, I survey the ancient citations of this result. The oldest extant reference, it appears, is from Ptolemy, where the ratio is stated simply as greater than 'triple plus ten seventy-firsts' [Heiberg 1910–1915, i 513.4–5]. The wording in the enunciation of Archimedes' prop. 3 (as stated above) appears to have been framed by an editor following a model from Theon.

earliest); for the *Dimensio circuli* in particular, the text has undergone considerable alteration through editorial reworking and scribal carelessness.[36] Thus, it would surely be perilous to make claims about Archimedes' fractional *notations* on the basis of this textual evidence. But the character of Archimedes' computation, including the numerical values here cited, would not have been significantly tampered with by the later copyists and editors. Thus, the extant figures can be taken as reliable witness to the sequence of his computation. In particular, the ploy of attaching fractional increments (like $^{9}/_{11}$), however denoted, would be indispensable to his procedure.

As far as notation is concerned, it is remarkable that in Eutocius' commentary on this proposition, where each of the square root values is checked by a fully worked out multiplication, the term 1838 $^{9}/_{11}$ is misconstrued as 1838 $^{1}/_{9}$ + $^{1}/_{11}$ [Heiberg 1910–1915, iii 253].[37] This must indicate a notation for the fraction $^{9}/_{11}$ that closely resembled a common unit-fractional expression like $\acute{9}\ \acute{11}$. Thus, we can have reason to date the use of general fraction notations no later than the 6th century. While this may not seem to place them back very far, it does refute the supposition that such notations are artefacts of Byzantine scholarship. Moreover, it supports the view that the earlier traditions stemming from Archimedes, Hero, and Diophantus also had access to some form of general notation for fractions. Without insisting that the extant manuscripts preserve those notations with complete accuracy, we may assume that the computations with fractions in their works, which we take up next, were recorded in some comparable form.

From the *Metrica* of Hero of Alexandria (1st cent. BC) numerous computations with fractions may be cited. For instance, in *Metrica* iii 3 the

---

36 My examination of the ancient versions of prop. 1 in Archimedes' *Dimen. circ.* indicates that the extant text is an adaptation from Theon's commentary on Ptolemy's *Almagest*: cf. Knorr 1986b; 1989a, pt. 3, ch. 1–3. This confirms widespread doubts about the extant text of this particular Archimedean work, although it proposes a far greater distance between Archimedes' original version and that extant than scholars have heretofore supposed. In a discussion of the numerical figures in the mss. of prop. 3, D. H. Fowler [1987, ch. 7.3a] emphasizes the textual corruptions and the great span separating the prototypes of the mss. from the time of Archimedes and cites these as factors complicating their use as indicators of the earlier notations.

37 Heiberg gives the correct figure in the body of the passage (line 7), but the incorrect figure is the basis of the computation in the appended working out of the computation (col. 3, lines 11 *et seq.*). In a note on this faulty computation, D. H. Fowler [1987, 244n] surmises that Eutocius' procedure is 'fudged'. But, examining the passage more closely, I have argued [1989, 522–523] that Eutocius' own text, which sets out correct figures, must be founded on a correct computation; the supplementary work sheet, which purports to lead to the same answer *via* incorrect figures, would be due to a later editor.

division of $100\,{}^{4}/_{5}$ by $10\,{}^{22}/_{65}$ is stated to be 9 2́ 4́ [Schöne and Heiberg 1903–1914, iii 148].[38] The method of division is not explained; but since the result is exact, a measure of contrivance is indicated. Again, in *Metrica* iii 8 the result of multiplying $12\,{}^{1}/_{14}$ by $5\,{}^{5}/_{26}$ and dividing by $7\,{}^{4}/_{7}$ is given as $8\,{}^{1}/_{4}$, here also without explanation [Schöne and Heiberg 1903–1914, iii 158].[39] One suspects that a method of raising terms is employed in cases like these, comparable to the modern school procedure.

We possess several extensive collections of problems in metrics, containing series of exercises usable in conjunction with the study of Hero's *Metrica*.[40] Here examples proliferate in which fractions are manipulated in ways distinctly comparable to the familiar modern methods. These sometimes invoke unit-fractions but, as with the papyri, in a merely notational, not a computational, mode. It is important to note that, whatever notations or dictions are used for expressing fractions, the scribes freely manipulate these as numerical terms. For instance, even if the ancient equivalent of what we write as 5/26 (as above) is expressed as 'of the 5 the 26th' or '5 26ths' or '5 26́' or even '5 26́ 26́' (where the duplication of 26́ indicates the plural),[41] and even if that term is there viewed only as the result of a division (namely, of 5 by 26), nevertheless, the quotient so found is treated as a numerical term, attachable to an integral term to produce a sum (the equivalent of $5\,{}^{5}/_{26}$), which can then be multiplied or divided by similar terms.

The six books extant in Greek of Diophantus' *Arithmetica* comprise almost 200 arithmetic problems whose solutions are as often fractional numbers as they are integers.[42] In book 2, for instance, in prop. 8 the solutions—for which Diophantus employs the term ἀριθμός, the standard Greek term

[38] The principal ms. (the Codex Constantinopolitanus, pal. vet. 1) dates from the 11th century. It is reproduced in photofacsimile in Bruins 1964, i.

[39] A modern procedure, exploiting cancellations, might run thus: $({}^{169}/_{14})\cdot({}^{135}/_{26}) \div ({}^{53}/_{7}) = 13 \cdot 135 \div 212 = 8\,{}^{5}/_{212}$; this could be reduced to very nearly $8\,{}^{1}/_{4}\,{}^{1}/_{36}$, whence Hero's answer $8\,{}^{1}/_{4}$.

[40] These are edited by Heiberg as the *Geometrica* and *Stereometrica* in Schöne and Heiberg 1903–1914, iv–v. Some examples of fraction computations from these works are discussed in Knorr 1982b.

[41] This would be in accordance with a common tachygraphic convention in Byzantine texts. On ancient fraction notations in general, one may consult Heath 1921, i 41–44; D. H. Fowler and Turner 1983; D. H. Fowler 1987, ch. 7. Fowler frequently remarks on deficiencies in the older standard accounts, such as Heath's.

[42] See Tannery's edition of Diophantus [1893–1895], and Heath 1910. The same prominence of fractional solutions is evident in those portions of the *Arith.* surviving only in Arabic: cf. Sesiano 1982, Rashed 1984.

for 'number' (that is, ordinarily, 'integer')—are $^{256}/_{25}$ and $^{144}/_{25}$. The solutions are fractional terms of this type in every problem thereafter in this book, through prop. 35, its last problem, save for props. 10, 19 and 23 (which result in mixed unit-fractional terms, like 72 $\acute{4}$), and 19 (which results in integral solutions).

It is clear that Diophantus' effort depends on a supple manipulation of fractions. This applies not only for explicit constant operands, but also for variables. For instance, in iv prop. 36 two indefinite fractional terms, $3p/(p-3)$ and $4p/(p-4)$ [lit.: '$3p$ in the part (ἐν μορίῳ) $p-3$', and so on] are stated to have the product $12p^2/(p^2+12-7p)$ [Tannery 1893–1895, i 288.9–10].[43] In the same problem their sum is worked out to be $(7p^2-24p)/(p^2+12-7p)$, with the following explanation:

> Whenever it is required to sum parts (μόρια), for instance, $3p$ pt. $p-3$ and $4p$ pt. $p-4$, the number [*scil.* numerator] of the part shall be multiplied into the alternate parts [*scil.* denominator of the other], e.g., $3p$ into the parts [denominator] of the other, *scil.* $p-4$, and again the $4p$ into the parts [denominator] of the other, into $p-3$. In this way the addition has made $7p^2-24p$ of the part [denominator] which is the product of the parts, *scil.* $p^2+12-7p$.[44] [Tannery 1893–1895, i 288.1–9]

Doubtless, the account is cumbersome. But this is an accommodation to the learner, who is being asked to extend arithmetic operations beyond the simple manipulation of definite terms to the case of unknowns. The rule here is only stated, it is not proved. Moreover, it appears as an appendage to prove (that is, to check) the correctness of the solutions derived in the main text of the problem. One may well suppose that here, as is suspected elsewhere in the *Arithmetica*, the added section is due to a later editor, and that Diophantus himself could assume such operations without comment.

The basic fractional operations thus appear to be taken for granted, while only the more elaborate extensions need to be explained. Since Diophantus is reported to have written a treatise on fractions (the *Moriastica*), it is possible that this work provided the basis for these passages of the *Arithmetica*.[45] Although the date of Diophantus is uncertain, the basic expertise in dealing with fractions can hardly be considered a novel feature of

---

[43] Here I denote by 'p' Diophantus' symbol for the unknown number.

[44] Diophantus' term μόριον (lit. part) here bears the sense of our 'denominator', while he uses the word ἀριθμός (lit. number) where we would write 'numerator'.

[45] On lost Diophantine works, see Heath 1910, 3–4.

his mathematical work: one may note comparable cases in book 6 (props. 12–14, 19, 21–22).

The *Greek Anthology* contains dozens of epigrams involving arithmetic word problems, among them the famous epitaph of Diophantus [Tannery 1893–1895, ii 60–61].[46] Most of these are accompanied by scholia which work out the solution, some in considerable detail. In one common type of problem a whole is diminished by denominated parts leaving a given remainder. For instance, in no. 2, the provenance of the gold used for making a statue of Pallas is identified thus:

> Charisios has given half the gold, Thespis an eighth and Solon a tenth part, and Themison an additional twentieth. The remaining nine talents plus his craftsmanship are the gift of Aristodikos. [Tannery 1893–1895, ii 44]

To find the whole weight of gold, the scholiast instructs us to find the least number having all of the stated parts, in accordance with the procedure in Euclid, *Elem.* vii prop. 39; the number is here 40; its $\acute{2}$ $\acute{8}$ $\acute{10}$ $\acute{20}$ (that is, $20 + 5 + 4 + 2 = 31$) when subtracted from 40 leave 9, as required, so our answer is 40. In the case of another remainder, say 6, he tells us to adjust the 40 in the ratio of 6 to 9, giving an answer of $26\ ^2/_3$.

Dating the epigrams in the *Anthology* and their scholia is difficult: a definitive compilation was made early in the 10th century, from elements of diverse date and provenance.[47] Tannery holds open the possibility that the grammarian Metrodorus (assigned to the early 4th century AD) was responsible not only for the arithmetic epigrams, but also for the scholia. It is certainly clear that any writer with a knowledge of the arithmetic techniques of Euclid and Diophantus was able to produce such a commentary.

For our present purposes the central point is the scholiast's use of Euclid. The procedure for finding the least number with specified parts [*Elem.* vii prop. 39] is a modification of the finding of the least common multiple of specified integers [vii prop. 36]. Why do these propositions appear in the *Elements*? As they bring book 7 to a close, one cannot suppose that they were needed for the proofs of later propositions. But the examples from the *Anthology* reveal how the parts-procedure is neatly tailored for solving unit-fraction problems, just as the l.c.m.-procedure has an obvious function

---

[46] Epigram no. 13 sets out as an arithmetic problem the finding of Diophantus' age at death: in effect, his age less its 6th, 12th, 7th, and half parts is nine years. The scholiast works out the answer as 84 years.

[47] On the editing of the epigrams, see Tannery 1893–1894, ii x–xii.

within the addition and subtraction of general fractions. Further, Euclid's inclusion of the parts-procedure, an ostensibly superfluous variant of the l.c.m.-procedure, can now be perceived as motivated from computational practice. A view along these lines has been suggested by Itard [1961, 128], noting possible associations with methods employed in the Rhind Papyrus [cf. Mueller 1981, 80, 114–115; R. Parker 1970, 8–10].

I believe this survey represents the range of the evidence, although dozens of additional examples of similar kind could be cited. What claims does it support as to the presence or absence of a general conception of fraction? First, I question whether it is justified to impose a strict division between late evidence (e.g., Graeco-Roman and Byzantine) and early evidence (e.g., pre-Euclidean and Egyptian), when the computational techniques are so uniform. If the 'modern' techniques are clearest in Byzantine texts, this need not preclude their use much earlier. Indeed, the examples from Hero and Archimedes, for instance, recommend assigning them a very early provenance within the Greek tradition. Second, I consider it more reasonable that a sophisticated theory like that in Euclid *Elem.* vii developed on a sturdy base of practical technique, than that, conversely, the abstract theory should have arisen spontaneously and preceded its application by some long interval. The technical distinctions we can discern may as well be assignable to pedagogical, as to chronological factors. The quaint persistence of unit-fraction techniques, for instance, is most strongly marked in the papyri and the metrical collections, a genre of writing pitched to novices. Even here, however, the unit mode is a cover for a more general computational technique. Doubtless, it was left to the expertise of the teachers to compensate for the gaps in the papyrus textbooks in explicating the general procedure.

The pre-Greek tradition in Egypt is represented to us in only a few documents, like the Rhind Papyrus, dating from over a millennium before the Greeks.[48] Despite their antiquity, these documents appear to foreshadow the general fraction technique. From the Hellenistic period, the demotic papyri give evidence of knowing the more general method, as R. Parker [1970, 9–10] has observed; and they must surely have received this through the native tradition, rather than by borrowing from the Greeks. This would plausibly assign familiarity with the general technique of fractions to the Greeks in the pre-Euclidean period, through their exposure to earlier Egyptian methods.

---

[48] The Rhind Papyrus has been edited by Peet [1923] and Chace and Manning [1927]. For surveys, see van der Waerden 1954, ch. 1; Gillings 1972; Robins and Shute 1987. Some examples of its fraction computations are examined in Knorr 1982b.

Sometime in the late 3rd or early 2nd century AD the Greek mathematical tradition came into a windfall, by receiving the Mesopotamian astronomical and computational techniques.[49] The sexagesimal place-system provided a flexible instrument for elaborate computations far beyond the capacity of the elementary fraction methods. But as far as numerical conceptions of Greek arithmetic are concerned, as the evidence from Hero and the papyri reveal, this exposure made essentially no difference. For the Greeks never absorbed these new techniques into their elementary lessons on arithmetic. Indeed, teachers like Theon of Alexandria (4th cent. AD) found it useful to refer to the conventional notions of fractions in order to explain the manipulation of sexagesimals.[50] The very fact that the Greeks could adopt the sexagesimal system for the more advanced scientific uses would appear to indicate the prior existence of an adequate conception of fractions. That is, they were already prepared to translate their techniques of manipulating terms of the form 'n of the mth parts' into the sexagesimal mode.

Operationally, the new procedures were radically different from the traditional ones known to the Greeks; to implement them, for instance, a whole new range of tabular auxiliaries had to be introduced, on the pattern of the ancient Babylonian computational system. But the introduction of sexagesimals need not have affected the underlying conceptual basis already established among Greek arithmeticians centuries before. Conversely, we should not assume that the persistence of unit-fractions in the popular arithmetics entailed any limitations in that underlying concept. In view of the silence of our sources on conceptual matters, we are compelled to treat the ancient concept of fractions as a sort of 'black box' associated with the technical procedures preserved in the texts. The evidence, in my view, would have to be far more extensive and explicit than it is to sustain

[49] For accounts of the Babylonian computational methods, see van der Waerden 1954, 37–45; Neugebauer 1957, 29–35.

[50] On sexagesimal computation in general, see Theon, *In Ptol. ad* i 10 [Rome 1936, 452–62]. In particular, Theon refers to the familiar Greek tradition, as represented by Diophantus, to explicate the operations on sexagesimal parts; for instance, Theon notes [Rome 1936, 453]:

> In the case of (fractional) parts of the unit... in the manner in which, according to Diophantus, the species are altered in the multiplications of the parts of the unit, for the reciprocal first power (ἀριθμοστόν), [e.g.] the third, multiplied into itself makes the reciprocal second power (δυναμοστόν), [i.e.] the 9th and alters the species, in the same way also here the parts of the unit alter the species.

Cf. Tannery 1893–1895, i 8: 'ἀριθμοστόν into ἀριθμοστόν makes δυναμοστόν'. Theon goes on to elaborate case after case of sexagesimal parts (e.g., minutes times minutes make seconds, minutes times seconds make thirds, and so on).

the hypothesis of a significant separation between the ancient and modern concepts.

4. The aim of Euclid's *Elements*

In the preceding sections we have taken up difficulties in the interpretation of Euclid's theories of proportion and number. Besides clarifying the meanings of particular passages, we have seen ways to exploit Euclid as a source for understanding aspects of the pre-Euclidean development of these theories. But the primary task of a literary analysis of a writing must be to explicate the meanings and objectives of the work as a whole. We undertake such an analysis for the *Elements* in the present section.

What sort of work did Euclid intend the *Elements* to be? Hirsch's account of genre will be useful here. The notion of genre, he proposes [1964, ch. 3], serves as focus for the external or public determinants of a text, complementary to the internal or private meanings of the author. Upon first encountering a text, we begin to construct its meaning on the basis of provisional views we have as to its genre, or type classification. Our idea of the genre embraces the expectations we have about the text, as we pursue our examination of it. At first our genre idea is quite broad; but it is progressively narrowed, being refined, modified, sometimes discarded and replaced by an entirely new genre idea, in the course of our reading. When we ultimately possess an adequate reading of the text, our idea of its genre (or, intrinsic genre) will be closely accommodated to this.[51] What is notable in this account is the reciprocity between the meanings attributed to a text and the ideas one has of its genre. The genre delimits at each moment the possible meanings one can assign to the text; but the genre idea is always provisional, subject to revision as it comes into conflict with new details and implications of the text.[52]

As we approach the particular case of the *Elements*, we will find that views of its aim and meaning are accompanied by implicit views of its genre. Conversely, establishing a credible view of the genre will assist one's inquiry into Euclid's aims.

The question of Euclid's objectives was already raised in antiquity. His commentator, Proclus (5th cent. AD), observes [Friedlein 1873, 70–71] that the aim (σκοπός) of the *Elements* can be viewed in two respects, relative to its research findings and to its instructional uses. In the former regard,

---

[51] For Hirsch's definition of 'intrinsic genre', see n12.

[52] Conversely, Hirsch [1967, 71–77] offers interesting examples where misconceptions of the genre can lead to persistent misconstruals of a text.

Proclus sees the culmination and goal of the work in its elucidation of the five cosmic figures (that is, the five regular polyhedra, elsewhere styled the 'Platonic figures').[53] But from the pedagogical viewpoint, he continues, Euclid aims to provide an introduction (στοιχείωσις) toward perfecting (τελείωσις) the learner's understanding (διάνοια) of the whole of geometry.

Proclus' first surmise, which is probably his own insight,[54] may be dismissed as the expected emanation of his own Neoplatonism [cf. Heath 1956, i 115]. To be sure, it is remarkable that so much of the *Elements* comes to bear on the constructions of the five solids in book 13—including results from the plane geometry of books 1–4 and 6, the solid geometry of book 11, and the theory of irrational lines of book 10. Even so, much of the contents of these books is unrelated to the solid constructions, while the substantial portion of the *Elements* contained in its other books would be wholly left out of account—the proportion theory of book 5, the number theory of books 7–9, and the exhaustion theorems of book 12. In Proclus' defense, one might observe that his position is hardly less credible than modern theories of Euclid's Platonism, based almost entirely on the definitions prefacing books 1 and 7.[55]

But in his second suggestion, about Euclid's pedagogical aims, Proclus takes the essence of the work to be its elaboration of complex theorems out of the simplest and most fundamental starting points. In this he foreshadows certain modern views, which emphasize the axiomatic architecture of the *Elements*. A conspicuous feature of the *Elements* is its deductive structure.[56] As a prototype of modern foundational efforts like Hilbert's *Foundations of Geometry*, it lends itself to a mathematical analysis of its axiomatic technique—e.g., what changes must be made in Euclid's scheme of axioms to make it sufficient for demonstrating the propositions in each

---

53 Proclus [Friedlein 1873, 23] also mentions the figures in the context of remarks on Plato's *Timaeus*. Referring to the same solids, Pappus once calls them 'the five figures of (παρά) the most divine Plato' [Hultsch 1876–1878, 352.11–12]. As a scholiast to the *Elements* observes [Heiberg and Stamatis 1969–1977, v.2 291.1–9], the Platonic association is due only to his inclusion of the solids in the *Timaeus*, for their discovery, he claims, goes back to the Pythagoreans and to Theaetetus.

54 Proclus qualifies his statement with 'I would say.' Heath [1956, i 33–34] notes, however, that such passages need not indicate Proclus' originality.

55 The most extensive effort to link Euclid with specifically Pythagorean and Eleatic precedents is Szabó 1969. I criticize some aspects of the argument in Knorr 1981b.

56 This is the focus of the analysis by Mueller 1981.

of the major branches of Euclidean geometry?[57] There has also been an extensive discussion of the related philosophical issue of Euclid's conception of first principles: What is it about Euclid's axioms and postulates that suits them to be the starting points for his deductive system?, in particular, How does Euclid's conception compare with Aristotle's prescriptions about the starting points of formal expositions of scientific knowledge?[58]

Implied in these views are two complementary notions of the genre of Euclid's *Elements.* In looking to the five solids as the focal point, that is, in defining the work in terms of its subject, Proclus conceives it as the exposition of a technical field of research. But when he, like the modern scholars mentioned, emphasizes its deductive system, the work moves into the category of philosophical exposition—the actualization, as it were, of some implied program of axiomatics. To these forms of technical exposition and philosophical exposition, we can add a third, also recognized by Proclus, the introductory textbook. In the following discussion, we will explore how the decision relating to genre affects our view of the goals Euclid pursues and his success at attaining them.

Proclus amplifies his second point on the aim of the *Elements* by considering the meanings of the term 'elements' (στοιχεῖα). The characterization he borrows from Menaechmus, Euclid's 4th-century precursor, is of particular interest, since it is one Euclid himself would have known. Menaechmus, as Proclus [Friedlein 1873, 72–73] reports, recognized two senses for the elements of a geometric system: in the relative sense, any proposition can be termed an element in the context of any other proposition whose demonstration assumes it; in the absolute sense, the elements are those simpler propositions (e.g., axioms) on which whole bodies of theorems depend. Proclus observes that under the former sense many propositions could qualify as elements, whereas under the latter sense the term will apply only to a specific, relatively small set of propositions.[59] One may note in Menaechmus' first sense of the term the reciprocal character of elements: two propositions can be elements of each other, if the proof of the one can be effected by assuming the other and conversely; this indicates a certain

[57] In his brief comparison of the formal approaches of Euclid and Hilbert, Mueller [1981, ch. 1] seeks, rightly I believe, to distinguish the ancient and modern views.

[58] Among many discussions of the relation of Aristotle's theory of deductive science and Euclid's procedure in the *Elements*, one may note Mueller's account in ch. 5, above; Mendell 1986, ch. 6; Hintikka 1981; and Heath 1949, 50–57.

[59] For further remarks, see Burkert 1959, 191–192.

fluidity in the order of proof, which would obtain in the field before the appearance of a definitive textbook treatment.[60]

The second notion of element is reminiscent of the position adopted by Aristotle in *Physics* i 1.[61] In the effort to organize a field of experience into a deductive science, in his account, we take complex wholes more knowable to us through sense perception and break these down into their simple components more knowable by nature; the latter are the elements or first principles which form the basis of the science. Implied in this remark is a two-part process in which, first, the analysis of phenomena leads to the discovery of the primary conceptions, and then the field is reconstructed deductively on the basis of these conceptions. On this view, Euclid's *Elements* might be viewed as a synthesis of the latter sort, where the fundamental principles, already discovered, can be stated at the outset and their consequences worked out as propositions, in systematic order.[62]

This sense of 'element' is linked with the modern views mentioned earlier, not only by virtue of its emphasis on deductive form, but also for an important nuance in its conception of the geometric field in relation to its principles. For they all imply an account of a formal system which assigns priority to its axioms and views its propositions merely as their deductive consequences. In the modern conception, in fact, the axioms define the system by specifying the entities it contains and their essential properties; whatever other properties can be deduced from these may be considered as already implicit within the axioms.[63] In effect, demonstration becomes a mechanical process, deriving consequences from the first principles in accordance with deductive rules. In the ancient mathematical terminology, this process is called synthesis, and it characterizes all the proofs one finds in Euclid's *Elements*.[64]

---

60 Barnes [1976] would take these remarks from Menaechmus to indicate the employment of circular reasonings in the proofs by mathematicians before Aristotle. This inference is, I think, too bold to be sustained on the evidence at hand.

61 On this passage, see Ross 1936, 456–458.

62 Other Aristotelian passages are cited in Heath 1956, i 116—*Top.* 158b35, 163b23; *Meta.* 998a25, 1014a35–b5: cf. also Heath 1949, 205–206. For additional discussion of the Aristotelian senses of 'element', see Mendell 1986, 492ff.

63 One may think of Aristotle's characteristic distinction between the potential and the actual, where in the present instance one would refer not to the existence of derived terms, but to knowledge of their existence: cf. *Meta.* 1051a30–33, and the discussion in Mendell 1984.

64 The principal ancient account of the method of analysis and synthesis is from Pappus [Hultsch 1876–1878, vii 634–636]. The modern literature on the method is extensive: see, for instance, Knorr 1986a, esp. ch. 8.2; A. Jones 1986, 66–70; Hintikka and Remes 1974; Mahoney 1968–1969; Gulley 1958; Robinson 1936.

This conception of the *Elements* as an inquiry about the first principles of geometry will undoubtedly sponsor interesting insights into the nature of formal studies. But it assigns too small a role to the technical content of the fields of Euclid's geometry, and thus misses a feature which, I believe, is critical for a historical understanding of his project.[65] In what follows, I will take up three aspects related to setting the genre of Euclid's work: its place as the paradigm of a particular type of technical treatise; the manner of Euclid's activity in editing his source materials; and the character of technical instruction toward which it was applied.

For these purposes, it seems more useful to pursue Menaechmus' first sense of the term 'element', namely, as any principle or theorem assumed within the proof of another. That is, *all* the propositions of Euclid's treatise are elements with respect to higher studies in geometry. As Proclus himself notes, this was precisely how Euclid was used by later geometers:

> the most fundamental and simplest theorems and most kin to the primary hypotheses are here [*scil.* in the *Elements*] joined together, taking the appropriate order, and the proofs of the other (theorems) use them as thoroughly familiar and arise from them. Just as Archimedes... and Apollonius and all the others seem to use the things proved in this very treatise, as agreed on starting points. [Friedlein 1873, 71]

Moreover, the title *Elements* was not restricted to Euclid's treatise. Proclus [Friedlein 1873, 66–67] denotes by this term geometric works produced by Euclid's predecessors Hippocrates, Leon, Theudius, and Hermotimus. Even if the term derives from Proclus' possible source, Aristotle's disciple Eudemus, rather than from the titles of the works themselves, we can infer its generic use in the 4th century to denote works of Euclid's kind.[66] From another of Aristotle's disciples, Aristoxenus, we possess a treatise in

---

[65] A similar distortion of emphasis, I believe, accompanies the familiar view that the ancients' geometric constructions played the role of existence proofs: cf. Knorr 1983.

[66] Burkert [1959, 193] takes *perforce* the designation 'elements' as the actual title of the pre-Euclidean works. Mendell [1986, 493] observes, however, that what these authors titled their works is immaterial; the salient fact is that Proclus' source, Eudemus, could designate them as 'Elements', whence the works themselves must have conformed to a Peripatetic conception of the term. In Mendell's view, Eudemus would surely have held the second of Menaechmus' senses for 'elements', namely, the simple components into which a complex can be resolved.

two books, the *Harmonic Elements*.[67] Apollonius [Heiberg 1891–1893, i 4] describes the first four books of his own *Conics* as 'falling into (the class of) elementary training' [πέπτωκεν εἰς ἀγωγὴν στοιχειώδη], in contrast to the contents of the last four books, of a more advanced character, which he terms 'supplementary' (περιουσιαστικώτερα); the commentator Eutocius [Heiberg 1891–1893, ii 176] cites the same work as the *Conic Elements*. Pappus [Hultsch 1876–1878, ii 672.12] denotes by the term *Elements* not only Euclid's work, but also treatises by other geometers, e.g., the *Conic Elements* by Aristaeus.[68] Archimedes once cites his *Elements of Mechanics* for a result we know from his *On Plane Equilibria* (i 8) [cf. Heiberg 1910–1915, ii 350.21]; several of his references [Heiberg 1910–1915, i 270.24; ii 268.3, 436.3] to the *Conic Elements* are likely to denote the works by Euclid or Aristaeus, precursors of Apollonius.[69] Interestingly, Archimedes once uses the phrase 'conic elements' to denote three specific *propositions* (i.e., *Quad. parab.* props. 1–3) for which he provides only the enunciations, since their proofs can be assumed from a treatise on conics, itself also called the *Conic Elements* [Heiberg 1910–1913, ii 266.3, 268.3]. Among later writers, e.g. pseudo-Hero [Schöne and Heiberg 1903–1914, iv 14.1, 84.18] and Diophantus, the term στοιχείωσις becomes synonymous with 'introduction'.[70]

These instances make clear that Euclid's *Elements* was the paradigm for a literary genre which embraced technical treatises extending beyond the specific field of Euclidean geometry. Proclus indicates this when he refers to the 'numerous compositions' falling into the category of elementary treatise [στοιχείωσις] in the areas of arithmetic and astronomy. In speaking of the general activity of producing such treatises, he remarks on the diversity of their editorial styles [Friedlein 1873, 73]:

> It is a difficult task in any science to select and arrange properly the elements out of which all other matters are produced and into which

[67] See Barker's discussion in ch. 9, below. Note that Aristoxenus' *Harm. Elem.* is discursive in style, thus in marked contrast to the rigorously deductive format of Euclid.

[68] Cf. also Hultsch 1876–1878, ii 552.4 (possibly referring to Theodosius or Euclid: cf. 553n), 608.2 (referring to Euclid's *Phaen.*), 660.19 (taken by Hultsch to refer to Apollonius' *De plan. loc.*). It is possible, however, that some of these passages are interpolations. For other uses of the term, see Hultsch's index *s.v.* στοιχεῖον.

[69] Note that two citations of Euclid's *Elements* at Heiberg 1910–1915, i 20.15 and ii 444.28 are certainly interpolations.

[70] A scholiast [cf. Tannery 1893–1895, ii 72.16–20] to Iamblichus refers to Diophantus' *Arithmetica* by this term. In Diophantus' preface, the initial materials are said to be elementary (ἔχοντα στοιχειώδως) [Tannery 1893–1895, i 16.3]. Of course, Proclus frequently refers to Euclid's work as στοιχείωσις.

> they can be resolved. Of those who have attempted it [Morrow 1970, 60: '*scil.* for geometry'] some have brought together more theorems, some less; some have used rather short demonstrations, others have extended their treatment to great lengths; some have avoided the reduction to impossibility, others proportion; some have devised defenses in advance against attacks upon the starting-points; and in general many ways of constructing elementary expositions have been individually invented.[71]

This passage has sometimes been taken to refer to treatments of elementary geometry;[72] but one would surely find remarkable the implied proliferation of ancient editions alternative to Euclid.[73] Further, Proclus is unlikely to have had access to pre-Euclidean versions. Proclus' source in the present discussion may be Geminus; but only Eudemus could provide Proclus information on the pre-Euclidean tradition, and Eudemus of course could not compare such versions with Euclid's treatment [cf. Heath 1921, i 114].[74] But as Proclus makes clear his interest in the expositions of 'any science', we must suppose that he is considering the whole range of technical treatises, not just those in geometry. Among extant treatises, reduction-avoidance happens to be characteristic of Menelaus in the *Spherics*, while the inclination toward lengthy proofs typifies Theodosius' *Spherics* in contrast with parallels in Menelaus and Euclid (*Phaenomena*) [see Heath 1921, ii 248–249, 263, 265].[75] The preface to one of the editions of Euclid's *Optics* forms

[71] Cf. Morrow 1970, 60, which I here follow. The characterization here of elements as that 'out of which all other matters are produced and into which they can be resolved' is reminiscent of Aristotle's passages on elements (in particular, the material kind of elements): cf. *Meta.* 983b8–11; *Phys.* 194b24, 195a16–19.

[72] Cf. Morrow 1970, 60n which explains Proclus' phrase 'of those who have attempted it' by the remark: '*scil.* for geometry'. Artmann [1985] begins with this very passage as the basis for developing his thesis that some of the proportion–avoiding proofs in Euclid's books 1–4 derive from a pre-Euclidean treatise. The difficulty, however, is that Proclus is not speaking here about pre-Euclidean precedents of the *Elements*, but rather of the whole ancient tradition of elementary mathematical treatises.

[73] Of course, there were many *commentaries* on Euclid, such as those by Hero, Geminus, and Proclus: cf. Heath 1956, i 33–45.

[74] We return below to the issue of the pre-Euclidean precedents for the proportion-avoiding proofs in the *Elements* [see n83, below].

[75] Similarly, Autolycus' *De sphaera quae movetur* is more abstract and verbose than the parallels in Euclid's *Phaen.*: cf. the specimens cited by Heath [1921, i 351–352]. For further discussion of these treatises, see Berggren's essay [ch. 10, below], Berggren and Thomas 1992, Knorr 1989b.

a sort of defense of its postulates, in that notions of the nature and transmission of visual rays are explained in a counterfactual manner: hypotheses different from those implicit in Euclid are refuted through consideration of associated physical phenomena.[76] Thus, Proclus' remarks on the variety of styles in this tradition of technical writing can be related to some extant works, even though most of the works he refers to must be ones that are now lost.

The *Elements* is thus the model, if not the first exemplar, of a particular type of scientific treatise, in which the content of the science is presented in a formal, systematic, and deductive manner. The exposition develops as a series of propositions in which the demonstration of each depends on those preceding and in its turn can serve toward the demonstrations of those following. The whole series is initiated by the statement of certain primary terms and propositions (the elements in the absolute sense), in the form of definitions and postulates, for instance, which in the context of the work may be taken as inderivable.[77]

In studying a treatise of the synthetic type, one will follow its deductive order. But the project of producing the treatise entailed the *discovery* of the appropriate deductive sequence and this heuristic phase will invariably proceed in the reverse order, that of analysis. In its usual sense in ancient geometry, 'analysis' refers to a method for finding solutions of geometric problems.[78] A targeted result (e.g., a construction satisfying certain specifications) will first be assumed as known; the quest for a derivation or proof will then bring forward certain other results, these in their turn still others, until one obtains a sequence linking the desired end result to ones

---

[76] The preface to one recension of the *Optics* is thought by Heiberg [1895, vii 144–154] to be based on introductory lectures by Theon. I discuss aspects of the preface in Knorr 1985b, sect. 9, and argue against Heiberg's assignment of the recensions in Knorr 1992.

[77] In the absolute sense, such inderivables would be postulates or definitions. But in advanced treatises, such as Archimedes' *Quad. parab.* and *Meth.*, one typically permits as initial assumptions the results established in more elementary treatises (such as the *Con. elem.* or *De plan. aequil.* i).

[78] Pappus [Hultsch 1876–1878, ii 634–636] distinguishes two types of analysis: the problematic (*scil.* analysis of problems) and the theoretic (that of theorems), and clearly he assigns priority to the theoretic type. In this I believe he has severely undervalued the significance of the analysis of problems: cf. Knorr 1986a, ch. 8.2. Hintikka and Remes [1974] also find the analysis of problems (constructions) to be the more fruitful domain for examination.

that are already known or admissible *per se*.[79] Although this method is usually employed in the investigation of individual constructions or propositions, a heuristic procedure of the same kind can be applied toward the organization of systems of constructions and theorems.[80] A nuance we have noted in Menaechmus' first account of 'elements' is that the deductive order is relative: within certain limits, the researcher can make the *choice* as to which results will be prior and which derived from them. The options will emerge as one pushes further into the analysis of the principal theorems and constructions of the field.

In the specific instance of the *Elements*, however, Euclid is not so much its composer as its editor. As Proclus informs us (following, it appears, the authority of Eudemus), other geometric compilations in the form of 'Elements' were produced in the century before Euclid, so that his treatise is a consolidation of several generations of geometric study. Euclid's sources must have provided expository models, not only for the proofs of individual propositions, but in some cases for the substance of major sections and even of whole books. It seems clear, for instance, that the structure of his proportion theory in book 5 existed in much this form among disciples of Eudoxus, and that similarly extensive prototypes existed for his number theory in book 7, his theory of irrationals in book 10, his exhaustion theory in book 12, and his constructions of the regular solids in book 13.[81]

---

[79] In analysis the examination pursues the deductive consequences of the assumed target; the formal synthesis reverses the logical order. In his account of the method, Pappus is ambiguous, at one time asserting that the reasoning in the analysis is deductive, but then describing it as a search for appropriate antecedents. Older accounts [e.g., Robinson 1936] attempt to reconcile the two views by emphasizing the convertibility of geometric propositions. Hintikka and Remes [1974] look toward the special nature of reasoning *via* geometric diagrams. Knorr [1986a, ch. 8] prefers a textual explanation, in which Pappus has merged a mathematical and an Aristotelian account conceived along contrasting lines.

[80] Burkert [1959, 195] suggests that the bi-directional character of research in geometry—the investigation of deductive consequences on the one hand and the search for prior principles on the other—gave rise in the 5th century to an application of the term στοιχεῖον (until then used only to denote a 'column' or 'file', as of soldiers) to designate the ordered sequence of geometric propositions. Knorr [1985] adopts an analytic strategy to explicate the development of the theory of irrationals in Euclid, *Elem*. x.

[81] van der Waerden [1954, 115, 123–124: cf. 1979, 352–353] has striven to assign the prototypes of books 2, 4, and 7 to the Pythagoreans, that of book 8 to the Pythagorean Archytas [1954, 112, 149, 153–155], that of book 10 to Theaetetus [1954, 172], and those of books 5 and 12 to Eudoxus [1954, 184–189]. Neuenschwander [1972–1973] sustains some of these claims on the Pythagorean provenance of substantial parts of books 1–4. But I think the effort to view whole Euclidean books as, in effect, mere transcripts of treatises written a century or

Presumably, Euclid did not merely transcribe his sources verbatim, however. Each of the books has a few key problems and theorems [e.g., *Elem.* i prop. 47, the 'Pythagorean theorem'], and Euclid must have been able to choose from among several variants for his own constructions and proofs. We can only speculate as to the manner of his examination of these materials, but one can readily suppose that it took the form of an analysis. That is, inspecting the preferred variants, Euclid would determine which prior results were necessary for establishing them; taking these up in their turn, he would determine what they required; and so on. Eventually, the backward sequence must terminate in results which can be accepted as primary, without further proof or justification. In this way, the elements or first principles, suitable as the basis of the corresponding synthetic exposition, would emerge as the last terms in the analytic inspection of the major theorems of each book. One may note, for instance, how the sequence of problems of construction in Euclid's book 1 traces back to the three postulates of construction which preface the book; or further, how the parallel postulate first enters within the proof of *Elem.* i prop. 29, and in fact is stated in precisely the terms that this proof requires [cf. Knorr 1983].[82]

One of the remarkable features of Euclid's formal style is his deferral of the methods of proportion until book 5. This commits Euclid to presenting congruence proofs for all the propositions on plane figures in books 1–4. At times this results in intricate congruence proofs where the use of the similarity of figures would be straightforward [cf. *Elem.* i prop. 47, iii props. 35–37, iv prop. 10]. Indeed, one would presume that simpler variants of the latter type were employed in Euclid's sources.[83] From the technical

---

more earlier drastically oversimplifies the likely process of transmission and editing. One will freely admit that the *content* of Euclid's propositions was, for the most part, entirely familiar among mathematicians by the middle of the 4th century and, for much of the more elementary material, far earlier than this. But the *organization* into treatises closely resembling Euclid's books was surely still in progress until very near Euclid's time and Euclid himself must be assigned a major role in establishing whatever stylistic unity one can discern in the *Elements* over all. Such an account is surely what is suggested by Proclus' survey [Friedlein 1873, 65–67] of the pre-Euclidean writers on elements.

[82] Note that the account in sect. 2 above of the development of proportion theory follows a similar pattern of analyzing proofs (i.e. those in the 'Eudoxan' style) back to prior principles which, in turn, become the basis for the synthesis of an alternative form of the theory (namely, that in Euclid, *Elem.* v).

[83] A very good *résumé* of these proofs and their alternative versions is given in Artmann 1985. As noted above, however, the evidence from Proclus [Friedlein 1873, 73] does not deal specifically—and perhaps not at all—with treatises from the early geometric tradition. Thus, we have no grounds for assigning the project of devising the proportion-avoiding proofs to any pre-Euclidean editor of Elements.

viewpoint, this is an entirely artificial project, since the results established in either event are the same. Why does Euclid divide the materials of plane geometry in this manner? Is it for philosophically interesting reasons, e.g., adherence to a principle of economy in setting up a deductive system? Or for the mathematically interesting reason of determining the precise domain accessible under specified postulates (e.g., the axioms of congruence)?[84] Yet the aesthetics of deciding between a simple proof which uses proportions and a complicated proof which does not would appear to be unclear. Moreover, Euclid is elsewhere not so concerned over formal niceties, as in his unexplicated assumptions on continuity in books 5 and 12,[85] or his appeal to geometric motions in book 13.[86] At best, Euclid's success in maintaining such formal restrictions would appear to be uneven.

On the other hand, the attempt to root Euclid's deferral of proportion theory in historical reasons does not face the immediate issue. It might be the case that the avoidance of proportions was recommended after geometers realized how the existence of incommensurable magnitudes rendered invalid the use of theorems established through an integer-based definition of ratio. But this state of affairs could have held only briefly, during the earlier part of the 4th century.[87] By Euclid's time Eudoxus had long since

Artmann is right, I believe, to see in these Euclidean proofs a subtle project: to establish as much geometry on as few assumptions as possible. But we can as well assign this effort to Euclid himself as to any of his precursors, and his motives for undertaking it could be other than purely mathematical (e.g., pedagogical), as is maintained below.

[84] One could pose similar questions about the motivation for restricting the means of construction to compass and straightedge: see n88, below.

[85] We have noted already Euclid's assumption of the existence of finite multiples of magnitudes greater than other given magnitudes. At several places he also assumes the existence of the fourth proportional of given magnitudes. Neither assumption is covered by explicit postulates.

[86] The definition of 'sphere' in book 11 and its applications in book 13 conceive this figure as a solid of revolution. Euclid could have defined it statically—as the locus of points in space equidistant from a given point—by analogy with his definition of circle in book 1. This is, in fact, the way the sphere is defined by Theodosius. Attempts to explain away his inconsistency, e.g., as occasioned by the specific exigencies of the solid constructions in book 13 [cf. Heath 1956, iii 269], seem motivated by the desire to save an interpretive principle (*scil.* the desire to avoid assumptions of motion) which did not actually bear on the ancients' view. It is clear, for instance, that Archimedes, Apollonius, and Pappus have no qualms in retaining and exploiting the generational conception of the sphere and other solids of revolution.

[87] It is doubtful, however, that any such dislocation of mathematics ever took place in antiquity. See the discussion of the alleged foundations crisis in the pre-Euclidean period in Knorr 1975, ch. 9.

resolved these difficulties. It would seem merely clumsy on Euclid's part to persist in the avoidance of proportions in his pre-Eudoxan sources (if indeed he did work directly from older sources, rather than contemporary editions), when the mathematical difficulties had been resolved.

An account along the lines of the pedagogical intent of the *Elements* seems possible: for Euclid may have judged that the simple notions of the congruence of figures constituted a manageable body of material for introductory purposes. The use of similar figures could extend the field and facilitate many of the proofs, and these could have been admitted on the basis of a naive conception of proportion. But Euclid's plan for the *Elements* includes the presentation of proportion theory in a fully rigorous manner. Set early in the sequence of books, instruction in the logical subtleties of the general proportion theory would distract from the geometric material, so that one is well advised to postpone this theory until after a basic block of geometry had been covered. Having made this choice, Euclid would be compelled to find alternative congruence-based proofs, however intricate those constructions might turn out.[88]

The strategy of avoiding proportions in the earlier books of the *Elements* is thus occasioned by the combination of the rigorous proportion theory with the whole field of plane geometry. Neither body of material, taken separately, would compel such alternative demonstrations. Presumably, none of Euclid's predecessors had attempted to compile such a large portion of the geometric field within the limits of a single treatise. Hence, our view carries the implication that the proportion-avoiding proofs were due to Euclid himself.[89]

In setting the genre of the *Elements* as a systematic geometric treatise, we thus perceive two different formats as it were, the research monograph and the introductory textbook. That a study of the elements of a field is the objective of a work of the latter sort is already attested in Aristotle, *Top.*

---

[88] One can account for the restriction to planar constructions (that is, circle and straight line) on similar pedagogical grounds: the use of other methods, such as *neuses* (sliding rulers), conics, mechanical curves, and the like, would require additional postulates and an appreciable body of lemmas, before they could be admitted into a formal work of the type of the *Elements*. The narrow base of the Euclidean postulates opens up a domain of constructions rich enough to make such additions unnecessary. Of course, the ancients studied the alternative constructions extensively [see Knorr 1986a, for a survey]. One might view Apollonius' *Conics* as an effort to axiomatize the field of conics on the Euclidean model. But it is unclear whether any of the other approaches received a comparably rigorous elaboration. Despite this, they were freely admitted into researches on advanced problems.

[89] Contrast Artmann 1985: cf. n83, above.

viii 3. As noted above, Apollonius divides his own *Conics* into two parts, the first half being elementary—that is, devoted to the systematic presentation of familiar results, presumably for the purposes of instruction—and the second half being an advanced supplement comprising new material. Most of the writings in the Archimedean corpus were produced as research works, as one can gather from the prefaces;[90] the first book *On Plane Equilibria*, however, fits better into the instructional category.[91] In the case of Euclid's work, it was in fact adopted as the standard textbook in its field.[92] This is evident in the manner of its citation throughout the later geometric tradition and the appearance of a substantial body of commentary on it, as by Hero, Apollonius, Geminus, and others.[93] Proclus [Friedlein 1873, 74], for instance, often alludes to its uses in teaching, as when he lists its advantages over other textbooks.

Indeed, for Proclus, the *Elements* defined the scope of an introductory course in geometry. Euclid's formal manner would be especially welcome within the curriculum of the Neoplatonic Academy, ultimately geared toward training in Platonic philosophy. But as a technical introduction, the *Elements* is surely remarkable—for its sophistication on the one hand, and its opacity on the other. For a presentation of the basics in geometry, one would find Hero's *Metrica* a more likely text.[94] In *Metr.* i, for instance, Hero sets out the different kinds of plane figures in order (triangles, regular polygons, circles, parabolas and ellipses, conical and spherical surfaces) with arithmetical rules for computing their areas, and likewise for the volumes of solid figures in book 2. (Book 3 is devoted to problems in the division of plane and solid figures.) In a very few cases derivations are provided (as for the circle-segment rule in *Metr.* i). But for the most part, the rules are only stated, together with details of the working of explicit problems; for formal justifications the student is referred to the appropriate writings by

---

[90] This does not preclude that some of his writings eventually found their way into school use. Archimedes' *De sph. et cyl.*, *Dimen. circ.*, *De plan. aequil.*, and *Meth.* are frequently cited by later commentators like Hero, Pappus, Theon, and Eutocius, who thus assume their availability to students of higher mathematics.

[91] Berggren [1976–1977] suggests that the extant *De plan. aequil.* i is an adaptation for school study. The manner of the origin of the extant text of *Dimen. circ.*, as a reworking of materials from Theon, would indicate the same for this work: cf. Knorr 1986b.

[92] A similar status would seem to apply for the *Data, Optics, Catoptrics* (whether or not this is an authentic Euclidean work), and the *Phaenomena*.

[93] On Euclid commentaries, see Heath 1956, i ch. 3–4.

[94] For the text, see Schöne and Heiberg 1903–1914, iii and Bruins 1964. For a survey, see Heath 1921, ii ch. 18.

Euclid, Archimedes, and others. The same concrete approach is adopted for an introduction to arithmetic problem solving in Diophantus' *Arithmetica*. Here one encounters sequences of problems, set in terms of explicit numerical parameters, with complete working out; formal justifications of any underlying arithmetic or algebraic relations are usually omitted, save for occasional references to companion treatises.[95] As another example of an introductory text, Ptolemy's *Almagest* covers the more advanced field of mathematical astronomy, where a thorough grounding in plane and spherical geometry is assumed [see Toomer 1984, 6]. While Ptolemy's exposition shares some features of Euclid's in that proofs of important geometric relations are often provided, it includes guidance in the more practical aspects of the field, like instrumentation, observations, numerical methods, tables, and so on.[96]

In contrast with these examples, Euclid provides no insight into the application of his theorems, nor does he take up any of the related practical aspects, like the nature and manipulation of instruments for the construction of problems. For all his theorems on prime numbers, perfect numbers, square and cube numbers, irrational lines, and so on, not one concrete example is provided of any.[97] Because he adopts the synthetic mode exclusively, the reasons behind the steps in his proofs and constructions—why, for instance, an auxiliary term is introduced or a particular proportion is used—are left unexplained. At times one is awed, even mystified, at the *dénouement* of an especially complicated proof.[98] Without the heuristic

---

[95] In some instances, Diophantus cites his own *Porisms* for the derivations of assumed results: cf. Heath 1910, ch. 5.

[96] On Ptolemy's procedures, see Pedersen 1974. Ptolemy takes pains to derive from observations the numerical parameters of his planetary models. But he offers perfunctory explanations at best to justify the specific basic geometric configurations themselves (e.g., eccenters, epicycles, equants, and so on). Presumably, the basic geometric options were fixed in the older technical literature, particularly, the work of Hipparchus, so that only the refinement of parameters needed detailed commentary.

[97] Numerical examples turn up in the scholia to the *Elements*. By contrast with Euclid, the arithmetic expositions of the neo-Pythagoreans, following Nicomachus (2nd cent. AD), are based almost entirely on specific examples. Here, general results must be inferred on the basis of incomplete inductions.

[98] Note, as particularly striking instances, the proof of *Elem.* v prop. 8, the indirect limiting arguments in xii props. 2, 5, 10–12, 18, and the solid constructions in xiii props. 13–17. The Euclidean proof of the 'Pythagorean theorem' on right triangles [*Elem.* i prop. 47] is frequently held up as a model of contrivance. In the case of problems, a reconstructed analysis mitigates the element of surprise, for it reveals, as the synthetic proofs do not, the reason for the critical auxiliary steps: cf. Knorr 1986a, 9.

insights that one would obtain from the analyses, Euclid's moves often seem arbitrary. Some parts of the *Elements*—most notably, the elaborate classification of irrational lines in book 10—evade description as introductory at all [see Knorr 1985]. In all, the student is drawn into a passive appreciation of Euclid's often imposing reasoning, rather than stimulated to develop active expertise in solving problems.

The presence of such advanced features in the *Elements* indicates that Euclid can already suppose the student's understanding of the basics of practical arithmetic and geometry, that is, properties and rules of the type set out by Hero. Although Hero cites Euclid and other authors in the formal tradition, that need not imply his students' prior exposure to these works; these could as well (indeed, more fittingly) be viewed as references forward to works of a more advanced nature for future study. Precedents for this more concrete, application-oriented mathematics are firmly established in the older Egyptian and Mesopotamian traditions.[99] The Greeks themselves, from Herodotus and Eudemus to Proclus look to practical contexts for the origins of mathematics, and modern scholarship tends to support their view of a transference of the older techniques to the Greeks sometime in the pre-Euclidean period.[100] The achievement of the Greeks in the 5th and 4th centuries, culminating with Euclid, would then lie in their provision of the rigorous deductive foundation of this geometric lore, not the creation of an abstract geometry *ex nihilo*.

The difficulty in assessing Euclid's aims in the *Elements* thus appears to result from what we would term a confusion of genres. For Euclid has fashioned his introductory textbook along lines which we more readily associate with research treatises. Instead of offering models of analysis, whereby the student would learn the arts of geometric inquiry, he provides the formal exposition of results in the synthetic manner. To be sure, the instructor would be free to supplement the text with suitable motivating

---

[99] For surveys, see van der Waerden 1954, ch. 1–3; Neugebauer 1957, ch. 2, 4.

[100] The Egyptian precedent is cited by Proclus [Friedlein 1873, 64–65: cf. Morrow 1970, 51–52]. The Mesopotamian precedent, although largely unnoted by the ancients, is standard in the current historical literature: cf. van der Waerden 1954, 124; Neugebauer 1957, ch. 6. I have proposed that the channel for transmission of Mesopotamian techniques to Greece in the pre-Euclidean period was Egypt during and after the Persian occupation (6th–5th cent. BC) [Knorr 1982b, 157]. Scepticism about the Mesopotamian element in early Greek geometry, however, has been expressed by Berggren [1984, 339]. To similar effect, D. H. Fowler [1987, 8, 285] maintains that the Mesopotamian characteristics evident in Heronian metrics entered the Greek tradition in the Hellenistic period, hence, well after the Euclidean manner had been established through the prior researches of the Classical period.

and explanatory insights, as he saw fit. But the *Elements* itself provides no guidance along these lines, and one perceives from the extensive technical commentaries by later writers like Pappus, Theon, Proclus and Eutocius, that the formal aspects of geometry dominated the course of university-level mathematics.

In effect, Euclid transformed the study of geometry and the other technical disciplines into a scholastic enterprise—the appreciation and criticism of standard texts.[101] While creative research of a high level was pursued in the immediate circles of the greatest figures, like Archimedes, Apollonius, Hipparchus, and Ptolemy, the tension between the aims of research and criticism emerge early. Already Archimedes can chide Dositheus and his Alexandrian colleagues:

> Of the theorems addressed to Conon, about which you continually write me to send the proofs, some I send to you in this book. ... Do not be amazed if I have taken a long time before issuing their proofs. For this has occurred through my desire first to give them to those proficient in geometry and committed to their investigation.... But after Conon's death, though many years have passed, we sense that none of these problems has been moved by anyone.[102] [Heiberg 1910–1915, ii 2.2–3, 5–10, 18–21]

One senses that the Alexandrian group had come to take greater pains over the assessment of proofs than the discovery of new results. By contrast, Archimedes' concern is to stimulate inquiry, as his praises of Conon here indicate.

Our view of the genre of Euclid's *Elements* expands as we move from the prefatory first principles—definitions, postulates, and axioms—and into the main body of problems and theorems. Initially, it may seem plausible that Euclid's aim is to articulate the absolute principles of geometric science, and to elaborate from these the content of the field. But Euclid withholds all commentary on the developing character of his system; the relation of any given proposition to the first principles is never an explicit issue for remark, but only the justification for each of the steps of its proof. Moreover, the presence and position of any construction or theorem are determined by what can be proved at that point; there is no indication

[101] One may observe that Alexandrian scholarship in the early Hellenistic period similarly transformed the study of other fields of learning, in particular, literature: cf. the surveys in Reynolds and Wilson 1968, Russell 1981.

[102] A similarly defensive tone, balancing the heuristic and apodictic aims of research, is evident in the *Method* [Heiberg 1910–1915, ii 428–430].

that what is absent is excluded on absolute grounds—e.g., that certain entities whose construction cannot be given do *not exist* within the system defined by the initial postulates [cf. Knorr 1983]. Euclid must be more pragmatic: he is not possessed of those algebraic techniques which can establish definitively which constructions fall within the Euclidean domain; he can only know which constructions have been worked out, and which (so far) have not.[103] Of those in the latter category, some may have been constructed *via* alternative means (e.g., cube duplication or angle trisection by means of mechanical curves or conics); but the *possibility* of effecting them *via* the Euclidean postulate remains for him unresolved. In view of this, Euclid's *Elements* could not be, even for Euclid, an exposition of the whole geometric field.

Thus, as we make our way through his treatise, we perceive how the structure is the vehicle for presenting a body of fundamentals. The structure, however, is not itself the subject of Euclid's interest. Euclid's meticulous attention to formal detail may well connect the *Elements* with sophisticated mathematical and philosophical inquiries into foundations. But this does not make it a treatise on foundations. The effort to interpret it as such a treatise qualifies as a philosophical critique of the *Elements*, but not as an *exegesis* of its own objectives.

One should consider remarkable—and perhaps unfortunate—Euclid's decision to adopt this formal geometric style in the context of a work intended as an introduction to higher studies. Insight into heuristic techniques is omitted, like the scaffolding scuttled upon completion of the edifice.[104] The *Elements* sets out the finished product for our contemplation; it is not a builder's manual. Yet Euclid must surely intend his work to serve as an introduction to the study of geometry—specifically, the formal, demonstrative type of geometry. Demonstrations in effect set out the causes which justify geometric procedures. What is striking is that Euclid considers this manner of presentation, namely, the formal exposition of finished results, to be the basis for instruction. That points to a fundamental difference between his views of research and pedagogy from our own. Euclid expects that

---

[103] For a discussion of the ancients' views on the classification and solution of problems, see Knorr 1986a, ch. 8.

[104] Cf. the related observation in Hirsch 1967, 78: 'Genre ideas... have a necessary heuristic function in interpretation, and it is well known that heuristic instruments are to be thrown away as soon as they have served their purpose.' Hirsch is here not expressing his own position, however; he goes on to argue that the notion of genre, specifically 'intrinsic genre', is indispensable for the articulation of meaning.

the learner will acquire expertise in geometry through the contemplation of its finished form, rather than through exercise in its production.

5. Conclusion

The scholarship on ancient mathematics invariably adopts an intentional vocabulary in framing interpretations. In discussing passages from ancient writings like Euclid's *Elements*, scholars present their views as if they were Euclid's own. Theorists of literary criticism have long recognized the difficulties that attend a naive conception of authorial meaning, however, and have generated a spectrum of positions on the admissibility of this concept. I have attempted to apply some ideas from E. D. Hirsch, Jr., whose defense of authorial meaning offers, I believe, a position more fruitful for the practicing historian than would the many varieties of literary scepticism. My aim has been, not to win a consensus for certain interpretations of my own on debated points about Euclid, but rather to show how Hirsch's insights can reveal the methodological assumptions implicit in the different views and thus inform the process of judging among them. His four criteria of legitimacy, correspondence, genre appropriateness, and coherence are especially helpful in determining whether a given view is likely to represent the ancient writer's meaning, or might instead be a criticism of it, that is, a projection of the text into the environment of the critic's concepts and concerns.

The standard views on Euclid's proportion theory (e.g., those cited from de Morgan and Heath) tend to read it in the context of modern notions of real number. In the definition of 'having a ratio' [*Elem.* v def. 4], it is maintained, Euclid's intent is one (or perhaps all three) of the following: to restrict ratios to pairs of homogeneous magnitudes, or to exclude non-finite magnitudes, or to incorporate incommensurable as well as commensurable magnitudes. All these views fit the technical demands applicable to any general theory of proportion. But each fares poorly as a rendering of Euclid's text: *legitimacy*—implicit meanings must be assigned to his terms, where there is no clear cause why he should not have made these meanings explicit, if such was indeed his intent; *correspondence*—aspects of the text are left out of account or appear superfluous (why, for instance, is the 'exceeding of multiples' set out as a reciprocal relation?); *coherence*—the definition is viewed in isolation from its textual position among the other definitions and propositions of Euclid's theory.

Like Heath, Mueller takes the intent of the definition to be the inclusion of incommensurables; but he transcribes it in a form different from that adopted by previous commentators. In his version, the definition is set firmly in the context of the comparison of equimultiples, characteristic of Euclid's form of the theory. As I have maintained, this effectively displaces all three of the former proposed views of the definition's intent and establishes an alternative one, namely to serve as a specific precondition for the definition of 'having the same ratio' which immediately follows. From this, one can develop a view of the origin of Euclid's fourth definition as a by-product of the recasting of a precursor version of the proportion theory of book 5. The resultant view, I believe, works well as a historical account of Euclid's meaning and editorial method. The standard views, while hereby displaced as accounts of Euclid's meaning, retain value as critical instruments. For we can now use them to judge Euclid's procedure in contrast to the standard modern theories.

With respect to their arithmetic theories, did Euclid and the other ancients conceive fractions as one does in modern mathematics—that is, did they intend by their terminology of fractions the same things we do? Fowler advocates the provocative thesis that they did not, that the unit-fractional mode adopted in the Egyptian and early Greek calculations acted as a barrier against formulating the more general notion of the fractional number. For instead of presenting the result of dividing an integer m by another integer n merely as the corresponding fraction (i.e., the equivalent of our $^m/_n$), the ancients habitually engage in further computations, casting the quotient as a sum of unit-fractions. Ostensibly, a textual analysis like that given in the preceding example would confirm the separation of the ancient and modern concepts. But here I perceive a difference, in that we are not dealing with a specific text set in a clearly defined textual domain, but instead, with a wide family of texts spanning the whole of Egyptian, Mesopotamian, and Greek antiquity. Thus, the criterion of *coherence* is ambiguous and its application circular: if we minimize our constraints as to what would be an acceptable fraction concept and insist only on an operational equivalence with modern fractions, then texts from later antiquity (e.g., Diophantus and Hero and the writers in their practice-oriented tradition of mathematics) would certainly qualify, and one would naturally date the concept back, in the absence of any clear signs of innovation in the late authors. But if we adopt the stricter sense advocated by David Fowler (where, if I understand his position correctly, one ought to accept fractions as forms of ἀριθμοί, thus conflating the categories of λόγος and ἀριθμός), then either the ancients never advanced to such a conception; or, if they did, this occurs only in late Hellenistic texts influenced by the assimilation of Mesopotamian sexagesimal methods and thus should be sharply marked off from the early arithmetic tradition of the Greeks.

The criterion of *genre appropriateness* raises a further difficulty. Our principal evidence for ancient practical arithmetic comes from the mathematical papyri. The two millennia from the Egyptian Rhind Papyrus to the Graeco-Roman papyri embrace a remarkably uniform tradition of school arithmetic, within which fractions are commonly handled in the unit-mode and applications of the general manner, while arguably present, are not expounded systematically as a separate technique. But should we expect otherwise in a genre of school writing consisting of solved examples, rather than exposition, proof, and commentary? But if we turn to the 'high tradition' of Archimedes and comparable writers of formal geometry, computations of this sort are assumed as part of the students' elementary training. Moreover, when notations for terms like 1838 $^9/_{11}$ appear in the formal tradition, are they merely the artefacts of Byzantine scribal conventions, or do they provide insight into the arithmetical expertise of centuries earlier?

Euclid's arithmetic theory in book 7 of the *Elements* falls within the latter genre of 'high geometry'. What its underlying concept of fraction is may be difficult to determine, since no such term appears there, beyond the undefined notion of measuring (μετρεῖν).[105] But it does deal centrally with ratios of integers, and we of course recognize how to establish an equivalence between ratios and fractions.[106] Further, the practical writers cite Euclid for the theory underlying their arithmetic procedures. One would naturally infer, for instance, that Euclid's problems on the finding of least common multiples [*Elem.* vii props. 36, 39] were intended to provide formal justification of techniques familiar within the practical arithmetic field. To be sure, our evidence does not directly sustain this view, nor would we expect it to. But I find it more plausible than the converse view that Euclid (or, more precisely, the theoretical tradition he consolidated) elaborated his arithmetic theory purely as an abstract exercise, while the later practical

[105] Already in *Elem.* vii def. 3, the relation of measuring (καταμετρεῖν) is assumed for defining 'part' (μέρος); cf. def. 5. The notion of one number's being measured (μετρούμενος) by another is exploited in the definitions of evenly even, evenly odd, oddly odd, prime, relatively prime, composite, and relatively composite numbers (defs. 8–15), even though a neater strategy could have been followed, applying the notion of part given in def. 3. The problems set out in vii props. 2–3 show how to find the greatest common measure of given integers; but neither 'measure' nor 'greatest common measure' has yet been defined. Similarly, one is assumed to understand the notions of measuring and measure as background for the definitions of commensurables and incommensurables in book 10.

[106] By this I mean that any operation on fractions can be re-expressed as a relation among integers.

writers serendipitously discovered its utility for their computations with fractions.

One of the contributions of criticism is to articulate the subtle implications inherent in our assumptions. Fowler performs this service in reminding us of nuances in our concept of fraction: we learn early and thereafter take for granted that 'm divided by n' is the fraction $^m/_n$, a term which can be manipulated with other such terms according to the familiar rules of arithmetic. But can one impute the same conception in ancient arithmetic texts? Fowler's negative position gives rise to the questions of when, by whom, and under what circumstances the general conception was introduced. Not only do our ancient sources provide no assistance toward answers, however, they seem unaware of any such questions. Like ourselves, they appear to take for granted the nuances implicit in their procedures for fractions.[107] It seems to me preferable, then, to take this silence seriously. The aspect of the fraction concept here at issue, the notion of the general fraction, is not a discovery in a simple sense, it would appear, but rather a concomitant of the basic notion of parts. To be sure, the persistence of the unit-fraction methods tends to obscure this general notion in many of our texts, but I would not take this to be conceptually significant. One might perceive a parallel in the persistence of 'English' standards in the United States today, despite the availability of the more efficient metric system. However much American learners might complain about the difficulties of the metric system, it is clear that no conceptual issue is involved, but merely the perpetuation of an outmoded technique. Doubtless, the ancients had comparable reasons of economic expedience for retaining the unit methods, long after the advantages of alternative procedures should have been evident. But this would not as such signify conceptual limitations.

The issue of *genre appropriateness* also illuminates one's understanding of Euclid's aims in the *Elements* as a whole. Euclid presents geometry according to a carefully worked out deductive structure—but his treatise is

[107] By contrast, it is clear when the commentators must assume that notions or techniques are not familiar to their readers. For instance, in his account of sexagesimal operations, Theon expounds the procedures at length, including detailed accounts of particular examples and full statements of individual cases (e.g., minutes times minutes, minutes times seconds, seconds times seconds, and so on): cf. *In Ptol. ad* i 10 [Rome 1936, 452–457]. This, of course, does not indicate the novelty of these techniques at Theon's time; they became available to Greek mathematical astronomers with the reception of Mesopotamian methods around the time of Hipparchus (2nd cent. BC) or earlier. But if techniques that are novel merely to the special group of learners entail such elaborations by Theon, all the more would actually new techniques receive such treatment near the time of their first introduction.

not *about* deductive structure. It is not in the genre of, say, Aristotle's *Posterior Analytics*, even though the juxtaposition of these two works may bring forth interesting details about the ancients' views on formal systems. As I have proposed above, the motivations underlying Euclid's arithmetic theory can be grounded in practice, and one can argue similarly for other parts of the *Elements*.[108] Indeed, much of its material on the measurement of plane and solid figures reappears in an arithmetical form suited for practical application in Hero's account of metrical geometry. Since this aspect of Hero's geometry may be seen to perpetuate the practical procedures of the more ancient Egyptian and Mesopotamian traditions,[109] it would follow that Euclid also developed his geometric theory on a comparable practical base. Ancient views on the term 'elements' link studies of this type to the introductory teaching of technical disciplines. It is, thus, plausible to associate Euclid's own intent in compiling the *Elements* with its actual use within the subsequent technical tradition, namely, as a basic textbook in geometry. But the *Elements* is not in the same category as the Heronian or Diophantine textbooks: Euclid appears to assume a practical grounding in the discipline, for which he aims to provide the appropriate formal demonstrations. In effect, the *Elements* is a treatise on the *causes* relevant to the geometric field; it offers the learner models of how to secure the results of geometry as deductive consequences ultimately rooted in certain notions (namely, the postulates and axioms) of figure and quantity. The learner is expected to gain expertise in geometric theory through the study of finished models of formal exposition.

Conceivably, Euclid himself was responsible for the decision to adopt the formal style in an *introductory* textbook. But the deductive form of geometric theory was a conception he owed to his predecessors. This emerged through the interaction of philosophical and mathematical specialists over the course of the 4th century. Although it is tempting to try to make

[108] The physical phenomena to which geometric propositions are related are often manifest in the cases of Euclid's *Optics* (and certainly the *Phaenomena* which utilizes the terminology of observational astronomy) and of Theodosius' *Sphaerica*. That parts of Euclid, *Elem.* i arose in the context of practical mensuration and instrumentation is noted by Proclus [cf. Friedlein 1873, 283, 352, for his remarks on Oenopides and Thales], while he claims that Pythagoras made mathematics an abstract study, whereas among the Egyptians and Phoenicians earlier it had been developed in the interests of commerce and surveying [Friedlein 1873, 64–65]. It is clear that the geometric constructions presented by Euclid and other writers draw directly from experience in construction with instruments, and many texts provide explicit information on practical execution: see Knorr 1983, 1986a.

[109] On the Greeks' debt to the older Mesopotamian and Egyptian traditions, see nn49–50, above.

explicit connections between Euclid's system and the epistemological pronouncements of earlier philosophers, particularly Aristotle and Plato, the situation was hardly this straightforward. Euclid had access to a variety of exemplars from the preceding generation of technical writers, and he was surely more likely to take his expository model from them than embark on a conscious effort to create a formalism satisfying the prescriptions of one or another philosopher. To the extent that Euclid is consistent with philosophical precursors, this can be assigned to their shared acquaintance with that technical corpus.

These three examples from the study of Euclid turn about a common methodological recommendation—that the historian of mathematics should give priority to the critical examination of the texts before undertaking a wider exploration of their philosophical and mathematical ramifications. This may sound too obvious to warrant special comment. But the combination of fragmentary evidence with a subject area readily associable with modern fields of mathematics and philosophy has made the study of ancient mathematics an arena for ambitious interpretation, where reconstruction overwhelms textual criticism. The result has been a striking use of intentionalist terminology in accounts so heavily dependent on the critics' special predispositions (mathematical or philosophical), that the ancient authors could hardly have actually intended what is claimed for them.[110] If the undesirability of that situation is now clearer and the potential of the alternative textual method evident, I shall have accomplished my purpose here.

[110] Ancient scholarship was hardly immune to the same charge. Russell [1981, 97] styles Hellenistic criticism as intentionalist, even among those Stoic and Neoplatonist writers committed to allegorical readings. In the ancient critics' view, then, Homer (to cite a specific example) actually intended the allegorical meanings they deduce from his texts.

8

Euclid's *Sectio canonis* and the History of Pythagoreanism

ALAN C. BOWEN

The treatise which has come down to us as the *Sectio canonis* or *Division of the Canon* consists in an introduction of thirty-three lines [Menge 1916, 158.1–160.4] and twenty interconnected demonstrations articulated in roughly the same way as those in Euclid's *Elements* [cf. Jan 1895, 115–116].[1] Beyond this most everything is in dispute. To begin, scholars debate the authorship of the *Sectio*. Those who deny or qualify the thesis that it derives from Euclid usually proceed by comparing it to treatises more commonly acknowledged to be Euclid's, and by pointing out supposed inconsistencies in the *Sectio* itself which are presumed inappropriate for a mathematician of Euclid's stature [cf., e.g., Menge 1916, xxxviii–xxxix]. None of the arguments, however, are particularly persuasive. In the first place, the critics tend to ignore the variety of logical structure and language evidenced throughout the Euclidean corpus, and to suppose that any ancient author writing treatises in the various sciences of his age would necessarily do so according to the same standards of expository style and precision.[2] Such an assumption fails when applied to the works in the Ptolemaic corpus, for example [cf. Neugebauer 1946, 112–113]. In the second place, the numerous inconsistencies 'discovered' in this treatise signify, in my view, a failure in scholarship rather than any serious problem in the document itself. Indeed, my main purpose in this chapter is to undercut these claims of inconsistency by setting out a new reading of the introductory part of this treatise.

[1] On the question of Euclid's date, which I put in the third quarter of the third century BC, see Bowen and Goldstein 1991, 246n30 or Bowen and Bowen 1991, section 4.

[2] For criticism of the case for authenticity based on linguistic data, see Menge 1916, xxxix–xl.

The learned debate about the provenance and nature of the *Sectio canonis* centers on five questions:

(1) What is the argument of the preface?,

(2) How does the preface bear on the subsequent twenty demonstrations?,

(3) What is the relation of the first nine demonstrations to the next nine?,

(4) Do the last two demonstrations, the very ones describing a division of the *canon*, belong with the preceding eighteen?, and

(5) Is the treatise complete as it stands?

The first is fundamental, since answers to the others all presuppose an interpretation of the preface. So, in what follows, I will concentrate primarily on the first question, though I will address a few remarks to the last. I will proceed, moreover, by way of a detailed analysis of the sequence of arguments comprising Euclid's preface to the *Sectio*, my aim being to suggest a reading of these arguments which joins them in a coherent, intelligible whole [section 2].[3] I emphasize that it is not my intention to argue that all other interpretations of the preface are wrong. For, not only would this be an improper category of criticism in the present case, it would belie my debt to these other interpretations and, in particular, to the nicely argued account offered by Andrew Barker [1981]. Rather, my purpose is to determine the minimum set of assumptions needed to present the preface as a credible, reasoned unity. And, in doing this, I will rely as much as possible on the internal evidence of the preface itself, and adduce assumptions from elsewhere only when necessary.

My basic contention is that the *Sectio canonis* elaborates in harmonic science the ontologically reductive thesis that all is number; and that once this thesis as it appears in the *Sectio* is properly understood, the most serious of the past worries about the structure and meaning of this treatise dissipate. In other words, if, as Barker suggests [1981, 15–16], the *Sectio canonis* shows above all how to analyze music precisely, it does this by displaying in detail how items in a specific domain, musical sound, are to be construed as number.

But, if this is correct, it would seem that we have replaced one set of problems about the *Sectio* with another concerning its alleged Pythagoreanism.

---

[3] Those familiar with this treatise may discern my approach to the second and third questions. The fourth, which is often raised in the context of reports by Proclus and Marinus of a *Musica elementa* by Euclid [cf. Menge 1916, xxxvii–xxxviii], and which was argued in the negative by Paul Tannery [1912, 213–215], has, I think, been well answered by Andrew Barker [1981, 11–13]. As for the fifth, it requires critical study of the entire treatise and introduces questions about technical writing in the various sciences which I must postpone for now.

For, according to Aristotle, one of the basic tenets of early Pythagoreanism is that all is number; and, as I understand him [see Bowen 1992], this means that numbers are what things really are. So, to conclude this chapter I will address the cluster of problems concerning Euclid, the *Sectio canonis*, and Pythagoreanism [section 3].

1. The preface to the *Sectio canonis*

Let us consider, then, how Euclid introduces the twenty demonstrations in the *Sectio canonis*. The Greek text reproduced here is taken from Menge's edition of 1916 with some slight changes in punctuation and the addition of sentence numbers in square brackets to assist textual analysis.

Euclid's *Sectio canonis*, pref. [Menge 1916, 158.1–160.4]

[1] Εἰ ἡσυχία εἴη καὶ ἀκινησία, σιωπὴ ἂν εἴη· [2] σιωπῆς δὲ οὔσης καὶ μηδένος κινουμένου οὐδὲν ἂν ἀκούοιτο· [3] εἰ ἄρα μέλλει τι ἀκουσθήσεσθαι, πληγὴν καὶ κίνησιν πρότερον δεῖ γενέσθαι. [4] ὥστε, ἐπειδὴ πάντες οἱ φθόγγοι γίνονται πληγῆς τινος γινομένης, πληγὴν δὲ ἀμήχανον γένεσθαι μὴ οὐχὶ κινήσεως πρότερον γενομένης,—τῶν δὲ κινήσεων αἱ μὲν πυκνότεραί εἰσιν, αἱ δὲ ἀραιότεραι, καὶ αἱ μὲν πυκνότεραι ὀξυτέρους ποιοῦσι τοὺς φθόγγους, αἱ δὲ ἀραιότεραι βαρυτέρους,—ἀναγκαῖον τοὺς μὲν ὀξυτέρους εἶναι, ἐπείπερ ἐκ πυκνοτέρων καὶ πλειόνων σύγκεινται κινήσεων, τοὺς δὲ βαρυτέρους, ἐπείπερ ἐξ ἀραιοτέρων καὶ ἐλασσόνων σύγκεινται κινήσεων, ὥστε τοὺς μὲν ὀξυτέρους τοῦ δέοντος ἀνιεμένους ἀφαιρέσει κινήσεως τυγχάνειν τοῦ δέοντος, τοὺς δὲ βαρυτέρους ἐπιτεινομένους προσθέσει κινήσεως τυγχάνειν τοῦ δέοντος. [5] διόπερ ἐκ μορίων τοὺς φθόγγους συγκεῖσθαι φατέον, ἐπειδὴ προσθέσει καὶ ἀφαιρέσει τυγχάνουσι τοῦ δέοντος. [6] πάντα δὲ τὰ ἐκ μορίων συγκείμενα ἀριθμοῦ λόγῳ λέγεται πρὸς ἄλληλα, ὥστε καὶ τοὺς φθόγγους ἀναγκαῖον ἐν ἀριθμοῦ λόγῳ λέγεσθαι πρὸς ἀλλήλους· [7] τῶν δὲ ἀριθμῶν οἱ μὲν ἐν πολλαπλασίῳ λόγῳ λέγονται, οἱ δὲ ἐν ἐπιμορίῳ, οἱ δὲ ἐν ἐπιμερεῖ, ὥστε καὶ τοὺς φθόγγους ἀναγκαῖον ἐν τοιούτοις λόγοις λέγεσθαι πρὸς ἀλλήλους. [8] τούτων δὲ οἱ μὲν πολλαπλάσιοι καὶ ἐπιμόριοι ἑνὶ ὀνόματι λέγονται πρὸς ἀλλήλους. [9] γινώσκομεν δὲ καὶ τῶν φθόγγων τοὺς μὲν συμφώνους ὄντας, τοὺς δὲ διαφώνους, καὶ τοὺς μὲν συμφώνους μίαν κρᾶσιν ἐξ ἀμφοῖν ποιοῦντας, τοὺς δὲ διαφώνους οὔ. [10] τούτων οὕτως ἐχόντων εἰκὸς τοὺς συμφώνους φθόγγους, ἐπειδὴ μίαν τὴν ἐξ ἀμφοῖν ποιοῦνται κρᾶσιν τῆς φωνῆς, εἶναι τῶν ἐν ἑνὶ ὀνόματι πρὸς ἀλλήλους λεγομένων ἀριθμῶν, ἤτοι πολλαπλασίους ὄντας ἢ ἐπιμορίους.

[1] If there were rest and lack of motion, there would be silence. [2] But, if there were silence and nothing moved, nothing would be heard. [3] Therefore, if anything is going to be heard, there must previously occur striking

and motion. [4] Consequently, since all musical notes occur when there is a certain striking, and since it is impossible that a striking occur unless a motion occurs previously—some motions are closer together but others are less close together; and the ones that are closer together produce notes higher (in pitch); but those less close together, notes that are lower (in pitch)—it is necessary that the former notes be higher (in pitch) because they are composed of motions that are closer together and so more numerous and that the latter notes be lower (in pitch) because they are composed of motions that are less close together and so less numerous; so that notes higher (in pitch) than what is needed reach it when lowered by subtraction of motion, and those lower (in pitch) than what is needed reach it when raised by addition of motion. [5] Wherefore, we should say that musical notes are composed of parts, since they reach what is needed by addition and subtraction. [6] But all things composed of parts are described in relation to one another by a ratio of (whole) number, so that musical notes must also be described in relation to one another by a ratio of (whole) number. [7] But some numbers are said to be in multiple ratio, some in superparticular ratio, and others in superpartient ratio,[4] so that notes too must be said to be in these sorts of ratio in relation to one another. [8] Of these [*scil.* musical notes] the multiple and superparticular are described in relation to one another by a single term. [9] In fact, we perceive some notes as concordant but others as discordant, and the concords as making a single blend out of a pair (of notes) but the discords as not. [10] Since these things are so, it is appropriate that concordant notes, being either multiple or superparticular, belong to (whole) numbers described in relation to one another by a single term, since they produce a single blend of sound out of a pair (of musical notes).

2. Analysis of the preface to the *Sectio canonis*

This introduction is, in fact, a series of five arguments establishing that

(a) prior striking and motion are required if anything is to be heard [1]–[3];

(b) the relative pitch of a musical note varies directly as the relative close-packedness or compactness of the motions constituting it [4];

(c) musical notes are composed of parts [5];

---

[4] If m and n are whole numbers, where $1 < n < m$, then ratios of the form m:1, $(m+1)$:m, and $(m+n)$:m are multiple, superparticular, and superpartient, respectively.

(d) two notes may stand in either multiple, superparticular, or superparticient ratio [6]–[7], and
(e) concordant notes are reasonably said to belong to those whole-number ratios which are predicated by a single term [8]–[10].

I suspect that this is sufficient to highlight the fact that the introduction to the *Sectio canonis* is peculiar. Indeed, the oddity of the locutions in these arguments, their sense, and how they fit together are real puzzles. And there is no way to solve them except by a careful study of what is actually written.

2.1 *First argument*

> [1] If there were rest and lack of motion, there would be silence. [2] But, if there were silence and nothing moved, nothing would be heard. [3] Therefore, if anything is going to be heard, there must previously occur striking and motion.

Though its structure is clear, it is not easy to see what this argument is about. Still, as we read on there are, I think, three alternatives to consider in deciding what moves and what is struck. The motion may be that of

(a) something which strikes a sonant body, a hand plucking the string of a lyre for instance; or
(b) a sonant body striking the ambient air, for example, the string of the lyre striking the air as it moves back and forth after being plucked; or
(c) the moving air which has been set in motion by the sonant body and strikes the ear.

2.2 *Second argument*

> [4] Consequently, since all musical notes (φθόγγοι) occur when there is a certain striking, and since it is impossible that a striking occur unless a motion occurs previously—some motions are closer together (πυκνότεραι) but others are less close together (ἀραιότεραι); and the ones that are closer together produce notes higher (in pitch); but those less close together, notes that are lower (in pitch)—it is necessary that the former notes be higher (in pitch) because they are composed of motions that are closer together and so (καὶ) more numerous and that the latter notes be lower (in pitch) because they are composed of motions that are less close together and so less

numerous (ἐλασσόνων); so that notes higher (in pitch) than what is needed reach it when lowered by subtraction of motion, and those lower (in pitch) than what is needed reach it when raised by addition of motion.

Here it is evident that not only must the striking or impact and motion precede the musical note, this motion must also be prior to the striking. In short, if there is to be musical note, there must first be motion which produces an impact which in turn produces the note. Yet, the story is now more complex, given that φθόγγοι (which I have rendered by 'musical notes')[5] are not only produced by motions, they are composed of them. In any case, if the φθόγγοι are to be *composed* of motions, it would seem unlikely that Euclid means to claim that the motions in question are (a) those of something which strikes a sonant body, like the hand's motion in plucking the string of a lyre, or (b) the motions of the sonant body, such as those of the sonant string to and fro. So, by elimination, it seems that the first argument concerns the motion of air as it strikes the ear. But this still leaves a problem: if the motions *constitute* the φθόγγοι or musical notes, it is difficult to see how the motions are to *precede* them.

This problem, however, is not insuperable. As our first hypothesis, let us grant Euclid a distinction between musical sound as heard (phenomenal

[5] The noun, φθόγγος, has a variety of attested meanings which include any clear, distinct sound—especially vocal sound, where this was primarily that of (male) voices and later extended to cover sound produced by any animal with lungs—as well as speech, musical sound, and sound in general. The tendency among scholars who have studied the *Sectio canonis* is to suppose that it here means 'a sound in general' and that the preface draws on ancient acoustical physics. I reject this for two reasons. First, I no longer see the point [cf. Bowen 1982] in elevating the sort of remark made in texts in harmonic science like the *Sectio* (or in others which attempt, for example, to explain hearing in terms of some philosophical theory of change and motion) to the status of an independent, acoustical physics: for, to do this without proper regard for the context of these remarks is to risk abstracting a domain of technical discourse which did not exist in ancient times, and to confound efforts to determine the sense and the history of the texts in question. In truth, regarding every discussion of sound as belonging to an acoustical theory makes as much sense as treating liver omens as part of some ancient veterinary science. Second, those who take φθόγγος to mean 'sound' arbitrarily introduce difficulties in explaining how the preface to the *Sectio* bears on the subsequent demonstrations—it is not surprising that they isolate the preface on the ground that it concerns sound in general, and view the first nine demonstrations as establishing truths about ratios without regard for musical phenomena [cf., e.g., Ruelle 1906; Mathiesen 1975, 237; Barker 1981, 1–3; Fowler 1987, 146]. Thus, to counter what I see as a gratuitous balkanization of the treatise, I propose, with equal justification *prima facie*, to start differently and to render φθόγγος as 'a musical note'.

musical sound) and musical sound as constituted of motions (objective musical sound), and let us suppose accordingly that the argument in sentences [1]–[3] of the preface is about the former. In other words, let us take the first argument to focus on conditions needed for the occurrence of musical sound as heard.

This hypothesis is plausible. In my view [cf. Bowen 1982], such a reduction of what is heard to objective, quantifiable conditions underlies the sequence of illustrations and observations made in the fragment from Archytas of Tarentum (who was active during the late 5th and early 4th centuries); and, indeed, this fragment bears interesting parallels to Euclid's preface.[6] More compelling is the fact that the notion of phenomenal musical sound is essential to the distinction of concords and discords in sentence [9] (note γινώσκομεν with the present participle construction), and that sentence [10] as a whole plays on the relation of phenomenal and objective musical sound [see section 2.5, below].

So far, then, it would appear that the motion mentioned in [1]–[3] and in the first two premisses of [4] occurs between the sonant body and the ear, and that this motion produces the musical note we hear by striking the ear. The second argument continues by way of an interjection, in which it is evident that the motion responsible for producing what we hear as a single musical sound is really a series of consecutive, discrete motions; and that the relative pitch of two musical notes as heard varies directly as how closely the motions in each series follow upon one another, that is, as the relative compactness (πυκνότης) of the series. Next, and most important, comes the conclusion that what we hear as a single note is in fact just the series of motions which produces it.[7] This is the force of 'because they (*scil.* the musical sounds as heard) are composed of (σύγκεινται ἐκ with the genitive) motions... '.

Several features of the argument in sentence [4] merit comment. First, that feature of phenomenal musical sound which most concerns Euclid is its pitch. The isolation of this feature is important. Though Euclid mentions the perception of a blending of concordant musical notes later in sentence [9], it is clear that he intends a blending of pitch. In short, this treatise prescinds from any other features of phenomenal musical sounds which one

[6] Curiously enough, Jan [1895, 132, 135, 146] adduces this same fragment in contending that Euclid's preface concerns the motions of a sonant body striking the air. Jan, however, neglects the claim that the φθόγγοι are composed of motions.

[7] For discussion of Boethius' treatment in his translation [Friedlein 1867, 301.12–308.15] of the *Sectio canonis*, of this reduction of phenomenal musical sound to a series of motions striking the ear, see Bowen and Bowen 1991, section 4.

might be disposed to view as contributing to their musicality (e.g., volume, rhythm, and timbre).

Moreover, given our hypothesis about Euclid's distinction of phenomenal and objective musical sound, it seems that not only does he focus on but one of the many salient characteristics of phenomenal musical sound, pitch, he takes this in turn to be nothing more than a series of motions that strike the ear. This is admittedly peculiar, but still intelligible. As I will explain more fully when we come to the problem of the relation between musical intervals and numerical ratios [see section 2.4, below], what we have here is the initial step in a reductive analysis of music as heard to relative number.

Next, it appears that, for Euclid, pitch is a relative phenomenon—he neither gives any hint that the pitch of a note is to be understood absolutely, nor, I maintain, is it necessary to the sense of this passage that pitch be construed as absolute. This is, of course, consistent with the absence of evidence from other sources that the ancient Greeks conceived of an absolute standard of pitch or that they possessed the means of measuring time so as to define one. Perhaps, the predilection evident in Greek scientific and philosophical documents for defining ratios only between quantities of the same kind explains this [cf. Euclid, *Elem.* v defs. 3 and 4]. In any case, it follows immediately that the numerosity or compactness of motions is not equivalent to frequency. In other words, Euclid does not assume here a vibrational theory of how sound propagates. For, although the various series of motions occur in time and are differentiated by the lapse of time between elements in each series, the series themselves are not to be quantified in relation to some unit of time. Thus, the compactness of motions is not the same as some number of motions *per* second, as Tannery [1912, 217] and Barker [1981, 8], for example, would appear to suppose.

If the compactness or close-packedness of each musical note is relative and not measured in relation to time, how then is it to be quantified? In the preface to the *Sectio*, it is clear that the higher pitch is assigned the greater number in the ratio of the musical notes, since the higher-pitched note is constituted of more motions. In order to quantify this, all one would need to know is that pitch varies inversely as the length of a sonant string or pipe. But this very assumption figures prominently in the last two demonstrations of the *Sectio canonis* [see, e.g., Menge 1916, 178.14–18]. So, for Euclid, it seems, if the compactness of the motions constituting a musical note at a certain pitch varies inversely as the length of the sonant string or pipe producing it, then, to quantify the relation between two musical notes *qua* pitches, one must measure the relative lengths of the strings or pipes producing them.

2.3 *Third argument*

> [5] Wherefore, we should say that musical notes are composed of parts, since they reach what is needed (τοῦ δέοντος) by addition and subtraction.

Again we have the eliminative reduction of phenomenal to objective musical sound, that is, the musical note as heard *qua* pitch to the series of motions that strike the ear. And as before, just as *relative* pitch is taken to be the primary or defining quality of phenomenal musical sound, *relative* compactness or close-packedness is to be the main characteristic of objective sound. What is added is the claim that each musical note so understood has parts, since it is constituted of motions to which motions may be added or subtracted. What sort of rationale might there be for this?

Consider the behaviour of a sonant string on a lyre. According to Archytas [cf. Bowen 1982], such a string strikes the air with each motion back and forth and sets the ambient air in motion like a projectile which strikes the ear causing one to hear a single sound at a pitch that varies inversely as the effective length of the string. It would, of course, be easy to elaborate this (in a way Archytas did not) by supposing that the pitch of the sound heard is determined proximately by the rate of the string's motion to and fro, and that this is inversely dependent on the string's effective length.[8] Since the string's motions to and fro are seemingly consecutive and discrete, it would seem plausible that the series of airy projectiles moving from the string to the ear is likewise consecutive and discrete, and that the relative numerosity or close-packedness of this series depends directly on the rate of the string's motions. Moreover, given that the the pitch of the note heard varies directly as the rate of the string's motion back and forth, it would follow that one may adjust the pitch by increasing or decreasing the rate of the string's motion. And, of course, one would do this by decreasing or increasing the effective length of the string itself. Thus, by identifying pitch as heard with the series of airy projectiles striking the ear, one gets the result that each musical note consists of discrete, consecutive parts subject to additive increase or decrease by decreasing or increasing the effective length of the string.

---

[8] Cf. the analyses offered by Adrastus in Theon [Hiller 1878, 50.11–21], Nicomachus [Jan 1895, 243.17–244.1; 254.5–22], and Porphyry's version of Heracleides' report of Xenocrates' remarks about Pythagoras [Düring 1932, 30.9–31.21].

Such an account of what underlies sentence [5] is admittedly conjectural. Its main advantage is that it adheres to, and is consistent with, what is actually written in the preface to the *Sectio*, and that it enables an intelligible transition from sentence [4] to sentence [6]. In any case, it is important to see that, were the relative numerosity or close-packedness of the series of motions constituting two musical notes to be understood and quantified in this way, there would be no need in addition to worry about the relative incidence of the component motions in pairs of series at the ear. Granted, one might well choose to develop some account of this for independent reasons; but it remains the fact that Euclid's writing of consecutive series of motions constituting musical notes as heard is by itself no warrant to suppose that the *Sectio* entails any views at all about how pairs of series impinge on the ear in relation to one another.

2.4 *Fourth argument*

> [6] But all things composed of parts are described in relation to one another by a ratio of (whole) number (ἀριθμός), so that musical notes must also be described in relation to one another by (ἐν) a ratio of (whole) number. [7] But some numbers are said to be in multiple ratio, some in superparticular ratio, and others in superpartient ratio, so that notes too must be said to be in these sorts of ratio in relation to one another.

From the conclusion that musical notes are composed of parts, Euclid now argues that such notes must stand to one another in whole-number ratios. Tannery [1912, 215–216: cf. Fowler 1987, 146] objects to the argument on the ground that it is simply not true that any two objects composed of parts need manifest a numerical ratio, and he concludes that a geometer like Euclid could scarcely have written this. Now, whether we should expect that Euclid would have written 'things composed of *discrete* parts' (i.e., 'pluralities') is a nice question. In any case, if I am right about the sense of the preceding sentences, this is what πάντα δὲ τὰ ἐκ μορίων in fact means; and so there is no real difficulty. Indeed, I suspect that Tannery is wrong to abstract this sentence from its context and to criticize it as though it were a universal proposition. As for what may be Tannery's assumption that an ancient author who writes in one scientific field will necessarily write with the same degree of precision on the same topics in another, I have indicated that this is not true of Ptolemy. Further, we should recall that the degree of articulation in the deductive structure of Euclid's *Elements* results in great part from its focus on problems of incommensurability, problems which require precise definitions for solution [see Neugebauer 1941, 25–26]; and we should realize that the varying sophistication in explanatory structure of the other sciences may likewise depend on the nature of their problems. In my view, given that the *Sectio* ostensibly presents a science of relations among the pitches of musical notes and is limited to the domain of commensurable magnitudes, Tannery's objection to this sentence in the preface is more captious than substantive: it is certainly no reason to deny Euclid's authorship of the *Sectio canonis*.

That the sort of harmonic science presented in the *Sectio canonis* is indeed limited to ratios of whole numbers follows immediately from two considerations already mentioned. The first is that pitch is to be understood relatively, that musical pitches are conceived only in relation to one another. The second is that in the *Sectio* one is apparently to quantify pitch by measuring string-lengths according to a common unit [cf. dems. 19–20: Menge 1916, 178.11–180.31]: such measurement by a common unit is an empirical process and will inevitably yield a ratio of whole numbers [cf. Bowen 1982, 96].[9]

As for the reference to multiple, superparticular, and superpartient ratios, there is no need in either grammar or sense to take this as an exhaustive tripartition. Were one moved to do this, however, it would follow that these three kinds of whole-number ratio are fundamental or basic, that the multiple superparticular and multiple superpartient ratios evident especially in the last two demonstrations are therefore derivative.[10] For my part, I prefer to suppose that Euclid mentions the three kinds of whole-number ratios he does and passes over the others because they are not germaine to the purpose of the preface [see section 2.5, below].

---

[9] If this is correct, we have an explanation for the fact that, when Greek theorists relied on ratios to analyze musical relations, they confined their attention to ratios of whole numbers. In a sense, then, this limitation is not arbitrary, though it is clear that not all the ancients understood it and that they may even have viewed it as a matter of convention. Adrastus [Hiller 1878, 50.14–16], for example, mentions ratios of incommensurable magnitudes and relegates these to noises or non-musical sounds. But I take this to be symptomatic of a somewhat specious logical completeness characteristic of much Peripatetic writing. For, according to Adrastus, if one assigns musical notes to ratios of whole numbers by quantifying speeds using some unit as a common measure, then one may assign noises to ratios of incommensurable speeds (presumably) by quantifying speeds using geometrical techniques and not a common measure. But see Barker 1984–1989, ii 214n16.

[10] If m, n, and p are whole numbers, where $1 < n < m$ and $1 < p$, then ratios of the form $(mp+1):m$, and $(mp+n):m$ are multiple superparticular and multiple superpartient, respectively.

Now, if the pitch of a musical note and the compactness of the series of motions that strike the ear are both relative, it follows that the fundamental musical phenomenon according to the *Sectio canonis* is the interval or separation (διάστημα) defined by two distinct pitches. In short, the phenomenon of music is not so much a sequence of pitches, as a sequence of separations defined by pitches.[11] Moreover, given the hypothesis of the

[11] Fowler [1987, 148] reiterates Szabó's claim [1978, 99–144: cf. Barker 1981, 13] that in the texts like the *Sectio* διάστημα signifies a '" distance between" or "interval" in a very general sense' [cf. Bowen 1984, 337–341], a claim which is perhaps one reason why he does not see that in the *Sectio* whole-number λόγοι (ratios) are what διαστήματα really are [cf. Bowen and Bowen 1991, section 4].

In any case, Szabó's claim rests on poor philology. As I have argued elsewhere [Bowen 1984, 340–341: cf. 1982, 95 and nn81–83], the root sense of διάστημα is 'separation'. Of course, the challenge is to characterize this separation and *one* way is to view it as a linear difference between pitches. But there are others and none is intuitively more correct. Indeed, Porphyry [Düring 1932, 90.24–95.23] suggests that the schools of harmonic science all start from the assumption that an interval is the separation of pitches, but differ as to how this separation is conceived. In particular, he reports that some think of musical intervals as differences (διαφοραί, ὑπεροχαί), whereas others say that they are whole-number ratios, and still others that they are continuous ranges of pitch defining τόποι (regions). Let us consider this further.

Pitch is a magnitude admitting a more and a less. The difference between two pitches may be likened to the separation of the endpoints of two line-segments which coincide and share a common origin. Now, there are three ways to describe this separation and each was adopted by some school of harmonic science. Some took the separation as the whole-number ratio specified by the magnitudes of the two line-segments: among these were the Pythagoreans and Euclid. Others defined the separation as the numerical excess of the greater line-segment over the less. Aristoxenus, who views theorists of this sort as his predecessors, calls them ἁρμονικοί; and for want of a better term we may follow him, though I must add that his use of the term may well be partisan—Aristoxenus so opposes Pythagorean theory that he denies it status as harmonic science and refuses to name any Pythagorean a ἁρμονικός or to allow that any was his predecessor [see Barker 1978a]. In any case, Euclid, the Pythagoreans, and the ἁρμονικόι all define the separation of two pitches by reference to their magnitude, the first two taking it as a ratio and the third as a numerical difference or excess.

But there is yet another way of looking at the separation of the endpoints of our two line-segments. Aristoxenus and his followers define an interval as the range of pitch between two pitches and stipulate that the identity of an interval is preserved as the magnitude of this range varies within boundaries which the ear, by attending to the melodic function (δύναμις) of the pitches, determines to be the limits of that interval. To apply this to our line-segments, then, the Aristoxenians think that the separation of endpoints is the range between them and hold that such a separation may preserve identity when this range increases or decreases in magnitude between certain limits determined on qualitative grounds.

eliminative reduction of phenomenal to objective musical sound, it also follows that each interval or separation is to be identified as a ratio of whole numbers. In effect, we have here what I have called an eliminative, reductive analysis of music *qua* system of relative pitches (intervals) to relative number (ratios).

Accordingly, it is, a mistake to suppose that Euclid's talk of adding and subtracting motions in sentence [4] means that musical notes are numbers and that the musical intervals defined by pairs of pitches are numerical differences. Hence, Düring [1934, 177] is, I think, wrong to maintain that Theophrastus' criticism of Pythagoreans for treating musical intervals as numbers [Düring 1932, 62.5–10: see Barker 1977, 3–5 for text and explication], that is, for confusing a ratio of two numbers with their difference [cf. Thrasyllus in Düring 1932, 91.14–92.8], should be read as directed against the Pythagorean tradition which (Düring thinks) the *Sectio canonis* retails.[12] For, not only does this misconstrue the *Sectio*, a document which may well not be Pythagorean, there is, so far as I am aware, no good evidence that any early Pythagorean was so benighted as to confuse notes evaluated relatively with those specified independently or absolutely. As I see it, Theophrastus' criticism is not directed against any real Pythagoreans at all: given Aristotle's scattered remarks about Pythagoreanism and the few fragments remaining of Philolaus' remarks concerning musical theory, I would say instead that Theophrastus' criticism is an assault on a straw man contrived on the basis of a literal reading of passages in Aristotle's *Metaphysics*.

2.5 *Fifth argument*

> [8] Of these (τούτων) the multiple and superparticular are described in relation to one another by a single term (ἑνὶ ὀνόματι). [9] In fact, we perceive some notes as concordant but others as discordant, and the concords as making a single blend out of a pair (of notes) but the discords as not. [10] Since these things are so, it is appropriate that concordant notes, being either multiple or superparticular, belong to (whole) numbers described in relation to one another by a single term (ἐν ἑνὶ ὀνόματι), since they produce a single blend of sound (φωνῆς) out of a pair (of musical notes).

---

[12] Ruelle [1906, 319] wrongly supposes that διάστημα in dems. 1–9 signifies a numerical difference: cf. Bowen and Bowen 1991, section 4.

These three sentences constitute a single argument which is in fact the culmination of the preface. But, though most will admit this, there is little agreement about what the argument really is.

The controversy begins with sentence [8]. What is the referent of the demonstrative in 'of these' (τούτων)? Some [e.g., Burkert 1972, 383–384; Barker 1981, 2–3; Fowler 1987, 144] think that it is 'numbers', i.e., that multiple and superparticular numbers are to be designated by a single term. Others [e.g., Jan 1895, 117–118; Tannery 1912, 218–219] suggest that the referent is 'ratios'. These views are in fact equivalent, since it is the same thing to talk of a multiple ratio and to speak of one number as a multiple of another; that is, λόγοι τοῦ ἀριθμοῦ are the same as ἀριθμοὶ πρὸς ἀλλήλους.[13] And so on either view, the problem is to discover what this single term is, because none is given in the text.

Jan [1895, 118] consults Porphyry [Düring 1932, 98.3–6] and proposes that the multiple and superparticular ratios are *potiores* or possessed of greater power (κρείττονες), because such ratios are simpler relations than the superpartient. Barker [1981, 2–3],[14] however, argues that there is in fact no general term for these ratios or numerical relations. Instead, he suggests that what Euclid alludes to is the linguistic fact that the Greeks expressed each multiple and superparticular ratio by a single term but used phrases for each superpartient. Such a thesis has the obvious advantage of explaining why no single term is given explicitly in the *Sectio canonis*—a problem which moved Jan [1895, 118–119] to posit a lacuna in the text—but like Jan's version, the resultant argument is not very convincing. After all, there is no compelling reason to connect the simplicity of multiple and superparticular ratios and the unity of concordance, or to connect linguistic practice (νόμος) in naming these ratios and the nature (φύσις) of concordant sound.

[13] Mathematically the same, that is: there is a difference between the two locutions which raises epistemological and ontological questions about the status of relations *vis à vis* their *relata*. When one says that some number is a multiple of another, one *relatum* may be treated as subject and the other as part of a complex predicate: e.g., p is a-multiple-of-q. In this account, the relation of p and q is to be seen as a property belonging to one *relatum* and specified in terms of the other. But, when one says that the ratio, $p{:}q$, is multiple, the relation of p and q is characterized first as a ratio, and then this ratio is qualified by the predicate 'multiple'. Hence, the relation is at least conceived apart from the *relata* exhibiting it.

More narrowly, the difference between the two locutions is that between treating music as sequence of musical notes and as a sequence of melodic intervals.

[14] Cf. Tannery 1912, 218–219; Ruelle 1906, 319; Burkert 1972, 383–384; Fowler 1987, 146–147.

There is not much to choose between these alternative accounts of the single term: where Jan focuses on the relative simplicity of the relation between the terms in multiple and superparticular ratios, the others adduce its manifestation in language. And were there no other possibilities, we would have to leave the matter here and content ourselves with a *Sectio* that simply falls apart just as it reaches its conclusion.

But let us look more closely at this final argument. As a matter of grammar, the referent of τούτων in [8] may, in fact, not be numbers or ratios but musical notes (φθόγγοι) [cf. Ruelle 1906, 319; Mathiesen 1975, 254n12]. So, though it is admittedly possible at first glance that the demonstrative τούτων refers to numbers [cf. [7]: τῶν δὲ ἀριθμῶν] or ratios [cf. [7]: ἐν τοιοῦτοις λόγοις],[15] let us suppose that it picks up the subject of the immediately preceding resultative clause (ὥστε τους φθόγγους ... ἀλλήλους). Accordingly, sentence [8] would mean that multiple and superparticular musical notes (that is, musical pitches *qua* series of consecutive motions) when taken in relation to one another form a single class of musical sounds.

Granted, this does entail that such musical notes belong to a special class of multiple and superparticular ratios. But it would now seem possible that the term for this class is musical and not necessarily some predicate appropriate to whole-number ratios as such. In other words, the *analysantia*, certain whole-number ratios, may have a predicate appropriate in the first instance to the *analysanda*, certain musical notes as heard.

Sentence [8] thus poses the question, What is this single term for multiple and superparticular notes? Since none is given explicitly in the text, there would seem to be two ways of seeking an answer. The first is to look elsewhere in other texts for a term satisfying the requirements of the argument in sentences [9] and [10]. This is the sort of approach taken by Jan and Barker, for example. The second is to consider the train of thought leading from sentence [8] to sentences [9] and [10] in order to see whether the term figures implicitly in the argument. (Of course, it is entirely possible that the term is simply unrecoverable, that there is an unbridgeable gap at this point in the logic of the preface.[16])

In considering the transition from sentence [8] to sentence [9], let us not forget that [8], as I construe it, is about objective musical sound. Suppose, then, that one musical note (*qua* series of consecutive motions) is 'taken

[15] Mathiesen [1975, 254n12] dismisses out of hand the possibility that Euclid is thinking of the linguistic fact that the Greeks use single terms to designate multiple and superparticular ratios.

[16] Such a break in logic, however, need not signify a lacuna in the text as it stands.

in relation to' a second. This means that these two notes manifest a whole-number ratio. When one regards the same two notes phenomenally, this whole-number ratio turns out to be the reality of the separation or interval (διάστημα) heard between the notes [cf. section 2.4, above]. In other words, the phenomenal counterpart of the claim that multiple and superparticular notes (*qua* series of consecutive motions) are described in relation to one another by a single term is that the intervals defined by these notes are determined by a single class of multiple and superparticular ratios. So the question about the single term is at the same time a question about a class of intervals or notes as heard.

Now, sentence [9] presents a distinction among phenomenal musical notes: those perceived as concords make a single blend of sound, whereas those perceived as discords do not. I emphasize that this distinction is not necessarily a dichotomy: contrary to the usual understanding of this passage, the text actually leaves open the possibility (a) that some melodic notes are neither concordant nor discordant, (b) that not every pair of notes perceived as a single blend of sound is a concord, and (c) that not every pair or notes *not* heard as a unified sound is a discord. (Note that those who assume a dichotomy quickly encounter difficulties in other parts of the *Sectio* which often they then use to impugn it [cf. n18, below].) Further, given that Euclid identifies phenomenal and objective musical sound, it would appear that the pairs of multiple and of superparticular musical notes (*qua* series of consecutive motions) mentioned in sentence [8] may either be concordant or discordant, that the single term said to designate these notes (objectively construed) may either be 'concordant' or 'discordant'.[17]

Sentence [10] continues as an inference from sentences [8] and [9]—as the phrase 'since these things are so' indicates—supplemented by way of two subordinating constructions. In effect, the inference in [10] is:

(p_1) since (pairs of) concordant musical notes are either multiple or superparticular

(p_2) since (pairs of) concordant notes are heard as a single blend of sound

---

(P) it is appropriate that (pairs of) concordant notes belong to ratios designated by a single term.

[17] Mathiesen [1975, 254n12] maintains that the term in question is 'concordant' and cites the same passage from Porphyry [Düring 1932, 98.3–6] which Jan adduces to show that it is κρείττων.

To unpack this and the final argument as a whole we need to determine the relation between sentences [8] and [9], and the subordinating constructions in [10] represented as premisses p_1 and p_2. It is obvious that p_2 recasts sentence [9]. So, does p_1 reformulate [8]? If we suppose it does, we do get the result that the term for multiple and superparticular notes (*qua* series of consecutive motions) is 'concordant'. Unfortunately, we also get an unproven conversion: saying that multiple and superparticular musical notes (*qua* series of consecutive motions) are concords (so sentence [8]) is not the same as saying that concords are multiple or superparticular. Thus, we should allow the phrase, 'since these things are so', some real significance and treat [8] as an independent premiss in the final argument of the *Sectio*. Accordingly, let us combine sentence [8] and p_1 as

(p_3) any pair of musical notes (*qua* series of consecutive motions) is designated by a single term, 'concordant', if and only if one is a multiple or superparticular of the other.[18]

---

[18] There are several features of this premiss to observe here. First is that the scope of p_3 is limited to the domain of phenomenal musical sound: there is no reason to suppose that Euclid countenances an unlimited plurality of concords on the ground that there is an unlimited number of multiple and superparticular ratios. For a clear statement of the issue and of the various positions discerned by Adrastus (who is much cited by Theon of Smyrna), see Hiller 1878, 64.1–65.9.

Next, there is the claim by Aristoxenus and later writers that the interval of an octave and a fourth (8:3) is a concord. But do such claims indicate that p_3 is false? Barker [1981, 9–10] maintains that they do, on the ground of p_2. In other words, he takes it for granted that Aristoxenus' assertion [Da Rios 1954, 25.17–26.1; 56.10–18] that the addition of an octave to any concord yields an interval which will be heard as a concord, is an accurate report of what Aristoxenus' contemporaries actually heard; and concludes that the *Sectio*, by virtue of p_2, is obliged to allow for this. But I think this concedes and requires far too much. To begin, unlike Barker I do not think that p_2 entails that every sound heard as a single blend is a concord: so, even if the interval of the octave and a fourth was heard as a single blend by Aristoxenus and his contemporaries, it does not follow for Euclid that it is a concord. (Nor, given that [9] does not state a dichotomy of intervals into concords and discords, does it then follow that it is a discord.) Further, Aristoxenus himself provides evidence [Da Rios 1954, 29.5–30.9] of disagreement in matters of musical hearing and of a tendency to extol music others find disagreeable. Indeed, my suspicion about Aristoxenus' claim regarding the interval in question is that it may well be a conclusion drawn from the (rather abstractly stated) principle that any octave added to a concord produces a concord. And, if this suspicion is right, then the claim about the interval of the octave and a fourth may in fact be wholly polemical. In any case, the real problem here is the use of Aristoxenus' testimony and, more generally, determining the relation between ancient harmonic science and musical practice. Solving this problem will, in the present instance, be extremely difficult: for, not

Next, there is the problem of the role of p_2 in sentence [10]. If the conclusion, P, is about objective sound, the inference in [10] becomes very puzzling, since p_2 concerns phenomenal sound. But, if P is about phenomenal sound, p_2 is essential.

I propose, then, to recast the final argument in sentences [8]–[10] (with redundancies) as follows:

(p_2) since pairs of concordant notes are heard as a single blend of sound [cf. [9]]

(p_3) given that any pair of musical notes (*qua* series of consecutive motions) is designated by a single term, 'concordant', if and only if one is a multiple or superparticular of the other

---

(P) it is appropriate that pairs of concordant notes heard as a blend of sound and being multiple and superparticular (*qua* series of consecutive motions) belong to (*scil.* are in reality) pairs of (multiple and superparticular) numbers designated in relation to one another by a single term.

The reader will notice that I have elaborated the conclusion, P, by spelling out (a) that the numbers to which multiple and superparticular notes belong are, in the first instance, themselves multiple and superparticular; and (b) that pairs of notes belong to pairs of whole-numbers in the sense that the latter are the reality with which the former are identified through reductive analysis.[19] Yet, this is not enough. The argument still needs an additional premiss,

---

only is there no independent evidence confirming that the Greeks of Euclid's time heard the octave and a fourth as a concord, it is clear that the harmonic science he presents is not intended to accommodate all of musical perception [cf. section 2.2, on pitch].

Finally, it seems to follow from p_3 that the tonic interval (9:8) is a concord, though, as is well known, Aristoxenus classifies this interval as a discord [cf., e.g., Da Rios 1954, 25.11–15, 55.12–56.5]. Whether Aristoxenus' views on the matter are a suitable basis for interpreting or criticizing the *Sectio* is a question that arises here too. Euclid, in any case, does not explicitly call this interval a discord, though the closing lines of dem. 12 [Menge 1916, 174.6–7]—if they are not interpolated—may be a problem, since they suggest that it is not a concord. Cf., e.g., Adrastus in Hiller 1878, 50.16–21.

[19] That the 'belonging to' locution signifies the final step in the reductive analysis is clearer given the language of dem. 1: cf. Bowen and Bowen 1991, section 4. See also section 3, below.

(p_4) characteristics of musical pitches uniquely determined by relations among musical notes *qua* series of consecutive motions derive from characteristics of the numerical relations which are the reality of what is heard.

Though this premiss does not appear in the text itself, it (or something like it) is certainly necessary on my interpretation of the sentences [8]–[10]; so, I introduce it here as my second hypothesis. p_4 is an adjunct of the eliminative ontological reduction that is essential (again, on my reading) to the *Sectio canonis*. In effect, p_4 isolates a subset of the predicates applied to music as heard (the *analysanda*) and asserts that these predicates hold because they apply above all to the numerical relations (*analysantia*) constituting what the sensible musical relations really are. Thus, sentences [8]–[10] set forth the argument that the musical notes we hear as concordant are, *qua* series of consecutive motions, multiple or superparticular and so are in reality multiple and superparticular ratios designated by a single term. And, given this much, it seems simplest to conclude that this single term is 'concordant' as well—and so I follow all who assume that the single term mentioned in sentences [8] and [10] is the same.

On this reading, then, the upshot of the final argument in sentences [8]–[10] is a justification of the thesis that concords belong to concordant numbers. This a result quite different from the usual claim that the point of the preface is to explain why the notes we hear as concordant are either multiple or superparticular [cf., e.g., Tannery 1912, 218–219; Ruelle 1906, 318; Barker 1981, 3], or to show that the study of musical notes 'should be assimilated into mathematics' [Fowler 1987, 146]. As I see it, the preface answers the question, Why are concordant notes concordant?, by proposing that concordant notes are heard as concordant because they are in reality concordant numerical ratios.

But what is the context for such a question and answer? Clearly, it is not Academic [but see Tannery 1912, 218]—at least, not as one might surmise given the question raised in Plato, *Resp.* 531c1–4, when Socrates asks which numbers are concordant and which are not and why in each case. Yet without some sense of the context, it is virtually impossible to assess the importance of the question or the adequacy of the answer, beyond determining the role of the preface in the subsequent theorems. So, since I have postponed the latter project to another occasion, I will turn now to the question of the context of the *Sectio canonis*.

3. Euclid's *Sectio canonis* and Pythagoreanism

While debate about the authorship of the *Sectio canonis* still continues, there is, in contrast, a consensus that this work is in the intellectual tradition we call Pythagorean [cf., e.g., Heath 1921, ii 444–445; Barbera 1984; Fowler 1987, 144]. The broad similarity between the preface to this treatise and a fragment [cf. Bowen 1982] of a work by Archytas on music is obvious and, though one may well doubt Jan's claim [1895, 146] that the source for the bulk of the treatise is Archytas, there is no denying that the *Sectio* retails one proof [cf. dem. 3: Menge 1916, 162.6–26] which is attributed to Archytas by Boethius (AD 480–524) in his *De institutione musica* [Friedlein 1867, 285.9–286.4]. Moreover, as Düring suggests [1934, 176–177], the opening sentence of the preface to the *Sectio* compares favourably with what Heraclides Ponticus (late 4th cent. BC) may be ascribing to Pythagoras in the first few lines of the fragment of his *Harmonica introductio* [Düring 1932, 30.7–8] preserved by Porphyry (AD 232–*ca.* 305).

But, regrettably, just as the debate about the authorship of the *Sectio canonis* may be ill-founded, so may this consensus about its philosophical character. There are, for instance, significant differences between the musical analysis in this treatise and the theory we may attribute to Archytas and which we find repeated in the works of Theon of Smyrna, Nicomachus of Gerasa (both second century AD), and of Boethius, for example. First, as I have already noted, whereas Euclid proposes to justify quantifying musical notes by means of the premiss that each pitch depends on (is) the relative numerosity of the series of consecutive, airy projectiles which strike the ear and produce what is heard as one sound, Archytas [Bowen 1982] maintains that pitch is determined by the relative speed/force of the airy projectile [cf. Archytas, Fragment 1.45–46]. Now, the same view as Archytas' (without the reference to force) is found in Nicomachus' *Harm. man.* [Jan 1895, 242.20–243.10] and in Theon's *Expositio* [Hiller 1878, 60.17–61.11]. Moreover, in Boethius' *De inst. mus.*, there is in book 1 an account fashioned after the preface of the *Sectio* [cf. Friedlein 1867, 189.15–191.4] which adapts it to Archytas' view, and in book 4 a translation of the *Sectio* that departs from the original on this very point [cf. Friedlein 1867, 301.17–18; Bowen and Bowen 1991, section 4]. What this all means is difficult to say.

Though Nicomachus is a Pythagorean and Boethius follows him in harmonic science,[20] and though Adrastus (according to Theon [Hiller 1878, 50.4–21]) attributes the sort of account found in the fragment from Archytas to the Pythagoreans, one should hesitate to say that is Pythagorean, if only because Adrastus [cf. Hiller 187, 61.11–17] also ascribes the same view to Eudoxus, who was not, so far as I am aware, regarded as a Pythagorean at any point in antiquity, and because Theon seems to be a Platonist. Indeed, the story is even more complicated.

Consider Fowler's proposal [1987, 145–146] to assimilate the *Sectio* to the Lyceum on the strength of *Prob.* xix 39 and a passage from Porphyry, *In harm.* attributed to Aristotle [Düring 1932, 75.14–27: cf. Barker 1984–1989, ii 98]. Now, in the passage from Porphyry, pitch is correlated with the speed of the motions striking the ear, whereas, in the *Sectio*, pitch is identified with the relative numerosity of these motions [cf. Barker 1984–1989, ii 98, 107n40]—as it is in *Prob.* xix 39 [cf. Barker 1984–1989, i 200–201]. So, it would seem that the thesis of the dependence of pitch on the speed of the motion striking the ear may not be peculiar to the Pythagoreans. In any case, Euclid and the author of *Prob.* xix 39—who is no longer thought to be Aristotle—are the odd men out in this group. Yet this hardly puts Euclid in the Hellenistic Lyceum. Not only is there no good evidence about the provenance of the compilation known as the *Problemata*, the preface of the *Sectio* only requires that relative pitch depend on (be) the relative numerosity of pairs of series of consecutive motions, a thesis which is intelligible and quantifiable as I have indicated, and which does not suppose or need the sort of talk found in *Prob.* xix 39 about the incidence of the pairs of series on the ear.

Furthermore, according to Aristotle, the Pythagoreans thought that all things are number and did not make the sort of ontological separation between appearance and reality found in the Platonic corpus. But, if the Pythagoreans maintained that number and numerical relations constitute the reality of all there is, then, it is interesting to observe that, for Euclid, though phenomenal musical notes are composed of series of consecutive motions which (therefore) stand to one another in numerical ratios, and though these series are said to belong to numbers, they are not said to be *composed* of numbers. In other words, Euclid appears to regard numbers as the reality of musical sound but—so far as I can tell from his language—he

[20] This is an inference based on the general character of the *De inst. mus.*, on the nature of Boethius' references to and treatment of Pythagoras and the Pythagoreans, and on how claims Boethius makes in his own voice (usually in the first person plural) compare with what he says of the Pythagoreans: cf. *De inst. mus.* i 9, ii 21–27, v 8.

does not treat them as a reality *constituting* what is heard. Indeed, Euclid leaves open the possibility of a different account of the relation between appearance (what we hear) and reality. Moreover, in the fragment from Archytas, the pitch of the sound is said only to vary as the speed/force of the motion producing sound at that pitch; it is not claimed that the pitch is *composed* of motion at this speed/force. Likewise, in the treatises by Nicomachus, Theon, and Boethius—all of which agree with Archytas in correlating pitch and speed—there is no such reduction of sound as heard to the speed of motion. Thus, again, Euclid stands alone: his account fits neither Aristote's outline of Pythagorean analysis nor the accounts given by such Pythagoreans as Archytas and the others.

Now I admit that such differences may only signify a divergence between rivals schools of the Pythagorean family. But, in the absence of independent evidence confirming this, we should not ignore the possibility that the *Sectio canonis* analyzes music from a standpoint, and for purposes, alien to Pythagoreanism. This means that we should resist the temptation to minimize these differences by carelessly lumping this treatise with other Pythagorean writings and, even worse, by interpreting all these texts in terms of one another.

The deeper problem in addressing the question of Euclid's philosophical allegiances such as they appear in the *Sectio canonis*, however, is that the modern, scholarly category of Pythagoreanism is not well defined in harmonic science, no doubt in part because the Pythagorean version of the science itself still eludes satisfactory interpretation. Most of the criteria currently used to classify a theory as Pythagorean are based upon ancient descriptions of the intellectual schools of thought. Unfortunately, when the ancient musical theorists do make remarks about their predecessors and contemporaries, they do not write as historians following the rules of evidence and interpretation which we now take for granted. Indeed, the most one should concede at the outset is that their classifications and criticisms of intellectual trends and so on *may* hold at best of the period and cultural context in which they were writing. Thus, for example, Andrew Barker [1978a] has argued that Ptolemy's characterization in the *Harmonica* of the controversy dividing the Pythagorean and Aristoxenian schools of musical theory does not hold of the fourth century BC. Yet, Barker [1978a, 1] still takes it for granted that 'a solid amount of what is attributed to these schools by such writers as Ptolemy and Porphyry quite genuinely goes back to the fourth century, to Aristoxenus on the one hand, and perhaps to Archytas and his followers on the other,' though this should be a matter for argument and proof if we are ever to get an accurate account of Greek harmonic science.

But surely, one may ask, can we not follow the ancients and suppose [cf., e.g., Barker 1981, 3; Fowler 1987, 144] that a theory is Pythagorean if it analyzes music by means of whole-number ratios and prefers reason to hearing in determining what is musical? Granted, these criteria appear to be adequate to the fifth and fourth centuries BC (albeit perhaps because we have so little clear, direct evidence of Pythagorean musical theory from this period). But, on the basis of these criteria, one might also conclude that the *Harmonica* by Ptolemy (*ca.* AD 150) is a Pythagorean text [cf. Barker 1984–1989, ii 270–271]. And this certainly does no good. For, not only does it conceal the profound differences in epistemology and argumentation which exist between the *Harmonica* and, say, the roughly contemporary *Harmonices manuale* by Nicomachus of Gerasa [cf. Bowen and Bowen 1991, section 3], it also ignores the fact that much of the material in Nicomachus' treatise may also be found in Theon's *Expositio*, a treatise which draws from Peripatetic sources (especially, Adrastus [cf. Hiller 1878, 49.6]) *inter alia* in order to elaborate what is needed to understand Plato. In short, these two criteria quickly prove inadequate to the complexity of relations between the ancient documents concerning music which we do possess.

Likewise, I see no reason to pursue Barbera's contention [1984] that the proper context for interpreting the *Sectio canonis* is the Pythagorean tradition which he thinks is defined by Theon and Nicomachus. Indeed, it begs the question. For, though Nicomachus presents his own work as Pythagorean, Theon makes little mention of the Pythagoreans except to point out where they agree with views he has already stated, and he introduces many of the same points as Nicomachus but as part of a general learning (some of it drawn from Peripatetic sources) that is propaedeutic to the study of Plato's writings. Thus, on what basis and how are we to decide whether the doctrine in question is Pythagorean? But this is the very question we started with. Further, if we follow Nicomachus and regard the doctrine as Pythagorean, should we also follow Theon and suppose that it was generally viewed as propaedeutic to Platonic philosophy? And what antiquity are we entitled to assign this doctrine in any case? But, until these questions, as well as others pertaining to the schools of harmonic science in the second century AD, are answered satisfactorily, there is little to be gained by using Nicomachus and Theon as authorities in interpreting a treatise written perhaps some four hundred years earlier.

In sum, the claim that the *Sectio canonis* is Pythagorean is, by rights, not a starting point but a conclusion; and the same holds of the too often repeated assertion that Euclid was a Pythagorean [cf., e.g., Menge 1916, xxxviii]. Moreover, the argument leading to this conclusion about the *Sectio* will be very arduous indeed. For, not only will it have to deal with this treatise itself, it will have to uncover plausible criteria of Pythagoreanism in harmonic science, criteria which may well differ from period to period. As matters stand now, we are not sufficiently informed to locate the *Sectio* in a Pythagorean context. But, until we are, we must resist the temptation to speculate by using it, for example, to elaborate the criticism of the Pythagoreans found in book 7 of Plato's *Republic* [cf. Barker 1978b].

Conclusion

The preface to Euclid's *Sectio canonis* has puzzled readers for more than two millennia. Even the ancients found it difficult, if the versions offered by Porphyry [Düring 1932, 90.7–23] and Boethius [Friedlein 1867, 301.7–302.6] are any indication: both Porphyry and Boethius omit the last argument. The main reason, as I interpret the treatise, is that by compressing the reductive, eliminative analysis at its core to the requirements of a deductive or inferential expository style, Euclid obscured his point. This is not, however, a criticism. It is very difficult to present an argument involving an eliminative, ontological reduction, when this reduction necessitates systematic ambiguity in the use of key terms (e.g., φθόγγος as 'the musical note or pitch heard' and as 'the series of consecutive motions that strike the ear producing a note at that pitch').

But if so, then harmonic science raised problems for Euclid not found in arithmetic and geometry. One has to be careful, then, in assessing criticisms of the *Sectio canonis* which take the *Elements* as a paradigm of style. As for completeness, let us observe that there are no hints in the manuscript tradition that the preface to the *Sectio* is part of a larger introduction. So, in this limited sense at least, what we have is complete. Yet, is the preface incomplete because it lacks the preliminary suite of definitions and so on that one would expect given the *Elements*? On balance, I would say that even in this sense the preface is complete. For, though the question itself, Why are concords concordant?, is unstated and the single term is implicit, what is written does constitute a very economical, compressed, and coherent answer; and to require a more elaborate account in which all is spelled out (for our benefit) seems unwarranted. Still, the contention that the preface is complete will not be demonstatrated satisfactorily in the absence of a reading of the entire treatise showing its unity and coherence, or without a thorough study of the other Euclidean treatises and of the corpus of texts in harmonic science that aims to discover the relevant criteria of exposition and argumentation.

9

Aristoxenus' Harmonics and Aristotle's Theory of Science

ANDREW D. BARKER

It is agreed on all sides that Aristoxenus was the giant of Greek musicology. His work in musical history and criticism was the point of departure for a host of informal essays on music by philosophers, antiquarians, and men of letters. Almost all the technical harmonic treatises of later antiquity drew heavily on the analyses set out in his writings: this is true even of authors in the distinct scientific tradition of 'mathematical' harmonics, writing under the banner of Platonism or Pythagoreanism. Aristoxenus himself insists loudly and often that nothing comparable in scope and sophistication had been attempted before his *Harmonica elementa*;[1] and though his reiterated claims to originality become irritating, they are undeniably true. It is not just that he was thoroughly acquainted with musical practice, acute in his observations, and tireless in the pursuit of detail. The crucial task of harmonics, as he conceived it, is to go beyond the essentially preliminary compilation of facts to their systematic coordination in a scheme of scientific understanding. He discussed, self-consciously, polemically and at length, the methods by which this understanding is to be achieved and the form it must take if harmonics is to be truly a science. His importance lies as much in his meta-musicological reflections and in the way he brought them to bear on the organisation of his material, as in any of his substantive doctrines about the musical facts.

Aristoxenus' conception of science and its methods exercised no noticeable influence in antiquity outside specifically musical studies. So far as I know, mathematicians, astronomers, medical writers, students of mechan-

[1] For the text I have used, see Da Rios 1954: all my references are by Meibom's [1652] pages and lines. See also Macran 1902.

ics, and the rest paid him no special attention.[2] This is not surprising, since—taken abstractly—his ideas were not new. They were borrowed almost without exception from Aristotle; and though modifications of Aristotelian positions can be found in Aristoxenus, it is principally his interpretation and application of his teacher's ideas that should earn him the attention of historians of science.

The influence of Aristotle on Aristoxenus has been studied from a number of angles in a recent book by Annie Bélis [1986].[3] Here I shall consider issues that arise out of just one aspect of the relationship, one in which, I suggest, Aristoxenus might be of considerable help to our understanding of Aristotle himself. It is notorious that none of Aristotle's own treatises offers itself, *prima facie*, as an example of the sort of science painstakingly described in the *Posterior Analytics*. It is sometimes argued that the *An. post.* should not be construed either as proposing a framework for scientific research or even as describing the form which a complete science should ideally take, but more modestly as articulating a blueprint for pedagogy, a way of organising scientific results so that they can effectively be taught [see esp. Barnes 1969, 1975]. But the *Harm. elem.* shows, I believe, that Aristoxenus drew directly on Aristotle's essay when discussing the methods by which his subject is to be investigated: hence, it was indeed possible for an associate of Aristotle to conceive the *An. post.* as offering a sound framework for something that can fairly be called a research programme.

Aristoxenus also treated the *An. post.* as conveying a description of the form of understanding that constitutes science, one at which the harmonic scientist, like any other, must aim; and the *Harm. elem.* seeks to articulate its harmonic truths in the pattern that the *An. post.* proposes, not just for pedagogic purposes, but because scientific understanding must itself have the structure that the shape of the treatise reflects. A careful study of the *Harm. elem.* will show that this treatment of the *An. post.* makes sense, and will clarify the notions of scientific discovery and scientific understanding that underlie the latter. At the same time it will be instructive to consider certain ways in which Aristoxenus found it necessary to modify the ideas of the *An. post.*, and certain difficulties that his project appears to encounter. The most important stumbling-blocks, I shall suggest, are not of his own making but are in fact inherited from Aristotle.

---

[2] An exception is Vitruvius, who announces his debt to Aristoxenus at *De arch.* v 4.

[3] Bélis' book reached me at a late stage in the preparation of this paper, and I have not been able to incorporate many reflections on it here. There are substantial points on which we differ, but it is a work from which much can be learned.

At the centre of our investigation will be a pair of closely linked Aristoxenian positions, one positive, one negative. I shall outline them briefly here and discuss them more fully in what follows. On the negative side is one of his rare departures from Aristotle's views. In the *An. post.* and elsewhere Aristotle identifies two sorts of harmonics, one empirical and one mathematical, and treats the former as subordinate to the latter. Empirical harmonics discovers only certain facts available to perception, and a list of unexplained and uncoordinated facts is not yet a science. The explanations, but not the data, are provided by mathematical harmonics. In more Aristotelian language, the empirical approach distinguishes and perhaps classifies the phenomena, but finds no ἀρχαί and generates no ἀπόδειξις. The ἀρχαί that stand as principles essential to the explanatory demonstration of statements describing the phenomena are principles proper to the mathematical branch of the subject.[4] This may well be an accurate transcription into Aristotelian terms of contemporary mathematical theorists' own view of their project. But Aristoxenus will have nothing to do with it. The ἀρχαί of his science, its coordinating and explaining principles, must—so he insists—be intrinsic to the domain of musical perception itself and not imported from the foreign territory of mathematics or quantitative physical acoustics. Aristotle, of course, engages in no very careful study of the two kinds of harmonics: he merely notes their existence for the sake of an example of the way in which one science may be subordinate to another, and apparently accepts the mathematical theorists' own estimate of their

---

[4] Mathematical harmonics, in Aristotle's sense, is exemplified in the pioneering work of Archytas [see Diels and Kranz 1951, i 428.15–430.12, 435.15–436.13; Bowen 1982], in a highly specialised application at Plato, *Tim.* 35b–36b, and later in such treatises as the Euclidean *Sectio canonis.* It treats pitch-relations as ratios between numbers (which may be conceived as attaching to physical variables such as speeds of movement: cf. Bowen 1982, and chapter 8 in this book). It represents harmonic structures, for instance, the octave-scale, as organised complexes of ratios, whose coordination may be explained by reference to a theory of proportions or to some other purely mathematical set of principles. In its application to acoustics, a focus of some interest in the Lyceum, such conceptions were used to account for phenomena including the correspondence of notes an octave apart and the concordance of octaves, fifths, and fourths: they are concordant *because* their ratios are of certain sorts. Aristotle accepts these accounts as giving at least a sketch of an appropriate explanation [see *An. post.* 90a18–23: cf. *De sensu* 439b31–440a3], even though it is not clear what reasoning is concealed in this 'because' [see n5 below].

relative ranking.[5] Aristoxenus rejects this view, as we shall see, for reasons drawn from the *An. post.* itself. He argues, in effect, that a consequence of taking the pronouncements of that treatise seriously is that the two existing forms of harmonic science cannot genuinely stand in the relation that Aristotle imagines. His intention is not, however, to elevate the work of previous harmonic empiricists from the humble station to which Aristotle had consigned it: their conception of the science, he believed, was as inadequate as that of their rivals was irrelevant. Nor had Aristotle failed to discern the merits of some other existing harmonic project. The real implication of the *An. post.*, for Aristoxenus, was that an entirely new harmonic science had to be framed which would absorb both the descriptive and the explanatory functions.

The positive side of the coin is Aristoxenus' insistence that harmonics must find its ἀρχαί through reflection on the phenomena revealed to musical perception, seeking forms of order intrinsic to the phenomena themselves as they are perceived (not in the ordering of an unperceived realm of 'causes', movements of the air, or the like, which physical acoustics might investigate). The ἀρχαί proper to harmonics articulate a φύσις, a nature or essence, that exists and is expressed in groups of heard sounds themselves in so far as they are melodically attuned, and qualify as instances of τὸ ἡρμοσμένον. These ἀρχαί describe the structures within which sounds are necessarily organised if they are correctly heard as melodic, since to hear a sequence as melodic is just to hear it as exemplifying the φύσις that the harmonic scientist seeks to describe. If a hearer is sufficiently attentive and carefully trained, he will come to realise that what the scientist articulates does indeed express the form a sequence of sounds must have when he

[5] Aristotle may have been persuaded by the apparent analogies between harmonics and his two other examples, optics and astronomy, particularly the latter. The Pythagorean treatment of harmonics and astronomy as 'sisters', articulated by Archytas [see Bowen 1982, 79–83] and reported in Plato, *Resp.* 530d6–9, derives from a conception of them as parallel studies of different forms of movement, audible and visible, or so I would argue. (Huffman [1985] gives reasons for rejecting a sentence in the text of Archytas which is important for my interpretation: I think his arguments can be answered but this is not the place to pursue the issue.) The achievements of Eudoxan astronomy may have encouraged the view that Archytan harmonics, already much the most impressive form of musical science, could be developed to rival it. In both these cases, and in optics, an abstract mathematical description of the phenomena, allied to purely mathematical principles, might plausibly be construed as constituting their explanation, as it is for a harmonic example at *An. post.* 90a18–23. Aristotle's antipathy to Pythagoreans is directed at their metaphysics and their use of harmonic theory in non-musical contexts: he does not criticize their development of it in its own sphere.

himself hears it as melodic or as being of a given melodic type. Similarly, the rules apodeictically derived from the ἀρχαί are also ones to which every alert listener would subscribe, on the ground that these rules make explicit criteria that were already implicit in his own experience. They are not surprising, unexpected constraints on what can count as melody: still less do they impose restrictions derived from principles in another realm, such as that of pure mathematics. The task of harmonics is to clarify and organise what educated perception implies, not to prescribe things that it will not autonomously accept. The apodeictic phase of the science serves to explain the rules that are implicitly accepted in ordinary practice by showing that they are not arbitrary or haphazard, but are coordinated expressions of a single nature or essence.

The immediate consequence of this approach is that Aristoxenus' harmonic science turns out to be a sort of musical phenomenology. It describes and classifies the phenomena according to the distinguishable ways in which they present themselves to perception, not according to classes whose members cannot be directly identified as such by the musician's ear. The audible appearances are not explained by reference to inaudible physical causes or mathematical principles, but by being displayed as aspects of a coherent nature that exists just in its audible instances. Audible melodicity is not an echo or a consequence of some other form or order that holds among the inaudible precursors of sound. The melodic is an autonomous form inhering in certain sequences of sounds under their aspect as objects of hearing and in nothing else. It is to be defined through a coordinated array of principles abstracted inductively from the audible appearances; and the rules governing what is and what is not an acceptable melody are accounted for when they are shown to be implicit in the pattern of organisation in which that form consists. In neither its descriptive nor its explanatory phase should harmonics call on data that are not presented to the musical ear or on categories that divide up the data on other than musical principles. Aristoxenus' reasons for taking this position, its implications, and certain problems it brings with it, will be explored in more detail below.

The *Posterior Analytics* and the structure of Aristoxenus' harmonic treatise

I have said that Aristoxenus' conception of a science corresponds closely to that set out in the *An. post.* A thorough evaluation of this claim would have to focus on the fine detail of what Aristoxenus does, in order to see how well the propositions of his treatise correspond to Aristotle's descriptions of the propositions of a science and their mutual relations. I shall do little at that level here. A more compassable project is to compare the explicit remarks made by each author on the subject of scientific method and on the conditions that an adequate science must fulfil, and here I shall draw attention to some obvious parallels. But the central investigation into which these closer ones must feed concerns the overall match between the designs proposed in Aristotle's essay and exemplified in that of Aristoxenus. This creates a difficulty, since the *Harm. elem.* as we have it is not a complete work, nor even, in my view, the remains of a single treatise. Issues about the relations between the parts of the surviving text have raised a good deal of scholarly dust:[6] I shall not attempt to sift it, but I must at least state the opinion on which what follows is premissed.

This is that books 2 and 3 belong together; that while the work of which they are parts was originally a good deal longer, and while the two books are not perfectly preserved even as parts, nevertheless they are enough to give a reasonably clear picture of the form which that treatise took. Book 1, on the other hand, is evidently in many respects (not in all) an alternative treatment of the project undertaken in book 2.

I must pursue these preliminaries a little further. Books 1 and 2 perhaps contain fewer genuinely equivalent passages than is sometimes assumed, but there are enough parallels and approximate repetitions to ensure that they cannot originally have been parts of the same finished work.[7] I take them to have performed roughly equivalent tasks in two different treatises. Further, though the first and second books differ little in what they claim to be facts about music, they do differ significantly in the conceptual resources which they bring to bear on the interpretation and organisation of these facts. Those deployed in book 2 are subtler and richer than those of book 1, and its author shows a markedly higher degree of methodological self-consciousness. (Some considerations that support these claims will be mentioned later.) For these and other reasons I have little hesitation in treating book 2 as the later essay, reworking in the light of greater maturity, and perhaps for rather different purposes, much of the material of book 1. The same criteria indicate that the affinities of book 3 are with book 2, not book 1, and hence encourage the belief that books 2 and 3 belonged to the same treatise. In what follows I shall be concerned mainly

---

[6] For a summary of opinions, see Da Rios 1954, cvii–cxvii.

[7] There are also several striking inconsistencies which cannot easily be resolved. But other scholars have taken different views, some fragmenting the work much more radically than I do, others declaring for its overall unity. See the survey mentioned in the previous note, and for a vigorous defence of a unitarian view, see now Bélis 1986, particularly 24–48: her position was already sketched in Bélis 1982, 450–451.

with the presumably later work represented by books 2 and 3, and shall draw only occasionally on book 1.

According to the *An. post.* [esp. 71b10–73a20] a science consists on the one hand of principles (ἀρχαί), and on the other of conclusions explained and deductively secured in the light of these principles. The pursuit of science, then, involves first the establishment of ἀρχαί, and secondly the demonstrative derivation of subordinate propositions. The establishment of ἀρχαί is not itself a matter of demonstrative proof (ἀπόδειξις): the scientist works his way up to principles from a starting-point in perception by a process that is in one sense or another inductive, and whose stages are outlined, rather enigmatically, in the last chapter of *An. post.* ii. The ἀρχαί include, beside the so-called common axioms, both what Aristotle usually calls ὑποθέσεις (and which I take to be propositions asserting that this or that exists or is the case), and definitions of primary entities or kinds within the relevant domain [see, e.g., 72a14–24]. Having arrived at these ἀρχαί through 'epagogic' reflection on perceptual experience, we then set them to work *as* ἀρχαί by identifying the special relations in which they stand to other propositions of the science, propositions which are not primary and which are scientifically understood only when they have been derived apodeictically from the appropriate ἀρχαί. The ἀρχαί, then, must be grasped as epistemologically and metaphysically prior to the facts expressed in these subordinate propositions, and as providing the explanatory ground for them (these points are summarily sketched at 71b19–22).

The ἀρχαί must also fulfil certain other conditions. Notably, they must be true and they must be immediate, not requiring explanation or demonstration in terms of anything else [see particularly *An. post.* i 2–3]. It is less than clear, epistemologically, how we can be sure in a particular case that either of these conditions holds: but the latter should be taken to mean that the ἀρχαί represent what belongs to the essence of things in the relevant domain, expressing what it is to be such and such. In the case of a class of perceptible things it is reasonably clear why, in Aristotle's view, the things that hold essentially of members of the class as such are not capable of being demonstrated from any higher considerations but must be grasped through a process of abstraction and coordination, a process directed at the data which our perceptual experience of them provides.

We have an understanding of the subordinate truths of a science only if we have demonstrated them: that is, we must not merely prove them by logical derivation from propositions known to be true, but we must derive them from principles that explain *why* they are true [see particularly *An. post.* i 2–3, 13]. The Aristotelian content of the notion of explanation is of course complex and takes us beyond the scope of the *An. post.* But in a general and abstract way, 'explaining why property P holds of subject S' here means 'showing that P holds of S *because* the essence of S (as expressed in some ἀρχή or ἀρχαί) logically requires it'. As a consequence, Aristotle argues, the conclusion of an ἀπόδειξις must hold of its subject *as such* in virtue of the essential nature of that subject (and not of some other thing or of the same thing differently conceived: see, e.g., *An. post.* i 4 and 6). And it is as a consequence of *this* proposition that Aristotle maintains the impossibility, except in certain very special kinds of case, of demonstrating ἐξ ἄλλου γένους μετάβαντα [75a38: cf. 71b22–23, and below]: a single science is delimited by a kind (γένος) that constitutes a single domain whose contents stand as the subjects of both the primary and the subordinate propositions. This requirement, as we shall see, has crucial work to do in determining the shape of Aristoxenus' harmonic theory.

But let us postpone consideration of that issue for the present and concentrate first on the general adequacy of fit between the *Harm. elem.* and the framework that Aristotle proposes. That there is some sort of rough correlation is reasonably clear: the two books we are principally considering seem to fall quite neatly under the two categories that the Aristotelian scheme demands. Book 2 articulates and discusses the ἀρχαί of the science: book 3 sets out a series of formal derivations from these ἀρχαί, demonstrations that this is a melodic sequence (while that is not) or that a melodic sequence in some one genus is subject to such and such conditions; and these demonstrations are at the same time explanations of why these things are so. Aristoxenus calls his derivations ἀποδείξεις and his use of the expressions is undeniably Aristotelian in intent.[8] He also makes several methodological declarations that distinctly echo the *An. post.* in both language and content. He alludes scathingly to other theorists who have

[8] Aristoxenus also refers twice to a section of his work as 'elements' or as 'concerned with the elements'. In i 28.34–29.1, a proposition ἐν τοῖς στοιχειοῖς δειχθήσεται: in ii 43.27–30, after introducing the main topics of harmonics but before elaborating his treatment of them, he says, μέλλοντας δ' ἐπιχειρεῖν τῇ περὶ τὰ στοιχεῖα πραγματείᾳ δεῖ προδιανοηθῆναι τὰ τοίαδε; and goes on to present methodological reflections that insist on a distinction between ἀρχαί and what follows from ἀρχαί. In the former passage, the στοιχεῖα evidently constitute part of the treatise: in the latter they seem to be the 'elements' of μέλος itself, some part of the harmonic scientist's project being designated by the expression τῇ περὶ τὰ στοιχεῖα πραγματείᾳ. It is not clear whether these parts of the treatise, or of the investigation, include the whole of its 'scientific' content (that is, everything after the discursive introduction), or whether they are restricted to the 'demonstrations' alone (that is, to the contents of book 3). Bélis [1986, particularly 34–48] takes a clear and strong line in identifying book 1 of the *Harm. elem.* as ἀρχαί and the other two as στοιχεῖα, but the problems are, I think, more complex than this treatment suggests.

asserted their propositions ἄνευ αἰτίας καὶ ἀποδείξεως: he himself, by contrast, will seek both to adopt (λαβεῖν) suitable ἀρχαί and to demonstrate (ἀποδεικνύναι) τὰ ἐκ τούτων συμβαίνοντα [32.29–33.1]. The ἀρχαί themselves, of course, cannot be demonstrated: τὸ γάρ πως ἀπαιτοῦν ἀπόδειξιν οὐκ ἔστιν ἀρχοειδές [44.14–15]. Again, every science that consists of several propositions must adopt (λαβεῖν) ἀρχάς ... ἐξ ὧν δειχθήσεται τὰ μετὰ ἀρχάς [44.3–7]. In passages like these Aristoxenus' debt to the *An. post.* could hardly be more obvious. He believed that he had succeeded in a task whose character his predecessors had not understood and whose necessity to harmonic science they had not even noticed, that of drawing the harmonic facts into an Aristotelian system of ἀρχαί and ἀποδείξεις. On this achievement he rests one of his major claims to originality and importance.

As in Aristotle's scheme, some of the ἀρχαί take the form of definitions: each of the seven μέρη of the science outlined in the early sections of book 2 involves a subject to be defined, and various other items are also defined along the way. The definitions initially offered, however, are explicitly schematic,[9] requiring more detailed articulation and differentiation as well as an enumeration of the more specific types of item falling under the subjects outlined. Not all Aristoxenus' elaborations of detail have survived: some that he promised may never in fact have been given.[10] What we do find are close accounts of some specific types of structure falling under the broader kinds that certain definitions sketch, notably his descriptions [i 21.37–24, ii 46.19–52.33] of tetrachordal divisions in the three melodic genera and some of their subordinate χροαί (nuances or shades). But this phase of Aristoxenus' project raises serious questions about his method. Though the analyses of these divisions might, not unreasonably, be given the status of definitions (definitions of species falling under a broader kind rather than more detailed definitions of the kind itself), the terms in which they are set make it very difficult to construe them as ἀρχοειδῆ, as principles proper to the science. On the other hand they are plainly not demonstrated in the technical sense, nor did Aristoxenus think they could be. There are problems here to which we shall return.

In addition to definitions, the ἀρχαί include what Aristotle calls ὑποθέσεις. Aristoxenus uses no corresponding noun; but he sets out a series of principles whose importance and primacy he vigorously underlines, each asserting that something is the case and each introduced by such words as

9 See, for instance, Aristoxenus' engaging appeal to his hearers at *Harm. elem.* 16.2–16.

10 Thus, at 36.17 he raises the question, What is a δύναμις?, as one that urgently demands an answer. But if he gave an answer, it left no trace in the writings of his successors and epitomisers.

ὑποκείσθω [in book 1, esp. 29.1–34], λαμβανέτω or θετέον [in book 2, esp. 54.7, 19]. As is appropriate to such primary propositions, they are quite ostentatiously presented with no attempt at proof or derivation. One of them he describes as 'the first and most indispensable of the conditions that bear upon the melodic combination of intervals' [τὸ πρῶτον καὶ ἀναγκαιότατον τῶν συντεινόντων πρὸς τὰς ἐμμέλεις συνθέσεις τῶν διαστημάτων: 53.33–54.1]. A little later he says of it, 'Let this, therefore, be posited as first in the order or principles: if it is not fulfilled, the harmonic attunement is destroyed' [θετέον οὖν τοῦτο πρῶτον εἰς ἀρχῆς τάξιν οὗ μὴ ὑπάρξάντος ἀναιρεῖται τὸ ἡρμοσμένον: 54.19–21]. It cannot be over-emphasised that the οὖν (therefore) in this sentence does not mark the conclusion of any sort of argument. The Aristotelian thesis that principles cannot be demonstrated is one that Aristoxenus wholeheartedly endorses: 'Anything that requires demonstration (ἀπόδειξις) is not ἀρχοειδές' [44.14– 15].

To the extent that it deals with principles, then, it is no criticism to point out that book 2 contains hardly anything in the way of arguments to support the musicological assertions it makes. (The same could be said of book 1.) There are arguments in book 2, but virtually all of them are methodological, and have to do with the way in which the subject is to be approached; they are not designed to establish substantive propositions of the science. The impressive and elaborate typology of melodic genera, the principles governing melodic succession, the enumeration of concordant intervals, the definitions of an array of harmonic concepts, these and all the rest are flatly asserted, not argued, rarely even supplied with supporting considerations. This is not a sign of either arrogance or incompetence: it is a necessary consequence of Aristoxenus' determination to take seriously the distinction between what can be demonstrated and what cannot [see 43.34–44.1].

The only proper approach to the latter is inductive, and it is important to be clear about what this implies. It implies that they cannot be established by any argumentative device capable of being presented in a written treatise. The function of the treatise, so far as they are concerned, is to systematise and draw to our attention what is implicit in our own experience, that is, in the perception by carefully trained and attentive listeners of certain phenomena as melodic. If we attend to our experience we shall *recognise* the authority of the principles that Aristoxenus articulates and the cogency of the distinctions marked by his definitions, as accurately mapping the framework within which our perception of melodies takes place.[11]

---

[11] Each of the primary propositions must be both ἀληθές and φαινόμενον: it must also be such as to be grasped (συνορᾶσθαι) by αἴσθησις as being among the πρῶτα of the various parts of harmonics [44.9–13: cf. 32.31–33.1].

The principles of harmonics are not recognisable as such *except* through an individual's reflection on his own perceptual experience.[12] No amount of reading treatises can give such experience and, hence, any attempt to establish their truth through the written (or spoken) word would be out of place and futile.

Similar considerations explain why Aristoxenus tells us nothing of particular examples of melody drawn from his own experience, on which his inductive generalisations might be founded. We might expect a treatise in empirical science to offer case studies or experimental reports at least by way of examples: Aristoxenus does not provide such data, though he occasionally mentions what he would expect his readers' experience to be under certain conditions. Whatever may be true of the other ancient sciences, I suggest that there are good reasons for the omission here. In most empirical sciences we typically assume—unless we affect a hyperbolic form of scepticism—that what the researcher observed at some time would have been observed also, other things being equal, by anyone else who had been there. Hence, the researcher's experience can stand proxy for our own, as a basis for the inductive extraction of principles. In Aristoxenian harmonics this is not so, not because Aristoxenus supposes that the assumption would be false but precisely because it is part of the task of harmonics to show that it is true. Little purpose would be served by descriptions of individual cases that Aristoxenus has observed, since we must not *assume* that we would experience them in the same way. Hence, he states his principles without supporting evidence: we ourselves must provide the grounds for believing that they hold universally, by finding that they are indeed implicit in our own perception of individual cases. To present examples in evidence, or any form of argumentation, would be to offer the illusion of support for the principles without the reality: that is something that the written word cannot provide.

Aristoxenus' text contains one superficially anomalous case, his 'argument' to the conclusion that the concord of the fourth is an interval spanning exactly two and a half tones [56.13–58.5]. But in fact this helps to prove the rule, since it is not offered as an argument in the relevant sense at all. That is, it does not seek to establish anything without recourse to the listener's own perception: it explains a practical method by which the student can satisfy *himself* of the truth of the proposition. He must work through a specified musical construction in practice, not on paper or in his head, and *listen carefully* to its results [see esp. 56.31–33]. If and only if the results are heard in a certain way, the proposition will have been

[12] See especially the contrast with geometry at 33.10–26.

established [56.33–57.3]. The method which Aristoxenus describes has its flaws,[13] but these do not affect the point I am trying to make. The intention of the passage is not to establish a musical proposition by argumentative means, but to show how perception may be used in order to judge whether the proposition is acceptable or not. To repeat: in a science based on the ideas of the *An. post.* it cannot be the function of a treatise to establish the truth of ἀρχαί. It can only *identify* them, on the basis of the author's own experience, in a form that allows us to recognise them in our own; and it can organise and coordinate them so as to bring out their interconnections and to make them available for use in ἀπόδειξις.

These reflections suggest something important about the way in which an Aristotelian scientist's work is to be construed. There is a sense in which Aristoxenus' writings do not by themselves constitute the science, the body of knowledge, of which he spoke. This is not to deny that these writings were, in their original form, as complete and accurate an account of their subject as any treatise could be: on this issue we do not need to pronounce. The point is that the treatise, no matter how finished a product it is, cannot itself be the knowledge which its author possesses and to which its readers may aspire. The thesis that knowledge is a property of minds, not of books, is not just a tendentious *a priori* dogma. There are good grounds here for saying that the book cannot even be a written representation of its author's knowledge, since there can be no such representation of the conditions that constitute it *as* knowledge. A treatise may enunciate the laws and principles on which its demonstrations rest, but it cannot incorporate the grounds that give them the status of scientific truths objectively established. The task of the written or spoken exposition is to guide others towards understanding, by articulating the truths that they must recognise if they are themselves to master the science of harmonics. Scientific understanding *establishes* these truths and the treatise does not. Then a science is not something that can exist in a book or a library or a data-bank: there is no impersonal corpus of scientific knowledge. A science must satisfy conditions that can be satisfied only a by mind that can draw on its own experience in confirmation of its claims: it is a δύναμις θεωρητική, a mental capacity or disposition, something that can be neither contained nor represented in the written or spoken word [cf. *Harm. elem.* 41.6–24].

Such a conception of a science is closely related to that of a τέχνη or practical skill (though there are differences, emphasised in 41.6–24.) The principles of such a skill may have been fully discovered and developed

[13] Exposed with acid precision by Ptolemy, *Harm.* 21.21–24.29.

long ago: they may even have been written down in works entitled, 'The art of such and such'. But plainly the art or skill does not exist in anything written: it is not even the sum of the propositions enunciated there, but consists of the capacities and dispositions of individuals who have mastered it. Similarly, a science is a mind's organised and systematic grasp of a determinate domain. What a treatise expresses can be knowledge only in so far as its principles are grasped by a mind that has recognised their truth; and that recognition depends on the mind's own experience, since their truth cannot be established by anything independent of that. Aristotle himself says something in this vein:

> Demonstration is not directed to external discourse, but to that in the soul... for it is always possible to raise objections against the external discourse, but not always against the internal.
>
> οὐ γὰρ πρὸς τὸν ἔξω λόγον ἡ ἀπόδειξις, ἀλλὰ πρὸς τὸν ἐν τῇ ψυχῇ ... ἀεὶ γὰρ ἔστιν ἐνστῆναι πρὸς τὸν ἔξω λόγον, ἀλλὰ πρὸς τὸν ἔσω λόγον οὐκ ἀεί. [*An. post.* 76b24–27]

In that case it seems misleading to treat the *An. post.* as offering only a framework for teaching, if this implies that it is not also designed to describe the form of a finished science and to commend certain approaches to research. In describing the conditions of ἀπόδειξις and the ways in which the ἀρχαί must be established, Aristotle is analysing the structure of a system of understanding that can exist only in a mind, not in a teacher's presentation of his material.

Again, since our grasp of a domain of experience becomes knowledge only when it is systematically ordered and grounded in the appropriate way, research in that domain must be, at least in part, a search for ways in which the phenomena of experience can be so ordered. Aristoxenus' success in finding principles and categories under which an empirical grasp of musical facts can be converted into an Aristotelian science is the core of his achievement; and it would be curmudgeonly to refuse his endeavours the title of research merely because they did not necessarily involve the discovery, or even the pursuit, of hitherto unsuspected first-order facts. The expression of the results of this research in writing ceases to be science and becomes pedagogy. No doubt it is appropriate to teach students who seek to become knowers by a method that brings out, as clearly as possible, the system of relations that must hold between propositions in their mind and between these propositions and their experience if knowledge is to be achieved. The treatise or lecture should therefore mirror the structure of the science so far as it can. But the structure of pedagogy, like that of research, is derived from that of the science itself as it may exist in a knowing mind and not the other way round; and as we have also seen, there are crucial aspects of the science, as Aristotle describes it and Aristoxenus pursues it, that the teacher's pronouncements have no power to represent.

2. The 'same domain' rule and Aristoxenian phenomenology

A scientifically perspicuous description of what I heard when I heard some sequence of sounds as a melody would draw attention to properties and relations that contributed essentially to its being heard as melodic or as being of some determinate melodic kind. The ἀρχαί of harmonic science are abstracted inductively from observations of phenomena to which such descriptions could be attached, clarifying, generalising, and coordinating them, but not introducing new matter from elsewhere. Subordinate rules are derived from principles deductively. It is considerations like these that underpin Aristoxenus' enthusiastic endorsement of what I shall call Aristotle's 'same domain' rule, according to which no ἀπόδειξις of what belongs to a subject in one domain can be derived from principles proper to subjects in another. In short, as Aristotle says, one cannot demonstrate ἐξ ἄλλου γένους μετάβαντα [cf. *An. post.* 75a38].

Aristoxenus lays great stress on this rule. To break it is ἀλλοτριολογεῖν [32.20: cf. 32.27] or εἰς τὴν ὑπερορίαν ἐμπίπτειν [44.17–18], and scientific understanding cannot come that way. In so far as a statement belongs to harmonics, it mentions nothing that falls outside the domain defined by the essence of the kind with which harmonics is concerned, τὸ ἡρμοσμένον and its species. Crucially, no laws describing the regular behaviour and properties of what is melodic as such can be demonstrated from principles expressing what is essential to things of another kind. Each principle of harmonic science, we are told, must be both true and φαινόμενον: it must be such as to be accepted as a primary principle by αἴσθησις [44.9–14].[14] That is, it must express what is essentially involved in something's presenting itself to our hearing as melodic, not offer descriptions referring to entities of another sort or even to sounds conceived under an aspect which is not that presented to musical αἴσθησις. It may not, then, impose rules based on a conception of sounds as movements of the air, differing in the rapidity of their transmission or in the frequency of the impacts that initiate them, since it is not *as* patterns of relative speeds or frequencies that sequences of sounds present themselves to the ear as melodic or

[14] Compare for instance *An. post.* i 6, particularly 74b24–6: the ἀρχή of a demonstration is τὸ πρῶτον τοῦ γένους περὶ ὃ δείκνυται· καὶ τἀληθὲς οὐ πᾶν οἰκεῖον. See also 81a38–b9.

unmelodic. Nor, for the same reason, can harmonic rules be founded in a representation of pitch-differences as ratios of numbers.[15] The entities, properties, and relations mentioned in the ἀρχαί must be exclusively ones to which precisely articulated observation-statements would already need to refer—statements saying just what it was about an experienced set of sounds that constituted it as melodic or as an instance of some specific melodic kind.

Aristotle grounds his 'same domain' rule in considerations of just these sorts, though of course without reference to the special domain of harmonics. The reasoning is essentially simple. Any putative demonstration that broke the rule would not display its conclusion, which asserts that some property belongs to entities of a given kind, as flowing from the nature of the entities themselves and, hence, as being explained by reference to that nature. The link between any entity's possession of that nature and its possession of that property would remain merely contingent, even if the pseudo-demonstration showed in the light of something else that the conclusion was indeed true.[16] If in fact the inherence of certain properties in things of kind K cannot be demonstratively explained from ἀρχαί expressing the φύσις of that kind, then either there is no such φύσις and K is an arbitrarily designated class of accidental aggregates, or else the properties in question do not belong to things of kind K as such but only under some description that applies to such things contingently (κατὰ συμβεβηκός).[17]

But Aristotle allows exceptions to the 'same domain' rule in cases where one science falls under another in a specified manner. There are cases where it is the task of one science to describe the perceived properties of phenomena in a specific domain and the relations between these properties, but that of another to explain why the properties belong to them and why they are so related [see *An. post.* 75b14–20, 76a9–15, 78b32–79a16]. The situation seems to be this. The 'empirical' version of a science identifies a range of properties that perception finds attached to subjects of a given sort. The 'explanatory' version then identifies these properties as special instances of a more generalised class of forms, instances whose perceptible character derives from the inherence of the forms in the kind of perceptible

---

[15] These issues are mentioned several times in book 1 in ch. 8–12, esp. 9.2–11, 12.4–32: see also ii 32.19–28.

[16] The statement at *An. post.* 75a38 that one cannot demonstrate ἐξ ἄλλου γένους μεταβάντα is introduced with the particle ἄρα, indicating that it is the conclusion of an argument and not a new assertion. The argument is contained in the whole of i 4–6. What I have said here is not even a summary or a paraphrase of those chapters, but may serve to indicate their general drift.

[17] See the forceful statement at *An. post.* 75a28–34: cf. for instance, *Phys.* 192b28–32.

matter under investigation, though they can, at least in principle, occur as forms organising matter in other perceptible domains. These forms remain what they are independently of the genus of matter on which they are imposed, and their nature and relations can be studied in abstraction from any such matter (though of course they cannot *exist* in separation from matter of some determinate sort). The prime examples of forms that are properly studied at this abstract level are the objects of mathematics, and in each case the explanatory version of the sciences that Aristotle mentions is a mathematical one.[18]

We can extract the skeleton of an example from *De sensu* 439b19–440a6, where Aristotle compares certain properties of colours and of sounds. He offers the view (though it is not clear whether he endorses it) that the same numerical ratios which characterise the relation between sounds that are perceived jointly in certain special ways, also express the relation between instances of the basic colour-types, dark and bright, when they are so related as to produce jointly certain perceptible results. He seems to suggest that the acoustic and visual phenomena explained in terms of the same numerical ratio may themselves be somehow analogous. The perceived property is of course different in each case, since one is seen and the other heard: but within its own perceptual domain, each is the analogue of the other.

> For the colours that are in the most well-ratioed numbers, like the concords in the other case, appear to be the pleasantest of the colours, ones like purple and scarlet and a few others of that sort (for which reason the concords too are few), while those that are not in numbers are the other colours.
>
> τὰ μὲν γὰρ ἐν ἀριθμοῖς εὐλογίστοις χρώματα, καθάπερ ἐκεῖ τὰς συμφωνίας, τὰ ἥδιστα τῶν χρωμάτων εἶναι δοκοῦντα, οἷον τὸ ἁλουργὸν καὶ φοινικοῦν καὶ ὀλίγ' ἄττα τοιαῦτα, δι' ἥνπερ αἰτιάν καὶ αἱ συμφωνίαι ὀλίγαι, τὰ δὲ μὴ ἐν ἀριθμοῖς τἆλλα χρώματα. [*De sensu* 439b32–440a3]

In each case, then, the subordinate, empirical science classifies the properties by the way they appear to eye or ear. They appear as colours or as sounds because of the character of the matter in which they are present. But the reason why they have their special perceived characteristics *within*

[18] For valuable accounts of Aristotle's views on these paired sciences, see Lear 1982 and Lennox 1986, particularly 31–44, which builds on Lear's approach. But while they give substantial help with the question how mathematics makes contact, in Aristotle's view, with the empirical subject matter of physics, I would argue that Aristoxenus' grounds for resisting Aristotle's position with respect to harmonics remain untouched. I discuss this resistance below.

any domain, and are related in their own special ways to other properties in that dimension (e.g., as concords stand to discords, or as primary and attractive colours stand to muddy intermediates) is revealed through some branch of mathematics. It is revealed when the perceived properties and relations are shown to be instances, in particular types of matter, of mathematical properties and relations, quantitative forms that can be abstracted in the same way from each.

It is striking that when Aristotle needs an example of such pairs of sciences, he finds it natural to turn to harmonics in its perceptual and mathematical guises (explicitly twice, and by implication in a third passage).[19] Yet Aristoxenus rejects with the utmost vehemence the suggestion that this relation holds, that the αἰτία of the phenomena which 'perceptual' harmonics classifies can be provided by mathematics or by mathematical acoustics [ii 32.18–28: cf. i 9.2–11, 12.4–32].

He does so not because he disputes Aristotle's theory of scientific explanation, but precisely as a good Aristotelian. The phenomena harmonics describes and classifies are the properties of groups of sounds heard as forming a melodic sequence: 'being melodic' is a property of sounds presented to the ear and exists nowhere else, and the properties that a melody has as such are necessarily and essentially those grasped κατὰ τὴν τῆς αἰσθήσεως φαντασίαν [8.23, 9.2–3: cf. 48.22]. It may be true, Aristoxenus seems to concede [cf. i 9.2–11, 12.4–32], that the pitches of sounds are in fact determined physically by their velocities, or by some other quantitative variable. It might even be open to him to accept, for instance, that the movements causally responsible for sounds an octave apart stand to one another in the ratio 2:1, those generating sounds a fourth apart are in the ratio 4:3, and so on, as the Pythagoreans had proposed, and both Plato and the acoustic scientists of the Lyceum agreed.[20] But facts like these, in Aristoxenus' view, could in principle do nothing to explain why certain sequences of intervals and not others present themselves to the ear as melodic. There are no mathematical reasons, or reasons within the province of physical acoustics,

[19] *An. post.* 78b32–79a16, 87a32–4: cf. 90a18–23. The other passages mentioned above [75b14–20, 76a9–15, 78b32–79a16] are also relevant. It is only at 79a1–2 that the two sciences are described as ἁρμονικὴ ἥ τε μαθηματικὴ καὶ ἡ κατὰ τὴν ἀκοήν: elsewhere they are ἀριθμητική and τὰ ἁρμονικά (or ἁρμονική).

[20] But to accept this would invite difficulties. Notably, since there is no mean proportional between terms in superparticular ratio, that is, in one of the form $(n+1):n$ [see Boethius *De mus.* iii 11; *Sect. can.* props. 3, 16, 18], it would be impossible to specify ratios of velocities belonging, for example, to notes an exact half-tone apart. Aristoxenus insisted, but mathematical theorists denied, that the tone and other superparticular intervals such as the fourth (4:3) can be divided into equal parts.

why there can be, for example, at most two notes between given notes a fourth apart in a single scale-system, or why no more than two *dieses* can be sung successively in the course of a melody, or why the lowest interval of a tetrachord between fixed notes is always smaller than the highest, and so on. Nor can such sciences explain why the boundaries of the genera lie where they do, or why the melodic structure called the πυκνόν cannot span an interval equal to or greater than half the magnitude of a concordant fourth. It is true that the boundaries of the genera and the limits of the πυκνόν can be quantitatively specified (a fact that raises difficulties of its own, to which I shall return). But the boundaries marked in this way correspond to no distinctions that are mathematically significant: from a mathematical point of view their placing seems quite arbitrary; and mathematical principles cannot show why melodically significant boundaries should lie there, rather than somewhere else. Thus, it is a fact of musical experience, according to Aristoxenus, that the sequence of two quarter-tones is heard as generically different from that of two intervals of one third of a tone, while the latter sequence differs only in χρόα, not in genus, from a sequence of two semitones [see, e.g., 50.22–51.1 with 48.21–26]. Nothing in mathematics would lead us to expect this result nor can mathematical laws explain it.

Indeed, *nothing* can explain it outside principles generalising or abstracting from the perceived data of harmonics itself. For Aristoxenus, harmonic truths are not to be explained by reference to something external to our perception of melodies. Rather, they are to be organised and understood solely in terms of the system in which they themselves appear. The task of harmonics is to reveal the structure of phenomena taken *as* phenomena, to show that for something to be melodic *is* for it to fit within a certain orderly system, and to describe that system's anatomy. It is not to show how the structure of the system is determined by something else, mathematical or physical, because there is nothing that so determines it. It is an independent essence, existing only in the realm of audible musical sound. It is not an ordering of entities that 'really' exist in some more fundamental domain, of which the heard sounds are just one aspect; nor is the ordering one that can be abstracted without loss from the 'material' of sound and transferred, even in principle, to another domain. The crucial relations can no more be abstracted from the domain of the audible than can the relation of sweet to bitter, for example, from the domain of taste.

Even among authors who espoused mathematical harmonics in Aristotle's sense, there were those who recognised a problem to be solved in this connection. It could not just be assumed that the forms which appear as properties of melodic phenomena are the very same forms as those with which the mathematics of ratio and proportion concerns itself: in hearing sounds as melodically related we plainly do not hear them *as* standing to one another in certain classes of ratio. That they are indeed the same forms is something that must be shown, a task that the writer of the *Sectio canonis* at least undertook, however unsuccessfully, in the introduction to the treatise [Menge 1916, 148–149: but see chapter 8 in this volume].

I do not mean to imply, of course, that Aristoxenus' stand on these matters is impregnable. Several points could be urged against him, of which two are worth mentioning here. First, he may have misconstrued the Pythagorean conception of the relation between audible phenomena and their mathematical counterparts. I shall not argue this question (though for what it is worth, I do not think that he was wrong), since some of his criticisms are in fact quite independent of the subtler nuances of interpretation that these relations may be given. In particular, he claims as a plain *fact* that there are significant differences between melodic systems, scalar systems, and so on, to which no comparably significant mathematical distinctions correspond, and that there are rules of melodic sequence whose mathematical counterparts could be derived from no rationally compelling principle: from the mathematician's point of view the rules must seem merely arbitrary. Now the Pythagorean ratios might be conceived as analyses of relations between physical events distinct from the sounds they cause. Alternatively, they might be conceived in various ways as characterisations of the mathematical form of the melodic relations between musical notes themselves. To Aristoxenus, however, such details are unimportant. Any audible relation between sounds, or any relation between the 'physical' causes of the component pitches, could no doubt be described in the language of ratios. But if the rules of melodic progression turn out to be purely accidental with regard to the principles of mathematics, or if in distinguishing between the mathematical counterparts of perceptually distinct melodic forms we are making classifications that are mathematically arbitrary and unintelligible, these translations into mathematical terminology will have achieved nothing. They will have brought us no nearer to the goal of explaining the musical phenomena or of elucidating the grounds of their coherence and order: this orderliness will, if anything, have been obscured under the 'accidents' to which mathematical descriptions draw attention.

Secondly, however, it might be argued that Aristoxenus is mistaken in supposing that the Pythagoreans intended, as he did, to analyse and coordinate the rules and distinctions implicit in ordinary musical experience: their real project was prescriptive rather than descriptive, concerned primarily with the excogitation of mathematical or metaphysical ideals. There would probably be some truth in this accusation if, for example, Aristoxenus' main targets were theorists of a more or less Platonist persuasion. But not all mathematical theorists were concerned solely with rational paradigms. I would argue that Archytas, for one, was deeply interested in the description, analysis, and coordination of the systems implied in current musical practice, as were such writers as Didymus and Ptolemy in later antiquity. Others, it appears, were not: examples would be Plato himself, at least in some of his moods, and later commentators such as Theon of Smyrna. Certain mathematical writers—Eratosthenes, for instance, and Theon's main source, Adrastus—seem to show traces of both approaches. There was no uniform 'Pythagorean' project with a single, unambiguous goal: mathematical analyses were offered by different authors for different purposes. But in so far as mathematical theorists did share Aristoxenus' ambition to describe and explain the credentials of melodic systems in familiar contemporary use, there can be little doubt that he had the better of the argument, for the reasons given above. His approach enabled him to identify and coordinate a far richer collection of musical forms and distinctions than the Pythagorean scheme of concepts could even describe, let alone subsume under mathematical principles. Even if Pythagoreans and Platonists could offer a mathematical account—an analysis and an explanation— of some properties essential to melodic systems, still a vast storehouse of properties and relations remained untouched.[21]

---

[21] The representation of pitch-relations as ratios was flexible enough to express many different forms of attunement or scalar series. In the fourth century, three distinct 'generic' versions were described by Archytas [Ptolemy, *Harm.* 30.9–31.18 = Diels and Kranz 1951, i 428.15–37], and these differ again from the 'Pythagorean diatonic' of Philolaus (if we accept as genuine the disputed fragment from Nicomachus, presented as the second paragraph of Diels and Kranz 1951, i 408.11–410.10) and Plato [*Tim.* 35b–36b], which is the same as that implied in *Sect. can.* props. 19–20. Different ones again were later offered by Eratosthenes and Didymus, and with impressive sophistication by Ptolemy: for all these, see Ptolemy, *Harm.* 70.5–74.3. But their mathematical principles, even Ptolemy's, could cope with relatively few tasks in the field of *explanation.* Attempts were made to account for the perceived difference between concord and discord [e.g., Porphyry, *In harm.* 107.15–108.21 = Diels and Kranz 1951, i 429.1–27; Menge 1916, 149.11–24: cf. [Aristotle] *De aud.* 803b26–804a9]. More importantly, the Archytan theory of means [Porphyry, *In harm.* 93.5–17 = Diels and Kranz 1951, i 435.19–436.13] was held by some authors to provide a rational basis for the construction of well coordinated systems; and it could then be used to explain one feature of a legitimate attunement that distinguishes it from improper ones—the former and not the latter divides the octave proportionally (most clearly Plato, *Tim.* 35b–36b: the same principles are also at work in Archytas' own divisions). Other authors found different principles to do similar work. Those adopted by Ptolemy have features that arise from reflection on the special characteristics of

Aristoxenus' concept of these essentially melodic properties and relations is closely linked to what he calls δύναμις. He has no corresponding adjective, but we can reasonably use 'functional' or 'dynamic' to describe the relevant properties of notes, intervals, and sequences. The notion of melodic δύναμις is the central pivot of Aristoxenus' approach to his subject in books 2 and 3. (The fact that it appears nowhere in book 1 constitutes the most significant difference between its ideas and those of the presumably later treatise.) It is too large a subject to be explored fully here, but a sketch is essential. The subject is best approached through the contrasts Aristoxenus draws between δυνάμεις on the one hand and purely quantitative features of notes and intervals on the other, particularly what he calls the μεγέθη (magnitudes) of intervals. A pitched sound may take its place in a melody by being perceived, for example, in the character of the note called λιχανός—that is, the note immediately below the note μέση which is the principal focus of the system. To be λιχανός is not to be a sound of any particular pitch. Nor is it even to be a sound standing at an interval of some definite size (μέγεθος) below μέση. That is, perceiving a note as λιχανός is not identical with perceiving it as a note at such and such a distance below μέση. Our perception of a note's melodic function, its δύναμις, is distinct from and may not even include a perception of the magnitudes of the intervals between it and other notes: in fact the note λιχανός may stand, according to Aristoxenus, at any distance from a tone to a ditone (inclusive) below μέση. Its being λιχανός and so its finding a genuinely melodic role depends only on its being perceived within the prevailing system as the note between which and μέση no other note can melodically be inserted.[22] (Μέση, of course, is also dynamically determined, and so is every other note in the system: all notes exist in their relations to one another, relations constituted partly by their sequential order, partly

superparticular ratios [see esp. *Harm.* i 7], whose privileged status was recognised before Archytas, though he gave it new emphasis [cf. *Harm.* i 13]: but they do not depend on the Archytan theory of means, relying instead on applications of several subtle and distinct conceptions of mathematical 'equality' [see *Harm.* i 7, 15, 16]. But none of the other essential features of melodic systems as Aristoxenus identified them could be accounted for by such purely mathematical expedients. Ptolemy is commendably frank about this, despite the 'rationalistic' ambitions declared in *Harm.* i 2. The point is most explicit in *Harm.* i 15, where he distinguishes sharply between 'principles of reason' and 'theses based on agreed perception', and insists that these are independent and equally indispensible starting-points for the derivation of properly ordered systems of attunement. On Archytas' own conception of the relations between mathematical principles and the data of experience, see Barker 1989.

[22] Aristoxenus discusses and defends his position elaborately in 46.24–50.14.

by relations of concordance between certain fundamental notes, and partly by the ways in which the actual locus of one note, within the boundaries of these concords, carries implications for the positioning of the others.)

To take another example, a chromatic sequence is not to be defined as an ordering of some particular set of intervallic magnitudes nor even as a disjunction of such orderings. A chromatic sequence is essentially one heard as having a certain melodic character, which it is the business of trained musical perception to recognise. That character is preserved no matter which of indefinitely many different sets of magnitudes its intervals may possess within determinate ranges: it is this character that constitutes its δύναμις as a form of melody, and it is only when we have recognised this character as definitive of the chromatic that we can begin to enquire which sets of intervallic magnitudes, in which melodic contexts, in fact present it to the ear [see 48.15–49.2]. Again, the so-called πυκνόν is not in essence a pair of intervals of such and such a size, or even one with a determinate range of magnitude. In fact all πυκνά do fall within a determinate range [50.15–19] but it is not this fact that defines them as such, nor is it in that character that they are perceived as πυκνά. A πυκνόν is to be defined as a pair of intervals that presents to perception a certain character of sound [πυκνοῦ τινὸς φωνή: 48.30], and it is in having that sort of sound, not in being of some size or other, that a πυκνόν plays its dynamic part in melody and affects the melodic character of what we hear. Conversely, two intervals of equal size may differ in δύναμις by differing in their locus within the system or in the genus of the system to which they are heard as belonging [see, for instance, 47.29–48.6]: this difference will determine distinct melodic roles for each of them and different possibilities for melodic continuation from them.

In general, if we identify the absolute pitches of sounds in a sequence or spell out the sizes of the intervals between them, we are not thereby giving an explication or analysis of the fact that they form a melodic series or that they possess some specific melodic character [see 40.11–24]. In hearing pitches as melody it is not these features as such that we attend to. Rather, their melodic character depends on their being heard in determinate dynamic roles that form a consistent pattern of reciprocal relations, relations which cannot be described in other than dynamic terms. The language of harmonic δυνάμεις cannot be translated into that of μεγέθη. A basic task of the student of harmonics is to learn to identify the melodic relations he hears under their dynamic categories, since it as δυνάμεις, not as μεγέθη, that they constitute the data that he must come to understand. Harmonics seeks to articulate the system in which the dynamic relations exist; to explore the implications, for the structure in which it occurs, of a note's being heard as λιχανός, a sequence's being heard as chromatic, a pair of intervals' being heard as a πυκνόν, and so on; and to show how these implications arise from a unified set of ἀρχαί that express the φύσις of melody. That dynamic relations have these specifically melodic properties involves and arises from their existence as elements in that φύσις which is articulated by the theorist as a systematically integrated relational structure.[23]

In the context of Aristoxenus' insistence on the primacy of perception, this entails that individual instances of the δυνάμεις with which he is concerned are themselves objects of perception. The δυνάμεις must be present as such to ordinary melodic experience and cannot exist only as concepts of reflection or theoretical constructs of the harmonic scientist. Harmonics studies the melodic: being melodic is a property only of sounds as heard, and to be melodic is to fall under dynamic descriptions of the sorts I have sketched. This raises problems of two sorts. First, is the claim the δυνάμεις are objects of perception an intelligible and plausible one? And secondly, is Aristoxenus consistent in maintaining it? One passage, at least, suggests otherwise.

The first question is perhaps tangential to an exposition of Aristoxenus' ideas, but it is worth a moment's attention. One might argue that the directly perceived character of a sound must be limited to such features as its pitch, timbre, and volume, that is, to features determined by the physical processes through which it impinges on the ear. The intricate pattern of dynamic relations that Aristoxenus conceives cannot exist in the physical event of sensory reception and, hence, a note's melodic δύναμις—which consists, precisely, in its place within that network of relations—cannot be part of our perceptual experience of it. The large issues raised by such argument cannot be pursued in any depth here. So I shall be dogmatic.

The restriction of direct perception to a grasp of the so-called proper objects of each sense, though traces of it may be found in philosophers as eminent as Aristotle, seems to me quite without foundation. We hear sounds as standing in certain relations to one another, and these include relations of melodic implication just as surely as they include relations of pitch and loudness: similarly, we see things in visual relations, not just of colour and size, but also of symmetry and pattern. Of course, our reception of these relational properties is a complex matter, and it would be legitimate to set aside the expression 'indirect perception' to describe it just so long as this description does not serve surreptitiously to insinuate that it is

[23] The striking orderliness (τάξις) of μέλος is emphasised at 5.23–4: compare the description of the φύσις τοῦ συνεχοῦς at 27.17–33. References to the φύσις, of μέλος and of μελῳδία and of τὸ ἡρμοσμένον recur frequently in this connection throughout the *Harm. elem.*

not really perception at all. In the case of sequences of sounds, such perception is conditioned by (at least) short-term memory, generating a complex within which the relations are grasped. Again, our capacity to engage in it can be heightened and sophisticated by training. Both of these are facts about which Aristoxenus is clear and emphatic [see 38.31–39.3, 33.1–26]. But there is no reason why they should impugn the status of these relations as inhering in what is perceived. A melody is something heard: its character as melody is not something intellectually constructed and imposed from outside our perceptual resources onto what is given to us only as a set of differently pitched sounds. To say that it is would imply that what makes something melodic is a structure discernible only by the intellect, a set of νοηταὶ αἰτίαι: such a Platonist or Pythagorean approach leaves it wholly mysterious how we can tell that something is a melody without the least recourse either to measurement and mathematical analysis or to the sophisticated investigations of Aristoxenian musicology. It would also fail to explain how Aristoxenian or similar conceptualisations can be recognised as accurate or inaccurate articulations of what we ourselves experience.

Whatever the truth about these issues may be, Aristoxenus is plainly committed in most of books 2 and 3 to the thesis that melodic δυνάμεις are indeed given to perception, and that it is in hearing sounds as related in these 'dynamic' ways that we hear them as forming a melody. But one passage can be read as denying this:

> The project depends on two things, hearing and reason. Through hearing we assess the magnitudes of the intervals, and through reason we study their functions.
>
> Ἀνάγεται δ' ἡ πραγματεία εἰς δύο, εἴς τε τὴν ἀκοὴν καὶ εἰς τὴν διάνοιαν. τῇ μὲν γὰρ ἀκοῇ κρίνομεν τὰ τῶν διαστημάτων μεγέθη, τῇ δὲ διανοίᾳ θεωροῦμεν τὰς τούτων δυνάμεις. [33.4–8: Macran emends τούτων to τῶν φθόγγων, but unnecessarily]

This suggests that only the quantitative aspects of intervals are detected by the hearing, while all grasp of δυνάμεις lies in the province of διάνοια (intellectual reflection). Now if that is what Aristoxenus meant and if anything in his procedures hung on it, it would be the ruin of his science. If the hearing can discriminate only intervallic μεγέθη, it can have no grasp of any of the major distinctions that are said to determine melodic form. But the immediate continuation of our passage makes such an interpretation impossible. At 33.32–34.10 it is said that we perceive (αἰσθανόμεθα) such things as the differences between genera, the differences between intervals of the same size lying in different parts of the system, the difference between two placings of an interval such that one, but not the other, creates a modulation; and all these are differences of δύναμις, not of μέγεθος, or of anything reducible to the quantitative.

So what do the sentences quoted mean? I think the answer is clear enough. Aristoxenus is not concerned here with the way in which we standardly perceive melody and melodic relations. He is discussing the special resources we must deploy in order to generate the analyses characteristic of harmonic science (πραγματεία, the project). To hear a melody is to hear sounds in certain dynamic relations: but to articulate what these relations are and how they fit together demands reflection (διάνοια). That seems clear and unobjectionable. On the other hand, we do not hear something as melody by hearing its notes as standing to one another at certain intervallic distances. But there *are* certain distances at which, as a matter of fact, they stand in any given case; and the identification of these distances has a part to play in harmonic analysis. Now *this* is not something that can be achieved by διάνοια: to say that it is would again be to lapse into Platonism or Pythagoreanism by supposing that given melodic relations *must* be associated with specific quantitative values for mathematical or other intellectual reasons.[24] We can discover what quantitative relations hold between melodically related notes only through perception, by devising a technique of auditory measurement that enables us to find what the quantities in fact are. But it must be emphasised—and Aristoxenus underlines it repeatedly—that though the ear is capable of these quantitative discriminations, and though their results help in the scientific articulation of harmonic structures, they are no part of the original perception of a sequence as melodic. Here is Aristoxenus in full flow:

> The fact that the perceptual discrimination of the magnitudes as such is no part (οὐδέν ἐστι μέρος) of the complete understanding [of μέλος] was stated in outline at the start, but is easy to see from what I shall say next: for neither the δυνάμεις of the tetrachords, nor those of the notes, nor the differences between the genera, nor, to put it briefly, the differences between the composite and the incomposite, nor the simple and the modulating, nor the styles of melodic composition, nor one might say anything else whatever, becomes known through the magnitudes as such. [40.11–24]

[24] See in particular Aristoxenus' contrast between harmonics and geometry [33.10–26] which he seems to have developed out of considerations like those discussed by Aristotle in *An. post.* i 12, *Phys.* ii 2. Aristoxenus' position is analysed, with this comparison in mind, by Didymus in Porphyry, *In harm.* 27.17–28.26 (esp. 28.9–19).

The knowledge and understanding of which Aristoxenus speaks depend on principles abstracted from what is essential to our perception of the melodic character of given sequences. Perception of magnitudes contributes not at all to this understanding. Hence, the perceptions that do so contribute are exclusively qualitative and dynamic. The perception of magnitudes as such, which we might call calculative perception, is a scientific resource that enables us to identify quantitatively the ranges within which given dynamic properties are found. It thereby gives some help in mapping the interrelations of δυνάμεις, but it has nothing to offer to the project of discovering and articulating the nature of these δυνάμεις themselves. Perceiving melodic δυνάμεις and perceiving intervallic μεγέθη are entirely distinct operations.

Let us review the gist of this part of our discussion. To hear a set of sounds as a melody is to hear its constituent notes as standing in certain intrinsically melodic relations, relations of the sort we are calling dynamic. Harmonics seeks to articulate precisely the nature of the melodic δυνάμεις and the ways in which they are related, to describe fully the anatomy of the system of relations that they compose, and to spell out the principles of behaviour by which these relations are governed. These principles express the nature of μέλος as such: from them follow special and subordinate rules such as those demonstrated in book 3. As a matter of method, the principles are abstracted inductively from a careful survey of perceived instances: they must be such that perception, not just the abstract intellect, will recognise their appropriateness and authority as ἀρχαί [44.11–13]. As a matter of metaphysics, it is because the principles are as they are, because the essence of μέλος is as it is, that we hear certain sequences and not others as melodic and discriminate perceptually in the way we do, for example, between the various generic forms. It follows from Aristoxenus' phenomenalism, which is grounded in his rigid adherence to Aristotle's 'same domain' rule, that nothing can stand as an ἀρχή and assert for instance that every melodic sequence necessarily has such and such a property unless an instance of the inherence of that property in a sequence would be recognisable as such by perception. More than that, it must be recognisable as such by *melodic* perception, not just by the sort of perception that I have called calculative: it must be recognisable as one of the properties we discriminate in the act of hearing something as a melody.

Now these properties, as I have tediously repeated, are dynamic properties: they are not, as such, quantitative features of intervals. In hearing melody, we may hear, for example, a sequence presenting the character of the enharmonic πυκνόν: and it is no part of such hearing to notice that its constituent intervals span exactly a quarter-tone each. Indeed, as Aristoxenus emphasises, the subintervals may not always be of just that size: the enharmonic character on which attention focusses is preserved so long as the pair of intervals falls somewhere within a certain range of quantitative variation. The half-tone πυκνόν, broken down into two quarter-tones, is merely his favoured instantiation of this enharmonic sequence [see, for instance, 49.10–21].

It would be natural to draw the conclusion that the relation between a sequence's perceived quantitative features and its melodic ones is wholly contingent. Further, if melodic properties are neither constituted by nor inferable from quantitative ones, then by the 'same domain' rule the latter should apparently not even be mentioned in any of the classes of proposition proper to the science—not in the observation-statements from which the ἀρχαί are inductively abstracted, nor in the ἀρχαί themselves, not in either the premisses or the conclusions of the apodeictic demonstrations. In that case assessments of the magnitudes and statements about them seem to have no place in the science. The fact is, however, that in Aristoxenus' actual text they recur continually. What role can such propositions have?

We shall look at his treatment of them in a little more detail shortly. But first we must consider briefly a stratagem that promises to provide a route between dynamic and quantitative propositions, a way of showing that they are after all essentially and not merely contingently related. I have tried to make sense of this stratagem elsewhere [Barker 1984, esp. 52–62], but I am increasingly doubtful about its credentials. The main effort is to persuade us that even if some of Aristoxenus' statements about magnitudes are methodologically anomalous, still there remains a large group that can be intelligibly accommodated into his science. Let us assume two things: (a) that the relations of concord and discord are properly functional or melodic (that, I think, is uncontroversial); and (b) that though melodic perception does not identify magnitudes as such, it nevertheless does have within its scope the relations 'larger than', 'smaller than', and 'equal to', as applied to intervals.

These assumptions will get us a surprisingly long way. I shall not describe the route in detail. But (i) they allow us to distinguish the three primary concords, the octave, the fifth, and the fourth, as respectively the third-smallest, the second-smallest, and the smallest of the concords presented to perception; moreover, (ii) they enable us to identify the octave as the sum of the fourth and the fifth; and (iii) to draw attention to the interval by which the fifth exceeds the fourth, and which is called the τόνος or tone. Finally (iv) Aristoxenus offers a method [*Harm. elem.* 55.8–58.5: cf. Euclid(?), *Sectio can.* prop. 17], proceeding through the construction of concordant fourths and fifths, for determining the size of the fourth relative to the tone (it is, allegedly, exactly two tones and a half); and by further, similar steps, we can assign magnitudes measured by the tone to all intervals that are a half-tone or its multiples.

Propositions correlating melodic phenomena with sequences involving intervals other than these remain anomalous. But *something* seems to have been achieved, for we have established links of a non-contingent kind between melodic properties on the one hand and magnitudes on the other: for instance, if an interval presents itself to perception as an instance of the smallest perceptible concord, then it is an interval that 'calculative' perception will assess as spanning two tones and a half. Some quantitative propositions are after all derivable from functional ones.

But this comforting conclusion cannot be allowed to stand, since it rests on an assumption whose methodological status is itself open to question. The assumption is made explicit by Aristoxenus at 55.3–11: this passage states that the magnitude of each kind of perceived concord, though not that of each kind of discord, is fixed and determinate, or so nearly so that it makes no difference. That is, there may be some variation in the size of interval capable of instantiating a given perceived type of discordant relation, but a given concordant relation can be instantiated in an interval of only one size or something very close to it indeed.

The development of artificially tempered tuning systems since the sixteenth century strongly suggests that this proposition is false.[25] But let us suppose that it is true and that we can check its truth against our experience. The trouble is that if it is true, it seems to be a truth of a wholly contingent sort. It is not implicit in our perception of something as a concord that this concord admits no variation in magnitude, nor, certainly, would Aristoxenus have supposed that it was: his statement of the proposition

> ἐπεὶ δὲ τῶν διαστηματικῶν μεγεθῶν τὰ μὲν τῶν συμφώνων ἤτοι ὅλως οὐκ ἔχειν δοκεῖ τόπον ἀλλ' ἑνὶ μεγέθει ὡρίσαι [55.3–6]

is noticeably tentative—it is no kind of necessary truth. The alleged fact might be verifiable through experience, but that experience must be built out of presentations that are not exclusively those of melodic properties as such.

---

[25] The problems that led to the development of tempered systems and the theoretical controversies that surrounded them are clearly and vividly described in Walker 1978.

Aristoxenus faces a dilemma. Either his propositions about tones, half-tones, and so on are genuinely concerned with determinate magnitudes (in which case they stand in no essential relation to our perceptions and conceptions of melodic phenomena as such) or they are shorthand expressions of relations that are properly melodic (in which case they only pretend to say something about determinate intervallic distances). Even in these most favourable cases, the attempt to build a necessitating bridge between functional and quantitative propositions must fail.

Quantitative propositions appear in Aristoxenus' work in various roles, but their task is primarily to establish correlations between specific melodic properties and particular magnitudes or ranges of magnitude. This is exactly what Aristoxenus' own principles seem to make methodologically suspect. Before we pursue the matter further, there are three preliminary points that need some emphasis. First, the problem is not (or not only) that quantitative propositions cannot be demonstrated from functional ones, or the converse. It would be entirely within the rules to establish their correlations inductively, as Aristoxenus generally seems to do. The problem is that the necessary information about magnitudes is drawn from observations that are not essentially melodic: they introduce facts drawn from a different domain.

Second, we are not concerned here with magnitudes that are problematic because they are not given to perceptual experience: they are not relative rates of vibration or anything of that sort. They are heard and identified by ear, but not as intrinsically melodic properties. Quantitative relations are never such that a set of notes perceived as standing in some such relation is thereby perceived as making melodic sense or sense of some specific melodic sort.

Third, I must underline the fact that Aristoxenus' deployment of quantitative propositions is far from being just incidental to what he is doing. The claims I have drawn on about the centrality of δύναμις and the irrelevance of μεγέθη to an understanding of it can be exemplified in a number of passages from about the middle of book 2 onwards. Yet this is hardly what we would expect from the way book 2 set out, after a short discursive introduction: here Aristoxenus seems to go out of his way to emphasise how crucial the ear's quantitative judgements are to the conduct of the science and how important it is that the student should be trained to make them accurately.

Thus, in the passage beginning at 32.18 we find those scathing remarks about people who try to proceed from αἰτίαι foreign to the domain (described as ἀλλοτριολογοῦντες and ἀλλοτριωτάτους λόγους λέγοντες), and about others who neglect the need for proper ἀπόδειξις. Aristoxenus, by contrast, will adopt ἀρχαί for his demonstrations that are all φαινόμεναι τοῖς ἐμπείροις μουσικῆς. Given all I have said, we would be surprised to find much emphasis on μεγέθη here. Yet he goes on at once, in another passage we have already reviewed, to assert that the practice of the science depends on two things, ἀκοή and διάνοια: through ἀκοή, he says, we assess (κρίνομεν) the magnitudes of the intervals, and through διάνοια we apprehend (θεωροῦμεν) their δυνάμεις. I have argued that this does not imply that δυνάμεις are inaccessible to perception, nor that hearing a sequence's melodic properties involves hearing as such the magnitudes instantiating it: but plainly it does assign some crucial role in harmonic science, if not in ordinary musical listening, to the identification of μεγέθη. By way of an example of the role in which such assessment appears, we may take his remark a page or so later [35.10–17] to the effect that his predecessors had neglected the task of identifying the point at which chromatic divisions of the tetrachord begin and enharmonic ones end. No sense can be made of this task unless it is that of picking out the *magnitude* of the smallest chromatic πυκνόν, a task that Aristoxenus does indeed undertake [see 50.25–51.1].

We cannot solve the problem by arguing that Aristoxenus changed his mind half-way through book 2. Propositions concerning magnitudes continue to appear throughout that book and in book 3. I have argued [Barker 1984, esp. 62] that the quantitative form of many propositions in book 3 can be taken *exempli gratia* and their import reinterpreted functionally, but even so the difficulty remains. Towards the end of the third book, in a long and ill-tempered digression, he argues [68.13–69.28] that harmonic science must be concerned in the first place with δυνάμεις, not μεγέθη, since its objects must be determinate.[26] In this sense δυνάμεις are determinate and μεγέθη are not: that is, there is a unique set of dynamic relations, but not of quantitative ones, by which any given kind of melodic phenomenon is to be defined. But in almost the same breath [69.22–28] he draws the conclusion that progressions of melodically successive intervals must be identified for just one χρόα at a time; and that claim makes sense only if the identification is quantitatively conceived. The dynamic properties he has in mind are common to sequences in more than one χρόα: that indeed is why they are determinate, in that definite general rules concerning them can be formulated. It is the magnitudes of intervals instantiating them that

[26] The emphasis on the determinacy of the object of scientific understanding and the terms in which the discussion is couched may echo Plato, *Phil.* 16c–17e: relations between Aristoxenus' treatise and the *Philebus* are discussed in Kucharski 1959. But the immediate source here may be Aristotle, specifically *An. post.* 86a3–7.

vary from χρόα to χρόα and cannot be specified for all, or for several, at once.

Despite all I said earlier, irreducibly quantitative propositions are as deeply embedded in the treatise as are dynamic ones. They apparently describe two different domains of perceptual experience. Now the ways in which harmonics can approach these domains are not parallel. Let us consider them first separately, and then in relation to one another.

First, δυνάμεις are objects of melodic perception. By reflection on them we proceed inductively to definitions of the δυνάμεις and rules governing their interrelations.

Second, quantitative assessments of intervals are also made by ear. But perceptions of intervallic magnitudes are not as such intrinsic to our experience of melody or melodic form. Hence, no rules of melodic progression or the like can be derived (inductively or otherwise) just from remembered perceptions of strings of μεγέθη. Quantitative hearing as such does not discriminate the melodic from the unmelodic at all.

But third, what Aristoxenus apparently thinks we *can* do is to establish, inductively, correlations between specified dynamic relations and the quantitative relations in which they are materially instantiated. It may turn out, and indeed it does, that a given dynamic relation can appear only in notes standing at distances within a determinate range of magnitude: yet there is nothing in the dynamic relations as such to ensure that this must be so.

Then, if we can formulate rules governing such correlations, what standing have they in the science? They form no part of a definition of the essence or φύσις of melody: they are neither ἀρχαί proper to the science nor propositions derivable demonstratively from ἀρχαί, precisely because they serve to span the gap between one domain, one experienced 'kind', and another. We may appropriately call them bridging rules but that is just a name: it should not be allowed to disguise the fact that Aristoxenus himself has no equivalent terminology, nor that they constitute an uncomfortable irregularity in the smooth surface of the scientific structure as he conceives it.

There are some slight signs, I think, that Aristoxenus was himself half-aware of a distinction between proper ἀρχαί and the propositions I am calling bridging rules. He does not articulate it; but the language in which he expresses propositions of the two sorts does something to encourage the belief that he approached them in rather different frames of mind, whether he knew why or not. If the distinction did indeed have an influence on his modes of expression, we would expect it to appear most clearly in his handling of such notions as necessity and necessary connection. We would anticipate that necessity-terms would appear in propositions dealing with relations between melodic δυνάμεις, but would be absent or modified in those correlating δυνάμεις and μεγέθη. Unfortunately, a survey of cases does not give unambiguous results, and I place little weight on the statistics reported below. In many cases relevant propositions are introduced only with unilluminating indicative verbs. In others there might well be dispute about the class—domain-bridging or purely dynamic—into which a given proposition falls: the differences are not always as clear-cut as my remarks may have suggested. Since I do not want to over-emphasise the significance of my findings, I shall present them very briefly.

First, we might anticipate that uses of the verb συμβαίνειν indicate something less than full-blooded necessity. There are some twenty-nine occurrences in books 2–3, of which five are casual and irrelevant. Of the rest, I would construe nineteen as expressing relations between μεγέθη and δυνάμεις. Only three plainly concern essential links between dynamic properties: two are doubtful.

Turning to the words most obviously expressive of necessity, ἀνάγκη and ἀναγκαῖος, these appear twenty-six times. Ten cases are irrelevant (three being quite informal, two mathematical, and five representing logical relations). None expressly links μεγέθη with δυνάμεις, though one may be accounted doubtful. Fifteen seem to indicate relations between δυνάμεις themselves.

These results look straightforward, but the amount of interpretation behind them is such that they must be treated with extreme caution. Nevertheless, they give a little tentative and provisional support to the hypothesis that Aristoxenus was not wholly unaware of the distinction I have made. To draw attention to the distinction, however, is not to answer the question how the domain-bridging rules are to be accommodated smoothly into Aristoxenus' enterprise: the problem is re-described rather than eliminated.

But perhaps we may get more help from Aristotle than from our own unaided wits. The problems we have encountered in Aristoxenus are already implicit, I suggest, in authentic Aristotelian views concerning the relation of matter to form; and Aristoxenus' two major categories, the quantitative and the dynamic, do seem to invite representation under these Aristotelian headings.

Thus, the melodic, defined dynamically, is evidently a formal essence. As such it requires matter of some specific kind for its instantiation and this must be the movement of the voice in the dimension of pitch. (Aristides Quintilianus helpfully describes κινήσις φωνῆς as the ὕλη μουσικῆς [*De mus.* 108.18].) Further, it turns out that any particular species of melodic sequence requires for its instantiation a set of magnitudes within some determinate range in this dimension, if not ones that are uniquely fixed at definite values. But the thrust of Aristoxenus' remarks about δυνάμεις is plainly that the *definition* of such a species of melodic sequence does not as such import any reference to the magnitudes that are its 'matter'—which is not, of course, to deny the Aristotelian view that there are some sorts of things which do demand such reference in their definition [see, e.g., *Meta.* viii 2–3]. The best statement of the sort of relation that may be conceived as holding between melodic essences and the appropriate μεγέθη is perhaps Aristotle's account of relations between form and matter in *Phys.* ii 9. The existence of suitable matter is necessary for the instantiation of a given form but is not sufficient for it, nor even a part or an aspect of it.

But what kind of necessity is this? Not, at any rate, one that can be displayed through the logical connections of apodeictic demonstration: Aristoxenus, at least, could certainly not regard it so, whatever may have been true of Aristotle. It is a sort of material, even contingent necessity. But what can that mean?

The matter of which a thing of a determinate natural kind is constituted is as such only potentially an instance of the kind. But though it is only potential, its being potentially *that* sort of thing marks it off from matter of other varieties or in other types of arrangement. Not just any type of material complex can be a tree or a bird or a thunderstorm: there is some sort of necessity in the fact that only *this* kind of matter is potentially *that* kind of thing.

Let us consider, briefly and impressionistically, how this relation is handled in Aristotle's reflections on organisms. An organism's body, conceived materially, is potentially an organism: its actually being a living organism, its possession of soul, is the appropriate 'actualisation of a body possessing organs'. The student of the soul, the person who investigates the activities in which only living things engage, may choose to study them in the abstract, in their essence and their relations to one another. His analysis of sensation, for example, may bring out the fact that it is the reception of form without matter, that a perceiver's various senses are not independent perceivers, that sensation is a precondition of φαντασία and φαντασία of thought, and so on.[27] But his enquiry will be incomplete if it fails to consider also the material conditions under which sensation is possible, to ask what organs are needed for seeing, hearing, and the rest, and to consider what material constitutions they require. No organism can see if it lacks eyes and nothing can be an eye unless it fulfils *inter alia* certain material conditions.

---

[27] These examples are all taken, of course, from *De an.* ii–iii.

In cases like this we are faced once again with a kind of natural necessity that cannot properly be presented as a necessity of logic, unless of course an account of an entity's matter is intrinsically involved in the definition of its essence along with the account of its form. We have noted that Aristotle recognises cases where this last condition is met, but in ones of the sort we are facing here the possibility is unhelpful. No doubt we could just stipulate that an account of the eye's material constitution must be included in what we are prepared to call its complete definition. Similarly, we could decide to include in a definition of the note λιχανός, along with its formal, dynamic properties, an account of the ranges of magnitude within which a λιχανός must stand in relation to other notes in order to be capable of playing the melodic roles that the dynamic properties identify. But the problem would only be disguised: the definition would merely place two sorts of description side by side without revealing how the kind so defined is *one* kind and not two whose classes of instances just happen to coincide or overlap [see, e.g., *An. post.* 87a38–b4].[28]

Aristoxenus' difficulty can then be stated adequately in Aristotelian terms; and if it threatens the coherence of his enterprise, Aristotle's own scientific projects in biology and elsewhere must be similarly at risk. It is all very well for Aristotle to say that the natural scientist must treat a thing's nature as comprising its matter as well as its form, though the latter is more important; or that the relation between matter and form is such that the form is essence and end, while the matter is what is necessary if the end is to be attained [see *Phys.* ii 1 and 9]. The fact remains that this necessity cannot be a matter of logic, and that the relation between a form and its material conditions cannot be the object of scientific ἀπόδειξις. It is open to Aristoxenus to treat his δυνάμεις as the correlate of Aristotelian formal essences and his μεγέθη as the correlate of matter. Hence, he can stand by his thesis that no contribution towards an understanding of the essence of μέλος is made by the perceptual grasp of intervallic magnitudes. Equally, since his science is a study of a class of perceptibles, not of abstract

[28] Lennox [1986, 34] puts the point clearly:

> The problem with offering a purely functional account of (say) man is that it gives the impression that his soul is only incidentally related to his body. But this is a false impression—a man is an animal of a certain sort, a specific perceptive being, requiring precisely structured organs to function properly. His psychological and physiological activities are simply the actual realisation of his specific bodily structure.

But the last sentence quoted is not a solution to the difficulty I am indicating, only a different way of expressing it. What is it, in point of scientific method, that can show the truth of such a claim?

or purely intelligible objects, it will be incomplete if it fails to discuss the ὕλη required for this essence's instantiation both in general (where it is the κινήσις φωνῆς κατὰ τόπον) and for each type of dynamic property that a melodic sequence can as such display (where it is a range of magnitudes or distances in the τόπος within which the voice can move). Essence and material condition, δυνάμεις and μεγέθη, can be correlated in a law-like way by inductive generalisations from experience. The accusation that this procedure either breaks the 'same domain' rule, or more simply just places the results of two different scientific programmes side by side without demonstrative justification, can be met if and only if δυνάμεις and μεγέθη are not after all two kinds (γένη), the inhabitants of two distinct domains, but are complementary aspects of just one. The study of muscle and bone, after all, seems to belong to the same science as does the study of animal movement; and so would the matter and the form of natural entities of any kind that are the subject of a coherent programme of investigation. If this is the direction in which we should be led, the result will be Aristotelian without a doubt; but it is one that leads to further difficulties. The 'same domain' rule, on which Aristoxenus takes so firm a stand, is no longer the clear and justifiable injunction that it seemed to be. In the case of domains, what is sameness?[29]

3. Essence and history

It is easy to accuse Aristoxenus of presenting prejudice in the guise of science. His conservative leanings are well known: in insisting that μέλος has a fixed nature, from which it follows that certain sequences are melodic while others are not, is he merely trying to shore up established practices against the perversions of modernism with the impressive but empty paraphernalia of pseudo-scientific argument? Can there be any justification for his thesis that τὸ ἡρμοσμένον is a real and objective φύσις, not just the invention of arbitrary human taste or whim? After all, it seems implicit in his procedure that what is perceived as melody *is* melody: Can he properly insist that some sequence is intrinsically non-melodic, if someone else asserts that to his ear it is part of a perfectly acceptable melody?

These difficulties can probably not be neutralised completely. It is striking, however, that while Aristoxenus is free with his abuse of predecessors who have failed in point of scientific method or who have given faulty

[29] This is a question I shall leave unresolved. If it has an answer, an extension of the analyses offered in Lennox 1986 may offer the best approach. The remainder of this paper is by way of a coda: it has something to say about the issues aired in this section but does not continue the argument directly.

quantitative assessments of certain intervals, and while he disparages performers who play out of tune and of connoisseurs who prefer saccharine, quasi-chromatic versions of the enharmonic to the 'noble' one that he himself admires, nowhere does he suggest that there are those whose ears perversely accept as melodic something that in fact is not. On the contrary, he goes to great lengths to accommodate all aesthetic preferences, including ones that he himself finds distasteful. There is a best form of the enharmonic, for example, but its excellence is not demonstrated within harmonic science (though Aristoxenus believes that its merits will gradually become clear to people who attend to it carefully and often): the nature of μέλος as such is neutral between melodies in good and in bad taste. It discriminates only what is a melody, whether good or bad, from what is not, and sequences of one melodic form or species from those of another [cf. 49.8–18 and 23.3–22].

Aristoxenus' overall enterprise, therefore, involves from the start two levels of judgement, distinguishing first what is melodic and secondly, within that class, what is admirable or fitting for given musical purposes. The sphere of harmonics, as he makes clear on several occasions, is restricted to the former [see esp. 1.18–2.2, 31.16–32.8: cf. [Plutarch] *De mus.* 1142f–1143e, 1144c–e]. Further, as we have already seen, there are at least two sorts of judgement associated with harmonics itself: there is the properly 'harmonic' investigation of dynamic relations and there is the assessment, by 'calculative' perception, of the magnitudes in which these relations are instantiated. Since each dynamic relation is capable of several different quantitative instantiations, which Aristoxenus shows signs of wanting to designate as aesthetically better or worse, it would seem that breaches of taste can occur in at least two ways. Legitimate harmonic relations can be instantiated in the less admirable of the magnitudes available to them as 'matter': alternatively (and this introduces a new dimension, beyond harmonics altogether), legitimate harmonic patterns may be combined and used in contexts and for purposes to which they are not suited. Deciding what is and what is not legitimately harmonic, however, involves no judgement of taste of either kind. But we may still ask what the grounds are on which Aristoxenus bases his confidence in the harmonic principles he asserts. If they stand on induction from his own experience and nothing else, the ground is exceedingly weak.

In the *Harm. elem.* we can detect hints of another source of confidence, the agreement of practical experts in the musical arts. It is fair to imagine that Aristoxenus exchanged views with such people and in particular that he took the trouble to find out how far the kinds of melody they aimed to produce corresponded, in their intention as well as in his ears, to the rules he believed he could discern. His purpose, after all, was not to enunciate a set of laws that prescribed obedience, but to articulate those that were presupposed by current practice.

But the core and the grounding of his ideas are most clearly revealed, I suggest, in his attitude to musical history. Very little of this appears in the *Harm. elem.*, but long fragments and paraphrases of his extensive writings on the subject have been preserved elsewhere, notably in the Plutarchan *De musica*. Here the manner in which his conservative prejudices are articulated is highly instructive. His aim was apparently to demonstrate the superiority of the music of an earlier period (up to and including the first decades of the fifth century), not just by asserting the excellence of the musical forms adopted by composers of those times, but by arguing that it was by deliberate policy that they restricted themselves to those forms.[30] The proliferation of elaborate new styles in the later fifth century and the passion for chromaticism in the fourth draw Aristoxenus' scorn. But his central thesis is that the melodic *possibility* of such styles was already implicit in the systems of the reputable ancient composers: it was by choice, not through ignorance, that they left them unused.

There are two points to be drawn from this. The first is the simple one that in proceeding by 'induction' to the articulation of ἀρχαί Aristoxenus did not start merely from his experience of contemporary practice. He took into account also what he knew—or thought he knew—of practices throughout Greek musical history. That by itself adds some weight to his findings. The second is the implication that even the earlier and simpler practices carried within themselves the seeds of later and more sophisticated ones: that is, even if earlier conventions restricted melodies to those based, for example, on the diatonic genus of scale, these conventions could not be understood except against the background of a system that embraced equally the possibility of the other genera. The nature of μέλος as a whole is then implicit in the simplest tune: for no ἀρχαί acceptable to perception can be found which would allow it to be understood, and

[30] This, at any rate, is the position taken by the speaker at [Plutarch's] dinner-party: it is a principal theme of *De mus.* 18–21, and of much of the discussion from ch. 28 to the end. But it seems safe to infer, from the nature of the Aristoxenian material he cites, that the theme was already there in his source. Thus, it seems likely, for instance, that the form in which the information on the σπονδειάζων τρόπος is presented [ch. 19] reflects that given it by the source: the line of argument in ch. 20 is plainly that of a fourth-century commentator: similar conclusions will apply in many other places. Notice also how the theme, still explicit in ch. 32, is merged without any sense of discontinuity into a wholly Aristoxenian discussion of the relation between harmonic science and other modes of judgement [in ch. 33–39].

its functional relations articulated, in isolated from the rest.[31] It is in this sense that the φύσις τοῦ ἡρμοσμένου, as expressed in the complex web of propositions that Aristoxenus sets out, is an objective and permanent reality. The musical conventions of particular times and places are partial exemplifications of it and can be made comprehensible only through an understanding of the whole.

This approach gives Aristoxenus' polemics against 'modern' composers a particular pungency. They have broken no laws of melody: they write tunes that are heard as melodies and so *are* melodies, and that are indeed much applauded by the vulgar. But their cheap and popular taste does not even have the merit of originality, of daring experiment. What they did was always there to be done: their achievement was only that of working out in practice and putting on display those aspects of melodic nature that earlier composers had deliberately and rightly avoided, deeming them to lack nobility or to be unsuitable for the social context in which their performances took place [cf. [Plutarch] *De mus.* 20–21, 28–30].

It was not Aristoxenus' intention, then, to use his technical writings on harmonics to defend his own conception of musical excellence. If it had been, we would no doubt have found in them 'proofs' that practices adopted by the composers he disliked are melodically improper, in breach of the principles of melody. In fact his attitude is the reverse: the legitimacy of these second-rate practices is implicit in the principles that underlie good melody. When he defends μέλος against the charge of being disorderly, of having no determinate nature, he does not do so by showing that modern manifestations of such disorder are outside the proper definition of μέλος: he argues that in fact they depend, for all their superficial confusion, on the same underlying system of order as does music of more traditional sorts. The discrimination of noble melody from meretricious rubbish is not within the scope of harmonics, though harmonics may provide some distinctions in terms of which judgements of taste may be set. What makes noble music melodic is no different from what makes a mere jingle so; and the melodic legitimacy of the individual instance or type, whether good or bad, can be understood only through its relation to the whole nature of μέλος which accommodates both [see [Plutarch] *De. mus.* 30–34].

This brings me to my final point, which is in a way the upshot of the whole discussion. The *Harm. elem.*, in my judgement, shows the principles of the *An. post.* at work, and brings out very clearly the conception of a

---

31 See, for instance, [Plutarch] *De mus.* 34 (esp. 1143e–f), a passage that is certainly derived from Aristoxenus.

science which that treatise implies. Its task is not conceived as the discovery of truths of fact hitherto unknown, but as the articulation of what is know and its ordering in relation to truths already implicit in what is known. A domain of experience is held together and show to be intelligible as a unity through the formulation of principles whose truth can be recognised by inductive reflection on experience. These, related to one another (not related by derivation from something standing outside of and 'explaining' the experienced content of the domain) form a description of a single essence or nature. From the principles expressing this nature subordinate truths are derived, but not as surprising new discoveries. Surprises, in fact, would be quite out of place. The ἀποδείξεις are designed to illuminate the known, not to uncover the unknown. They show how particular and familiar facts in the domain are implicit in the web of relations constituted by the primary essence, rather than being essentially disconnected items of experience related only casually to one another. Harmonics, as Aristoxenus envisages it, does not expel from the melodic domain anything we supposed that it included or import any novelties on theoretical grounds: it seeks to display the pattern within which our actual melodic experience falls and so to draw our experience into the sphere of our understanding.[32]

[32] I should like to express my gratitude to the organisers of the IRCPS conference in Pittsburgh for the opportunity to present an earlier version of this paper. Comments made by participants on that occasion have been very valuable: my thanks are due especially to Alan C. Bowen, James G. Lennox and Alexander P. D. Mourelatos. The shortcomings of the product are of course my own.

10

The Relation of Greek Spherics to Early Greek Astronomy

J. L. BERGGREN

My subject is the history of a tradition in the science of spherics that developed from the fourth to the first centuries BC. Spherics is a name which goes back to antiquity for that science whose subject is the mutual relations of arcs and angles formed by circles on a sphere. This science as it is found in texts of the fourth century has been variously described as 'sufficient for the astronomy of its time' [Heiberg quoted in Hultsch 1906], as 'fumbling attempts to obtain some quantitative and geometric insight' [Neugebauer 1975], and as the kind of literature that Plato thought typical of real astronomy [Mueller 1980]. In this paper I wish to examine these and other views of this tradition of spherics in the light of the ancient texts themselves and of the recent writing on this subject, in order to explore the relation of this tradition to the mathematics and astronomy of its own time.

All who have written on spherics have recognized its intimate relation with certain problems in ancient Greek astronomy. These problems, like all those which are at once beautiful and difficult, called forth a variety of solutions; and I should like to begin with a summary review of these problems and their ancient solutions in order to set ancient spherics in its proper context. I begin, then, with Ptolemy's *Almagest*.

After discussing some philosophical, physical, and mathematical preliminaries to the study of astronomy, Ptolemy devotes the end of book 1 and all of book 2 to solving a set of problems concerning spherical arcs and angles. The importance of the solutions of these problems for both astronomy and geography, is evident from the following sample:

1. Find the height of the Sun above or below the equator (its declination, δ) and its distance east or west of the equinoxes (right ascension, α) corresponding to any position of the Sun in the zodiac.
2. Given one of the three quantities—the maximum length (M) of daylight, the altitude (ϕ) of the north pole above the horizon, or the ratios of equinoctial and solstitial shadows to the length of the rod casting them—find the other two.
3. Find the variation in length of daylight during the year, given any one of the three quantities just listed.
4. Find the angles between the ecliptic and such great circles as the horizon or meridian.

Ptolemy's solutions to these problems are the earliest recorded that use the trigonometric methods developed between the time of Hipparchus of Bithynia (150 BC) and that of Menelaus (AD 100).

However, Vitruvius in *De arch.* ix 7 tells us that already by his time in the late first century BC the problem of the length of daylight had been solved by the geometrical method of the *analemma*, and Pappus in *Coll.* iv prop. 40 cites a (now) lost work on this method by Diodorus of Alexandria, an older contemporary of Vitruvius. Otto Neugebauer [1975, 301] suggests that Hipparchus, who lived more than a century before Vitruvius, may have used an *analemma* to determine the arc of a parallel of declination from a setting star of known declination to the point of the same declination on the meridian. Further, we have it on the authority of Synesius of Cyrene (*obit ca.* AD 415) that a century before Diodorus, Hipparchus described the method for solving problems about the sphere now known as stereographic projection. We also know from Vitruvius, *De arch.* ix 8.8–14 that by his time this method had given rise to an instrument, the anaphoric clock, which provides a solution to the problem of finding the length of daylight for a given position of the Sun in the ecliptic.

Not all efforts to solve the sorts of problem Ptolemy addresses relied on geometry, however. At about the same time as Hipparchus, the Alexandrian scholar, Hypsicles, in his *Anaphoricus* exploited a number of fairly weak assumptions about relations between arcs on the ecliptic and those on the equator in order to solve arithmetically the problem of the length of daylight by beginning with the maximum length of daylight as the sole datum. The methods utilize the linear increase and decrease of numbers; they are very different from the geometrically-based methods we have been describing, and their origins are to be found in Mesopotamia. However,

Hypsicles' computations worked with much smaller ecliptic arcs than we find in related, earlier works; and his clear recognition that the same procedures, further refined if necessary, could be employed in any of the seven *climata*, showed that arithmetic methods were entirely capable of producing what King [1987] has called universal solutions in spherical astronomy.

Thus, the mid-second century BC marks the beginning of a period when two important geometrical methods developed, the one based on trigonometry and the other on the *analemma*. It also marks the end of a period of development of arithmetical methods. There is, however, another geometrical tradition, one in which none of the writers we have mentioned participated, despite the fact that it began probably two centuries before the time of Hipparchus and continued as a source of new treatises for another century after him. This tradition comprised a body of theorems on solid geometry which pertain to the sphere and whose principal interest lay in their relevance to astronomy. It is known to us through the following works: *De sphaera quae movetur (The Rotating Sphere)* and *De ortibus et occasibus (Risings and Settings)*,[1] both by Autolycus of Pitane, who flourished in the latter half of the fourth century BC; *Phaenomena* by Autolycus' contemporary, Euclid; and *Sphaerica, De diebus et noctibus (Nights and Days)*, and *De habitationibus (Habitations)*,[2] all by Theodosius of Bithynia, whose probable *floruit* of 100 BC would make him a younger contemporary of his countryman, Hipparchus. None of these writers makes explicit reference to any of the others; and all presuppose not only basic theorems of solid geometry such as one finds in Euclid's *Elements* xi, but theorems on arcs and angles of the sphere which are not found in the *Elements*. Apparently, then, none of these works stands at the beginning of this tradition.

Indeed, the often-remarked relation of the entire theory of spherics to some of the astronomical problems listed at the beginning of this paper shows that the mathematics in these treatises originated after the time when Greek astronomers began to try to derive explanations of observed phenomena as well as predictions from the model of a spherical Earth fixed at the center of a rotating cosmos. As to when this was, Goldstein and Bowen [1983] argue that ancient Greek astronomy may be profitably studied by dividing its history to 300 BC into two periods. The first of these two phases is an ancient tradition characterized by a concern with calendrical matters. In this tradition the dominant astronomical activity

---

[1] Our study concerns only the first of these works; for the relation of the two books of the latter work to each other, see Schmidt 1949.

[2] I have used only the first of these three works in this study.

was the composition of *parapegmata* in which the phases of important stars (that is, their first and last morning and evening appearances) were given in terms of some calendar along with meteorological phenomena that could be expected to accompany these astronomical phenomena. It was an activity that attracted men of considerable distinction, as Ptolemy's inclusion of Meton, Euctemon, Democritus, and Eudoxus, in his list of parapegmatists testifies. Indeed, Ptolemy mentions these predecessors in his own *parapegma*, book 2 of his *Phaseis*, which is by far the acme of this tradition. It is almost a corollary of the intent of this kind of literature that there is no mention of eclipse-phenomena or planetary matters. In fact, it was concern with eclipse-phenomena and the dimensions of the cosmos that marked Greek astronomy after 300 BC.

Goldstein and Bowen also point to other activities, contemporaneous with the ancient tradition of *parapegmata* and relevant to the history of astronomy. Among them were the numerological speculations of the Pythagoreans and their idea of explaining the heavens by a ἁρμονία (*harmonia*) of whole numbers. Indeed, as Aristotle reports, it was the primacy of numbers in their science and the way in which numbers seemed to reflect a moral order based on the properties and ratios of ἁρμονία which led the Pythagoreans to suppose 'the elements of numbers to be the elements of all things and the entire heaven to be a *harmonia* and number'.[3] Also important, according to Goldstein and Bowen, were the cosmological speculations of the Presocratics, since they introduce elements that were later to become basic parts of scientific astronomy. Thus, some time before Eudoxus, the image of a sphere of stars rotating around a concentric, spherical Earth had been proposed; and such an image is found in the cosmological-moral tradition leading up to works like Plato's *Republic* and *Timaeus*. Indeed, in recent studies, Charles Kahn [1970, and in this volume] has argued that one of the principal achievements of Presocratic speculation is the construction of 'a cosmic model, including a spherical heaven, a spherical Earth, and a geometrical account of celestial motion'.

However, Goldstein and Bowen argue that this world-picture, held by some Presocratics, was not yet a mathematical model, in the sense of an explicit mathematical analogy between physical domains (both idealized mathematically) that served as a basis for computation, or at least comparison, of magnitudes. They believe that with the work of Eudoxus the second phase of early Greek astronomy began and that the distinguishing feature was what they call the two-sphere model, so named because

[3] Aristotle, *Meta.* 985b23–986a3 [quoted from Goldstein and Bowen 1983, 333].

it placed a spherical Earth at the center of a spherical cosmos which rotates daily around an axis passing through the Earth's center. Although components of this model can be found in various Presocratic texts, what was new was the exploitation of the model with its reference circles, including the horizon, equator, and ecliptic as well as those below, to provide a *mathematical* explanation of phenomena associated with the risings and settings of stars and the length of daylight. It is this that differentiates Eudoxus' tradition from that exemplified by the Presocratic and the Platonic writings mentioned above, and it is this that gave rise to the science of spherics.

The mathematical utility of the model lay in its division of (the celestial) sphere into five regions concentric around the poles: the torrid region between the tropics, temperate regions on either side of these, and finally the frigid zones around the poles. Although these names reflect climatological characteristics, the origin of the zones is astronomical. The tropics are bounded by the parallel circles defining the northern and southern limits of the Sun's annual motion, and the boundaries separating the temperate from the frigid regions are defined by the circle of always-visible stars and the circle of always-invisible stars, circles dependent on the latitude of the observer. All these circles and the regions between them were then transferred to corresponding circles on the Earth bearing the same names.

Eudoxus and his successors elaborated this model to include not only a sphere for the fixed stars but spheres for the Sun, Moon, and planets; but this system of homocentric spheres lasted only until astronomers realized it could account neither for the retrograde motion of Mars nor for the variation in apparent sizes of such luminaries as the Moon and Venus. After that time the general theory had no more impact on mathematical astronomy; however, the two-sphere model itself became part of the professional astronomer's stock in trade and the science of spherics took the place it occupied in the education of the astronomer throughout antiquity. Since, however, not all recent writers have seen the Platonic writings as being so distinct from the mathematical tradition initiated by Eudoxus, we must here take some account of Plato's idea of what spherics should be, as indicated in a well-known passage, *Republic* vii 528e–530c.[4] Here Socrates rebukes Glaucon for his notion that simply looking at stars is somehow the same as gaining knowledge through higher, rational speculation. Glaucon accepts the rebuke and, on asking how Socrates thinks the study of astronomy should be reformed, he learns that man must use his reason to conceive 'the true realities—the real relative velocities, in the world of pure

[4] The following account and all translations are taken from Cornford 1951.

number and all perfect geometrical figures, of the movements which carry round the bodies involved in them'. The real astronomer, we are told, will admire the observed sky as a geometer might admire 'diagrams exquisitely drawn by some consummate artist like Daedalus'. But,

> when it comes to the proportions of day to night, of day and night to month, of month to year, and of the periods of other stars to Sun and Moon and to one another, he will think it absurd to believe that these visible material things go on for ever without change or the slightest deviation, and to spend all his pains on trying to find exact truth in them.

After Glaucon assents, Socrates concludes that the genuine study of astronomy proceeds as does that of geometry, by problems, and proposes to leave the starry heavens alone.

What Plato meant by this is sufficiently illustrated in the following section of the *Republic* where there is a discussion of ἁρμονία, and Socrates and Glaucon lament the folly of those who in seeking to understand it 'waste their time in measuring audible concords and sounds one against another'. Such people 'do not rise to the level of formulating problems and inquiring which numbers are inherently consonant and which are not, and for what reasons'. Presumably, such principles when applied to the study of astronomy as recommended earlier, would produce a calendar like Philolaus' where numerological considerations forced a 59-year cycle in which each year had $364\frac{1}{2}$ days. This absurd number was chosen only so that the number of months in the resulting cycle would be a Pythagorean number for the Sun, 729, where $729 = 27^2 = 9^3$ and 27 is the number for the Moon, and 9 the number for the Earth [Neugebauer 1975, 619].[5]

In a recent paper, Ian Mueller [1980] considers the passages in the *Republic* cited above and argues that Plato's assimilation of astronomy to geometry and harmonics to arithmetic is 'not unreasonable' given 'certain Greek scientific texts which, I believe, make clearer the kind of astronomy and harmonics Plato has in mind in the *Republic*'. The texts to which he refers are Theodosius' *Sphaerica*, Autolycus' *De sphaera quae movetur*, and Euclid's *Phaenomena*.

In regard to astronomy, two problems arise from Mueller's arguments. One of these concerns the interpretation of Plato's intent in the passages quoted and the other concerns chronology. I shall discuss them in turn. First, as concerns intent, I proposed earlier that what Plato meant when

[5] For other examples of numerology in Greek astronomy, see Neugebauer 1975, 630–631, 659–660, and 693.

he talked about 'real' velocities and 'perfect figures' is sufficiently illustrated by his remarks on ἁρμονία, where we are urged to study not the consonances we hear but numbers which are inherently harmonious; that is, Plato's intent is uniform in the two sections I have mentioned. If this is correct, then, it is hard to see in the geometrical and kinematical idealizations of the three treatises mentioned above, the spatial analogue of the sort of numerological speculations Plato suggests for harmonics. That is, if Plato really had texts like those of Theodosius in mind, he could not have intended the same sort of thing in his comments on astronomy as he did in his remarks on ἁρμονία. Second, there are serious chronological problems in assuming that texts like those of Autolycus were available at the time Plato wrote his *Republic.* Under the current view of the composition of Plato's dialogues, the *Republic* was written before the year 370 BC. But comet-observations reported by Aristotle [see below] suggest that the two-sphere model was probably introduced after 372 BC and before 340 BC. Thus, there is every reason to believe that the introduction of the mathematical model, which must have antedated the three texts Mueller addresses, did not occur until some time after Plato had written the *Republic*; consequently, these three texts could not have been the genre of literature that Plato had in mind when he wrote the *Republic.* Thus, I prefer to take Plato at his (or at least Glaucon's) word, when Glaucon asks Socrates, 'How do you mean the study of astronomy to be *reformed*, so as to serve our purposes?' [emphasis added]. It is a reform of astronomy that Plato advocates; he is not describing a current genre of literature. In fact, Plato underlines how far his ideal is from current practice, at the close of that section when Glaucon says, 'That will make the astronomer's labour many times greater than it is now.'

If, then, the mathematical two-sphere model was introduced too late in the fourth century to have any effect on the composition of the *Republic*, when was the model invented? In this regard Goldstein and Bowen refer to Aristotle's report that

> in the archonship of Nicomachus (*scil.* 341 BC) a comet appeared for a few days about the equinoctial circle (this one had not risen in the west), and simultaneously with it there happened the storm in Corinth. That there are few comets and that they appear rarely and outside the tropic circles more than within them is due to the motion of the sun and stars.

That Aristotle [*Meteor.* 343b1–7] refers neither of the other comets he mentions (427/26 and 373/72) to any reference-circle increases our faith that this one detail was not added in later times and that it thus supplies us with a *terminus ante quem* for the introduction of the mathematical two-sphere model, namely, 341 BC. A *terminus post quem* is, to stay on the conservative side, 372 BC, the date of the latest unspecified comet-sighting; and so we have an interval of approximately thirty years within which the mathematical two-sphere model was developed by Eudoxus. (Goldstein and Bowen provide evidence that Eudoxus not only used this model in his astronomy but actually invented it.)

There is good evidence, moreover, that the science of spherics developed fairly rapidly. Indeed, two of the three treatises that will concern us were written by two contemporaries who were active, it seems, at the end of the fourth century. I mean, of course, Euclid and Autolycus, whose treatises on the subject of spherics are the earliest we have.[6]

It is often said of Autolycus' *De sphaera quae movetur* that it represents the earliest extant Greek mathematical treatise; but one must agree with Neugebauer that there is so little hard evidence to separate Autolycus and Euclid chronologically that all we can say with any confidence is that their treatises were all written at roughly the same time, probably in the second half of the fourth century BC. Even Germaine Aujac [1984], who places Autolycus and his writings some thirty years before Euclid, agrees that in any case Autolycus was not a source for Euclid; and there seems to be unanimity on the central point that both Autolycus and Euclid rely on an earlier work for the basic theorems of the subject. This earlier work must have appeared between the years of, say, 360 and 320 BC; of its content and range we may form some notion indirectly from the writings of Euclid and Autolycus.

We turn, then, to the nature and apparent purpose of these treatises. As early as John Philoponus, who in the sixth century of our era provides the first reference to the title of Autolycus' work, *De sphaera quae movetur*,[7] commentators have seen the difference in character between this treatise and Euclid's *Phaenomena*, for Philoponus points out that Euclid's work is the more 'physical' of the two in that it considers not only motion, a distinguishing concern of ancient physics, but οὐσία (substance) as well, that is, the Earth and the stars. (It also mentions by name all the principal astronomical circles on the sphere.) Furthermore, Euclid's treatise

---

[6] Autolycus' *floruit* is securely dated to the last quarter of the fourth century BC: see Mogenet 1950, 5–7 for details. Regarding Euclid's date, we have nothing more secure than the claim that he lived in the period between Aristotle and Apollonius. A later date would only strengthen the arguments I will give subsequently about the purpose of Euclid's work.

[7] For the text and a summary, see Mogenet 1950, 160.

is preceded by a long introduction whose purport is that the (mathematical) two-sphere model applies to our world; and it begins with definitions of 'horizon', 'meridian', and 'tropic circles', while such purely geometrical objects as tangent circles on the sphere and angles between great circles on the sphere are not defined. I would conclude that in defining circles of astronomical and physical importance and in passing lightly over purely geometrical objects, Euclid is telling his readers what are the really important ideas in his treatise relative to his intent in writing it. Indeed, he does not mislead the reader: *Phaen.* prop. 1 (*The Earth is in the middle of the cosmos and occupies the place of the center in relation to the cosmos*) puts us in no abstract, geometrical setting but in our own cosmos. The proof uses a diopter pointed at the beginning of Cancer rising and is obviously intended to put an important physical image in the reader's mind. The *structure* of the proof, however, reflects in all details the structure of a Greek geometrical proof; and I conclude from this that Euclid intends us to take the proof as seriously as any of his proofs in the *Elements*. That we have trouble doing so reflects our tastes and not Euclid's, nor those of his time. Indeed, Galen tells us, 'Euclid showed in Theorem I of the ≪Phaenomena≫ in a few words that the earth is in the midst of the cosmos, as a point or center, and the students trust the proof as if it were two and two is four' [quoted from Neugebauer 1975, 748]. In any case, we know from proposition 1 onward that we are dealing with a demonstrative science whose subject-matter is astronomical phenomena.

These phenomena are further defined by the following propositions. The first part of proposition 2 states that a great circle through the pole is, in one rotation of the sphere, twice at right angles to the horizon; and the second part concerns the angles made by the ecliptic during its daily rotation, with the meridian and horizon—the latter angles being of importance for the phases of a star. Euclid's remark that 'this has been shown' has often been taken to refer to an earlier treatise, but it seems unlikely that out of the large number of results on spherics which his treatise presupposes Euclid should have chosen only this particular one to cite. Following this are four propositions [*Phaen.* 3–6] on star-risings, of which we may take proposition 5 as typical: *Of stars on the circumference of a great circle that cuts the always visible circle the ones more to the north rise earlier and set later.* The method of proof in both *Phaen.* props. 4 and 5 is to project the stars from one parallel circle to another by means of great circles that are rotations of the horizon, thereby reducing the argument to the case when the two stars lie on the same parallel circle. (The parallel circles are not, however, identified as such in Euclid's treatise, though they are in Autolycus' proposition 8.) As this technique is applied, there is abundant appeal to visual evidence from the diagrams,[8] and the lemmas needed to establish the necessary assumptions are never stated.

The same technique as that employed in *Phaen.* props. 4 and 5 is fundamental to Autolycus' *De ortibus et occasibus.* In the *Phaenomena*, rotations of the horizon are used to find points on the day-circle of one star that rise and set simultaneously with a star on another day-circle. In Autolycus' work, on the other hand, rotations of the horizon are used to find points on the ecliptic that rise and set simultaneously with a given star. When such points are located, then the points 15° behind and ahead of them mark the points that determine the four phases of the star.

The next two propositions, *Phaen.* props. 7 and 8, deal with arcs of the horizon where the whole zodiac or individual signs rise, and would have been of interest in the theory of lunar eclipses, that is, in what was called the *prosneusis* of eclipses.[9] The inclusion of such a theorem could indicate that when the treatise was written some of the elements of the theory of eclipses were in place. However, it also provides an introduction to the idea of ortive amplitudes, that is, of the angle on the horizon north or south of the East-West line indicating the point where the Sun rises, and may equally well have been included for that reason. As is the case with the whole treatise, the results here are entirely qualitative. Aujac [1984, 100] has pointed out that a good example of the more geometrical language employed by Autolycus occurs in *Phaen.* prop. 7, where Euclid relies on the theorem, *That the zodiac rises and sets at all places of the horizon between the tropics*—the tropic circles are assumed to be at least as large as the always-visible and always-invisible circles. Compare this with Autolycus' statement of the same theorem in *De sph.* prop. 11:

> *If in a sphere a great circle, which is inclined to the axis, demarcates* [ὁρίζων] *the visible (half) of the sphere and the invisible, while some other inclined great circle is tangent to larger circles than the horizon is tangent to, then it makes its risings and settings along the whole arc of the horizon between the parallel circles that it (the other great circle) is tangent to.*

Perhaps Autolycus includes this proposition in order to introduce the idea of the Sun's ortive amplitude.

---

[8] For a discussion of the principles on which the diagrams in extant mss of Greek texts on spherics appear to be drawn, see Neugebauer 1975, 751–755.

[9] '*Prosneusis*' refers to the angle formed by the ecliptic and the great circle passing through the centers of the Moon and the Sun (in the case of a solar eclipse) or the Moon and the center of the Earth's shadow (in the case of a lunar eclipse): see Neugebauer 1975, 141–144.

With *Phaen.* props. 9–13 we come to what appears to be the central purpose of the treatise, a consideration of the rising-times of the signs of the zodiac. The notion appears in proposition 9 but is nowhere defined, so a word of explanation may be useful here. The rising-time of some arc of the zodiac is simply the time it takes that arc to rise over the horizon; and one may measure it either in equinoctial hours or in equatorial degrees (since the equator rises over the horizon at a uniform rate), where the relation between equinoctial hours and degrees is $1^h = 15°$.[10] Since in the period from sunrise to sunset of a given day, 180° of the ecliptic must rise, one may use rising-times to compute the length of daylight at a given locality on a given day by finding the rising-times of arbitrary semicircles on the ecliptic.

The first two propositions on the subject, *Phaen.* props. 9 and 10, provide qualitative results concerning the rising-times of semicircles,[11] while the next four, *Phaen.* props. 11–13 and the lemma, concern rising- and setting-times of equal arcs placed variously. The notion of a rising-time was obviously considered well understood, in contrast to the basic spherical apparatus which was so carefully defined at the beginning of the treatise. One imagines an audience already familiar with the idea of rising-times from linear arithmetic schemes, but which still needed to be schooled in the geometrical elements of the science of spherics.

Phaen. prop. 12 is a good example of the theorems demonstrated:

> *Equal arcs of the semicircle that follows Cancer set in unequal times. Those nearest the tropics set in the greatest times while the following set in lesser times. Those nearest the equator set in least times while those equidistant from the equator rise and set in equal times.*

The evidently qualitative nature of this result and those accompanying it should not, however, mislead us into thinking that they are of little use; for, in fact, these theorems justify all the inequalities on rising-times that Neugebauer [1975, 712] shows to be sufficient for deriving arithmetic sequences for rising-times under either Babylonian System A or B. Indeed, the inequalities Neugebauer uses are just those Hypsicles (who was a contemporary of Hipparchus) presents for the linear scheme in his *Anaphoricus*,

---

[10] In making this definition I do not mean to imply that Euclid or his contemporaries had defined equinoctial hours. The expressions used in the Greek text may be translated as 'in the time that arc X rises arc Y has also risen' or 'arc X rises in more time than arc Y', where X and Y are arcs of the ecliptic. Nothing is said of the units in which time is measured.

[11] If one knows the rising-time for any semicircle, one can find the day-length for any day of the year.

and the scholiast to that treatise saw clearly that the theorems at the end of the *Phaenomena* were just what were needed to justify the assumptions Hypsicles stated [see De Falco and Krause 1966, 41–45]. I maintain, then, that the *Phaenomena* represents an effort to provide a more systematic basis for those intuitively plausible symmetry-considerations needed to justify linear schemes for rising-times. This would at least explain both Euclid's casual reference to the subject as well as his qualitative treatment of it.

Hypsicles lived some one hundred fifty years after Euclid; but his *Anaphoricus* was certainly not the first treatise of its kind, as is shown by his mention of 'those who occupy themselves with rising-times'. According to Neugebauer,[12] the earliest methods for calculating the length of daylight were schemes which used linear interpolation to compute, month by month, the length of daylight. These schemes began with the length of the shortest day and increased this by a constant amount each month, an amount calculated so as to arrive at the correct value for the length of the longest day. These schemes are found in ancient Egypt in a Ramesside papyrus (12th century BC), in the Hibeh Papyrus and the so-called Eudoxus Papyrus (both containing views from the 3rd century BC); and they survive well into the Middle Ages. However, the more sophisticated arithmetic approach to the problem, which sums up the rising-times of ecliptic-arcs (these being taken to be in an arithmetic progression), was used not only by the Babylonians of the Seleucid period but also by Epigenes (perhaps *ca.* 250 BC) in Alexandria and pseudo-Berossus (1st century BC: cf. Kuhrt 1987, 36–44). Neugebauer suggests that a lunar tablet dating from about 400 BC, which gives length of daylight as a function of the Sun's position on the ecliptic, may represent a transitional stage between linear schemes for length of daylight and an approach which uses linear schemes for rising-times and then sums these to obtain length of daylight.

It seems, then, that the method of calculating the length of daylight by rising-times was introduced around the end of the fourth and beginning of the third century. I suggest, therefore, that Euclid's *Phaenomena* reflects the kind of mathematics that was needed in contemporary astronomy. In this I am following Heiberg's evaluation of the *Phaenomena* as a treatise sufficient for the astronomy of its time [cited in Hultsch 1906, col. 1048]. While I agree with Neugebauer [1963, 530b] that 'Euclid and Aristarchus... demonstrate the inadequacy of traditional mathematics to cope with spherical astronomy and trigonometry at the end of the fourth century', I would suggest that Euclid at least was not trying to cope with spherical astronomy or trigonometry but rather was aiming to give the

[12] For the following remarks, see the extensive discussion in Neugebauer 1975, 706–733.

student the theoretical background to understand methods that had been developed in astronomy.

In any case, with the inequalities of rising-times given in *Phaen.* props. 11–13, Euclid is in a position to finish the treatise with a comparison in *Phaen.* props. 14–18 of the time it takes equal arcs of the ecliptic to leave the hemispheres above and below the horizon. (I see no reason to doubt that this material is genuine, despite the fact that the last two propositions are found only in recension B.) The idea of leaving (viz. changing: ἐξαλλαγή) a hemisphere is defined at the end of the introduction to the *Phaenomena* as the passage of an arc of the ecliptic from its first point being on the eastern horizon to its last point being on the western horizon. Such a concept would arise in the calculation of rising-times as follows.[13] When the Sun rises, it is at a certain point P of the ecliptic which is on the horizon. Since the Sun always moves slowly westwards along the ecliptic, it happens that, when the rotation of the heavens brings P to the western horizon, the Sun is not yet on the horizon because it has traversed a certain arc PQ on the ecliptic (on the order of half a degree) in a direction opposite to that of the daily rotation of the heavens. Thus, the arc PQ must leave the hemisphere for the Sun to set, so the true length of daylight is the time it takes PQ to leave the visible hemisphere.[14] Since this is just the rising-time of the arc from P to $P + 180°$ added to the rising-time of the $1/2°$-arc diametrically opposite PQ, we may calculate a good value for the true length of daylight by interpolating linearly between the values for rising-times which Hypsicles gives at intervals of 1°.

Thus, again, one may see in Euclid's *Phaenomena* a geometrical account of topics that may have been known from the arithmetic methods of his day. Although there is no evidence that points to such fine calculations in Euclid's time, it is still true of ancient mathematics—and more so than of later times—that 'absence of evidence is not evidence of absence'; and the hypothesis of a relation with the arithmetic methods of the time at least gives some point to an exercise which otherwise seems rather mad—worrying about the rising-times of arcs on the order of $1/2°$ when one is, in fact, unable to determine even approximately the rising-times of whole signs.

Quite different from Euclid's *Phaenomena* in character and subject is Autolycus' *De sphaera quae movetur*. I have already quoted the remark

[13] This is pointed out in Schmidt 1943, a study from which I have derived great benefit. It is a pity that it has never been published.

[14] I am surprised to find nowhere in the literature a paradox of the Achilles-and-the-tortoise type, since one is suggested by the fact that, by the time Q gets to the horizon, the Sun has again moved away from it.

by Philoponus on the more geometrical nature of this treatise, so we may proceed immediately to a brief account of the subjects it treats. To begin, there is no long preface as with Euclid; instead Autolycus starts by defining the uniform motion of a point [but see Aujac 1979, 42nn1, 4] and then turns immediately to a series of twelve propositions which treat the following topics:

1–3 Generation of parallel circles, which are perpendicular to the axis, and similar arcs by points on the surface of a uniformly rotating sphere.

4–6 Cases when no points, all points or some points rise and set. Introduction of the *sphaera obliqua*, that is, the sphere for an observer not at the equator.

7 In a *sphaera obliqua* points rise and set on the same parallel and all parallels are equally inclined to the horizon.

8 Great circles tangent to the same (parallel) circles as the horizon are rotations of the horizon.

9 The co-risings and co-settings of stars in a *sphaera obliqua*.

10 In a *sphaera obliqua* a rotating circle that passes through the poles is only twice perpendicular to the horizon.

11 Where, on the horizon, does a circle tangent to circles larger than the horizon rise and set? [See my earlier comments on *Phaen.* prop. 7.]

12 If a fixed circle always bisects a moving circle, neither circle being perpendicular to the axis or passing through the poles, then each of these is a great circle.[15]

It is apparent from this list of theorems that this work, unlike the *Phaenomena*, is not dedicated to a goal more specific than discussing various phenomena arising in the *sphaera obliqua*. Certainly, some of the same topics are broached, for example, the perpendicularity of circles through the poles to the horizon and the arcs of the horizon which the ecliptic circle passes by during its daily rotation. There is also a discussion of the risings and settings of stars, a discussion which begins with rare cases and then turns to the *sphaera obliqua*. But, on the whole, one senses that this

[15] In his preface Euclid cites a weaker version of this theorem, in which the moving circle is given as a great circle, in order to show that the horizon is a great circle. His wording is slightly different and he neglects the necessary restrictions which Autolycus states here. It is certainly the sort of result one would appeal to in trying to apply the abstract model of spherics to observed phenomena.

treatise, with its smattering of topics—all fundamental but none pursued very far—is the sort of text that one would have a student read prior to reading the *Phaenomena*. The half-physical, half-abstract nature of this treatise is clearly exemplified by the fact that although the horizon and the always-visible and always-invisible circles are mentioned, there is only an allusion to the ecliptic and tropics in propositions 11 and 12, where these circles are described rather than named.

From the mathematical point of view, there are several features of both the *Phaenomena* and *De sphaera quae movetur* worth mentioning. First of all, on a formal level, the Euclidean treatise defines astronomically important circles whereas Autolycus defines uniform motion. Moreover, the similarity of the structure of the proofs of the propositions in both treatises to that familiar to us from Euclid's *Elements*—πρότασις, ἔκθεσις, διορισμός, κατασκευή, ἀπόδειξις, and συμπέρασμα—shows that both writers were writing within a mathematical tradition, whatever the degree to which physical notions entered. Finally, the formal incompleteness of these treatises, both of which quite casually cite results they need as apparently well known, is evidence that these treatises are but individual patterns in a larger mosaic: both writers are evidently working in a background of familiar results and methods, not only in the special area of spherics but generally in solid geometry as found in book 11 of the *Elements*. Both writers may have aimed to train readers in special topics, but neither wrote for beginners in geometry.

Turning now from the formal aspects of the treatises to the mathematical methods they employ, let us consider Hultsch's suggestion [1886] that the theorems from Theodosius' *Sphaerica* which Euclid and Autolycus refer to as known and available for their use, together with the geometrical results used to establish those theorems, may be taken as the kit of mathematical tools available to writers on spherical astronomy at the time of Euclid and Autolycus. According to Olaf Schmidt [1943, 11–12], however, the story is not so simple. Certainly, if a result proved by Theodosius is cited word for word by Euclid in the *Phaenomena*, one may assume that the result in that form was part of the mathematics available to Euclid; but this does not at the same time justify arguing that, because we know how Theodosius proved the result, we may extract from the theorems used in this proof other theorems Euclid must have known.

Schmidt's example of how such an assumption can mislead concerns proofs using tangent circles on the sphere. Neither Euclid nor Autolycus defines these circles, but it is apparent from such proofs as *De sph.* prop. 6, which shows that the parallel circles touching the horizon are either always visible or always invisible, that Autolycus considers two circles as tangent if they have only one point in common; and there is no reason to suppose that Euclid's conception was any different. On the other hand, at the beginning of *Sphaerica* ii, Theodosius defines two circles to be tangent at a common point if the line through that common point and in the plane of each circle is tangent to each circle. Theodosius then uses this to establish a group of propositions, *Sphaer.* ii props. 3–5, which together could be called the Fundamental Theorem of Tangency, and whose import is that two circles touching at a point are tangent if and only if that point and their poles lie on a single great circle. Admittedly, the last of these three, *Sphaer.* ii prop. 5, is used without proof by Euclid [*Phaen.* prop. 2] and by Autolycus [*De sph.* prop. 10]; however, in light of the fact that neither of these treatises contains any hint of the Theodosian definition of tangency, it would be unwise to assume that the source Euclid and Autolycus used for the theorem proved it from the same definition of tangency that Theodosius used.

Another important group of theorems that is basic to the theory of tangency is the group *Sphaer.* i props. 13–15, the import of which is that if G is a great circle and S a small circle on a sphere, then the following statements are equivalent: (1) G bisects S, (2) G is perpendicular to S, and (3) G contains the poles of S. The proofs of these propositions are built on several previous propositions of book 1, which utilize *Sphaer.* i props. 1, 7, 8, and 9, and these propositions are, in turn, used in the construction-problem, i prop. 21 (*To construct the pole of a given circle on the sphere*) and in i prop. 17, which is itself used in the proof of i prop. 21. *Sphaer.* i prop. 15 is also used in Autolycus, *De sph.* props. 5, 6, 7, and 10, while *Sphaer.* i props. 13 and 15 are both used in Euclid, *Phaen.* prop. 2.

However, it is instructive to compare the different ways in which *Sphaer.* i props. 13 and 15 are used by Autolycus and Theodosius. The latter uses them in the proof of the first part of what we have called the Fundamental Theorem of Tangency, namely, in *Sphaer.* ii prop. 3, which states that two intersecting circles are tangent if the point of intersection and their poles lie on a single great circle. His argument is that, by *Sphaer.* i prop. 15, the two intersecting circles are perpendicular to the great circle joining their poles, so that the line in which their planes intersect will be perpendicular to the great circle. Also, by *Sphaer.* i prop. 13, the great circle bisects each of the intersecting circles so that each circle intersects the great circle in a diameter. Accordingly, the line of intersection of the planes of the two circles is perpendicular to each diameter and is, therefore, tangent to each circle. By definition, then, the circles are tangent.

In contrast, Autolycus, when he must prove in *De sph.* prop. 6 a particular case of *Sphaer.* ii prop. 13, namely, that the horizon and greatest always-visible circle are tangent, begins just as Theodosius does but ends in a very different way. Autolycus starts by remarking that, since the meridian contains the pole of the horizon, it follows [cf. *Sphaer.* i prop. 15] that the meridian is perpendicular to the horizon and bisects it. Now, however, he takes a different tack. He observes that a section of a circle (the meridian) is upright on the diameter of a circle (the horizon) and is divided into unequal parts at the north pole. Next, as we remarked earlier, Autolycus applies the result later proven by Theodosius in *Sphaer.* iii prop. 1 to show that the greatest always-visible circle is tangent to the horizon. It is significant, I think, that although *Sphaer.* iii prop. 1 uses the notion of tangent circles, Theodosius' proof uses only the Pythagorean theorem and some elementary results in solid geometry concerning one plane's being perpendicular to another—in other words, Theodosius' proof makes no appeal to his definition of tangency. This, then, is one result in Theodosius' *Sphaerica* which is relevant to tangency and whose proof could go back to a pre-Euclidean source. In short, it would appear that *Sphaer.* iii prop. 1 forms part of the ancient theory of tangents, but that at some time after Autolycus someone saw how to apply *Sphaer.* i props. 13 and 15 to establish a new theory of tangents.

Theodosius himself uses *Sphaer.* ii prop. 5 in the proof of ii prop. 13 which introduces the idea of disjoint semicircles of great circles on a sphere that are tangent to the same parallel circle. This notion is used by Euclid, *Phaen.* props. 4–7, 12, and 14 to prove that certain arcs are equal. It is also applied in Autolycus' *De sph.* prop. 8 which states that 'great circles tangent to the same circles as the horizon will, when the sphere is rotated, coincide with the horizon'. Here Autolycus uses precisely the same terminology of disjoint semicircles as found in Theodosius.

On the other hand, *Sphaer.* ii prop. 13 introduces a sequence of propositions, *Sphaer.* ii props. 13–16, which provides constructions [ii props. 14 and 15] and theory [ii prop. 13 and its partial converse, ii prop. 16] relevant to comparing arcs on parallel circles. The centerpiece of this group, which finds important application in the *Phaenomena*, is *Sphaer.* ii prop. 15 on the construction of a great circle tangent to a given parallel circle and passing through a point between that parallel and another which is parallel and equal to it. As the table of the logical structure of *Sphaer.* ii shows [see Table 1], this proposition is well embedded into the structure of book 2 and makes use of the Theodosian theory of tangent circles. Euclid, however, appeals to this theorem on three different occasions [*Phaen.* props. 4, 5, and 12], and of crucial importance to these appeals is the result proven in *Sphaer.* ii prop. 13 concerning arcs of two parallel circles cut off by nonintersecting semicircles. Autolycus, too, states the enunciation of *Sphaer.*

ii prop. 13 word for word in his treatise. Here again it seems plain that the results of *Sphaer.* ii props. 13–16 represent[16] part of a pre-Euclidean chapter on the subject, which some writer later than Euclid put on a different basis.

	1	2	3	4	5	6	7	8	9	10	11	12	13	14	15	16	17	18	19	20	21	22
1																						
2																						
3																						
4		•	•	•																		
5				•																		
6		•	•		•																	
7						•																
8			•			•																
9																						
10																						
11																						
12																						
13					•					•	•	•										
14			•																			
15			•						•			•										
16										•			•	•								
17									•			•										
18																	•					
19																		•				
20									•													
21																						
22					•				•			•									•	
23					•				•													•

Table 1. The logical structure of Theodosius, *Sphaerica* ii

A • at row *m* in column *n* means that the proof of proposition *m* uses proposition *n*. For example, proposition 4 in book 2 relies on propositions 2, 3, and 4. This table does not show that ii prop. 1 uses i prop. 10; that ii prop. 2 uses i prop. 10; and that i prop. 15 is cited in the proofs of ii props. 3, 9, 10, and 21.

Further evidence of progress in the treatment of spherics between the time of Euclid and that of Theodosius consists in the much more sophisticated treatment of the angles of inclination between the ecliptic and horizon as it is discussed in *Sphaer.* ii prop. 22.[17] We are still dealing here, it is

[16] See Aujac 1984, 104–105 for details and a convincing discussion.

[17] According to Theodosius, one plane is more inclined to a base plane than is another plane if it makes a smaller angle with that base than the other does.

true, with angles between planes and not with angles on the surface of the sphere; but, even given this, Theodosius' treatment of the way the angle in question varies monotonically from a maximum to a minimum, together with its discussion of symmetries, goes far beyond the beginning steps in the solution of the problem that we see in Euclid's *Phaen.* prop. 2.

Theodosius' method of proving the result is of some mathematical interest in that it measures the variation in one magnitude, the angle between ecliptic and horizon, by that of another, namely, the height of the pole of the ecliptic relative to the horizon. The standard phrase for 'height of the celestial pole' is ἔξαρμα τοῦ πόλου, which refers to the arc of the great circle through the zenith and the pole contained between the pole and the horizon. The phrase does not occur in the *Sphaerica*, however: in comparing two positions of the pole of the ecliptic, Theodosius describes the one pole as μετεωρότερος than the other, and measures the elevation of the pole by the perpendicular from it to the horizon. The result is that in going from greater elevation of the pole to greater inclination of the ecliptic, the arc mentioned earlier is introduced secondarily and at the expense of some complication in the proof of *Sphaer.* ii props. 21–22, as a quantity that varies monotonically with the pole height and that allows one to pass from it to inclinations of planes.

Any study of the mathematical methods of ancient spherics would be incomplete, however, without some consideration of the methods used in the theorems requiring some construction.[18] *A propos* of such theorems in Theodosius' *Sphaerica*, Schmidt [1943, 13–14] sees in them evidence that those who worked with spherics in an astronomical context used constructions on a solid sphere. This is no doubt correct, but there is more to be said on this point.

Schmidt's principal arguments concern the group *Sphaer.* i props. 16–21 which provides the basis for the construction of a great circle through two given points and for finding the poles of a given circle. However, the proof of i prop. 18 assumes that one can draw lines on a circle within the sphere, and this theorem serves in i prop. 19 which requires one to find the diameter of a given sphere. Moreover, the very first construction-problem, namely, i prop. 2 (*To find the center of a sphere*), is obviously not solved by construction on the surface of the sphere; indeed, one wonders what use it would be for working with a solid sphere. It seems to me that i prop. 2 is an instance of a theorem that occurs in the *Sphaerica*, and, more to the

[18] The question of the status of construction-problems relative to that of theorems in Greek geometry has been discussed in Bowen 1983; the question of the motivation for constructions is discussed in Knorr 1983. These are recent considerations of issues raised in Zeuthen 1896.

point, was developed in the first place, because it was thought that such a proposition belonged in an elements of spherical geometry.

Theodosius' view of construction-problems, however, goes deeper than a simple desire to do problems on the sphere analogous to those one wants to do in the plane; and this is shown by the solution of *Sphaer.* i prop. 20, which requires the construction of a great circle through two given points on a sphere. He begins the proof as follows:

> Let A and B be the two given points on the surface of the sphere. It is required to draw the great circle through A and B. When A and B lie diametrically opposite it is clear that arbitrarily many great circles can be drawn through A and B, so let it be supposed from now on that A and B are not diametrically opposite.

Now if Theodosius' object had been to show astronomers how to do useful constructions on a solid sphere, he would hardly have omitted a case which is at least as useful as any other. Again, if the object were simply to write a treatise containing what a treatise on geometry ought to contain, then one would expect to see all cases treated. In fact, what the proof of this problem suggests is that one function of some of the constructions was to guarantee the existence of certain objects.[19] In this case the existence of a great circle through two diametrically opposite points would have been clear to any reader who had understood *Sphaer.* i prop. 6 and the meaning of 'diametrically opposite'.

Finally, Theodosius' motives for including construction-problems are well illustrated by the proof of problem *Sphaer.* i prop. 21, which requires finding the pole of any given circle on the sphere. The existence of poles is guaranteed by *Sphaer.* i prop. 8 together with the existence of the perpendicular to any given plane at any point on it (assumed, in any case, in problem i prop. 2); yet Theodosius does not give such a proof. His procedure for the construction is certainly such that one could carry it out on the surface of the sphere. Yet, if it were intended simply as a 'how to' recipe for the novice astronomer working with the solid sphere, it is

[19] Zeuthen [1896] suggested that this was the motivation for construction-problems in Greek geometry. Knorr [1983] argues, on the other hand, that such a purpose accounts for only a very few of the extant constructions and is largely a backward projection of mathematically trained 19th-century historians of then-current concerns onto an ancient canvas. Knorr's arguments are generally convincing, but my point here is that there is internal evidence in the proof of *Sphaer.* i prop. 20 that one of the motives for including the theorem was to prove the existence of great circles satisfying certain conditions. That other motives were also operative is evident from the construction-methods Theodosius uses.

curiously incomplete, since it assumes, without explanation anywhere in the treatise, the bisection of arcs of circles. Of course, the existence of the midpoint of an arc is clear from the considerations of continuity, as is also the case with the existence of a fourth proportional to three given arcs, something Theodosius assumes in the proof of iii prop. 10. Given that Theodosius says nothing more about the midpoint, it may well be that continuity was what he was tacitly appealing to; but constructive proofs are also at hand with the material at his disposal, so it seems we are in no position to argue that Theodosius held any one attitude uniformly towards all his constructions.

Such then are the mathematical methods of spherics in the fourth century BC, partly as they are found in the texts from that century and partly as we have reconstructed them from a text of the first century BC. The goal of these methods was to explicate astronomical phenomena and their origin lay in the mathematical two-sphere model of Eudoxus. That the explanations were entirely qualitative should not surprise us given the state of astronomy in fourth-century Greece, since astronomers only had available to them at that time a collection of data based ultimately on rough observation and qualitative estimates.

On the other hand, interacting with this material in a complex way were arithmetic methods of varying degrees of sophistication. We have seen how, for example, in the case of the length of daylight, all such schemes were based on the idea of interpolating a sequence of numbers between the annual minimum (m) for the locality and the annual maximum (M). The different levels of sophistication lay not in the values for M but in:

1. the rules according to which the sequence of numbers was chosen,
2. the density of the derived sequence of numbers within the interval $[m, M]$, and
3. the integration of one sequence with another in a scheme such as that of the *climata* which would, so to speak, cast a numerical net over the whole οἰκουμένη (inhabited world).

This is but one example of how the precision of the ancient exact sciences lay not in the exactitude of their observations, which was poor by modern standards [see Aaboe and Price 1964], but in the fact that the scientists developed or used mathematical methods to turn, if I may overstate matters slightly, observational dross into scientific gold. Thus, it is entirely consistent with the qualitative, approximate character of ancient observational data that the first attempts to geometrize the world-picture should themselves display a similarly qualitative character. In fact, the subject of spherics produced exact results only when it was combined, by means of trigonometrical tables, with numerical procedures which had long preceded it. But that belongs to the history of trigonometry and is another story.

11

The Definition, Status, and Methods of the Medical Τέχνη in the Fifth and Fourth Centuries

G. E. R. LLOYD

The debates on the definition, status, and methods of the medical τέχνη that took place in the fifth and fourth centuries BC offer us a remarkable opportunity to explore a whole range of issues. In part these are sociological and concern the relationships between the various groups who laid some claim to the title, ἰατρός, or who were involved, in practice, in one or other aspect of one or other rival or complementary tradition of treating the sick. In part they also relate to epistemological and methodological questions concerning the nature of a τέχνη, its defining characteristics, and the type of knowledge it presupposes. Although I shall be concerned chiefly with the second group, it would be foolish to attempt to deal with those questions in isolation from those that come under the first heading. Certainly in many Hippocratic texts besides the three treatises I shall be using as my principal evidence (*On the Art, On Regimen in Acute Diseases, On Ancient Medicine*), there is a recurrent preoccupation not just with defining medicine and setting down its methods, but also with how the doctor is to be distinguished from the layman, and again how from imposters, charlatans, or doctors in name alone.

It is well known that apart from the medical writers represented in the Hippocratic Corpus (themselves a highly heterogeneous set) several other groups were also involved in healing the sick. There were, for example, the 'midwives' (μαῖαι) concerned, of course, with much more than that conventional translation might suggest. Again, the collection, sale, and administration of drugs, especially herbal remedies, were chiefly the province of those called ῥιζοτόμοι (root-cutters) and φαρμακοπῶλαι (drug-sellers). On occasion, as texts such as [Demosthenes] *In Neaeram* lix 55–56 confirm, the women of the household went and bought drugs and administered them themselves without, it seems, calling in an ἰατρός of any kind. There were also those whose discourse on diagnosis and cure drew heavily on the categories of the divine and the supernatural; but again *they* can hardly be treated as a single group, since there are important differences between on the one hand the itinerant purifiers and sellers of charms and incantations whom we hear about from Plato, for instance, as well as from *On the Sacred Disease*, and on the other those who practised temple medicine in the increasingly well-established shrines of Asclepius and other healing gods or heroes.

Now plurality of medical traditions is far from unique to classical Greece: it can be paralleled in ancient Egypt and in ancient Babylonia to look no further afield in ancient, let alone modern, societies. But what is more exceptional is the overt and explicit attacks by one group or another. The point must not be exaggerated. As I have recently had occasion to stress elsewhere, not *every* Greek group of healers attacks or criticises every *other* one. Some of the relations between more or less clearly demarcated groups are marked by a certain tolerance and a mutual, though not necessarily symmetrical, respect. We do not find Hippocratic writers of any kind engaging in polemic with the midwives as a group, for instance, or with those whom the author of *On the Diseases of Women* i 68 calls the women healers; rather, as I tried to show in *Science, Folklore and Ideology* [Lloyd 1983a], there is some cooperation between them. On the other hand, some of the boundaries between one group and another *were* highly contested. The onslaught on the itinerant purifiers in *On the Sacred Disease* is the best known example, but we should be careful not to assume that it was always those represented in the Hippocratic Corpus who did the attacking (even though much of our evidence comes from them). It is clear enough from Aelius Aristides that, in the second century AD at least, temple medicine often set itself apart from and indeed criticised other styles of healing; and there are signs of that developing already much earlier in the fourth century BC in the inscriptions from Epidaurus.

Quite *why* in Greek society there are these explicit attacks is, to be sure, a highly complex and controversial question which I shall not attempt to reopen here. But *that* they existed is both clear enough and important. Evidently, when some of the frontiers or boundaries were a veritable battlefield, it was not going to be enough for those within a particular group to sit back and assume that sufficient *raison d'être* for their style of activity came from the very existence of the group itself. Self-justification was the order of the day and many Hippocratic treatises revert to the point, even when they are not devoted to it entirely (as *On the Art* is). Of course, self-justification could take many forms. In some Hippocratic works (such as the *Oath, Decorum*, and especially *Precepts*) the writers show themselves chiefly concerned with the moral respectability of their fellow-ἰατροί. Several of the deontological treatises are products of the Hellenistic period, and *Precepts* in particular shows clear signs of influence from Hellenistic epistemology. Yet—whatever its date—it certainly illustrates how methodological recommendations could be combined with advice about bedside manner, about morality, about economic aspects of medical practice, and so on. The author is as anxious about doctors getting a bad name for themselves for ostentation, avarice, and lack of φιλανθρωπίη, as he is about their doing so through faulty medical practice.

Attacks on other healers sometimes involve the exposure of an entire group with whom the author certainly does not identify himself, as when *On the Sacred Disease* sets out to demonstrate the fraudulence of the purifiers. But the accusations of bad practice, even charlatanry, can also be levelled at those whom you recognise—or others do—as your colleagues. Often in the surgical treatises, for example, criticisms of bad surgical practices involve no more than an explanation of the real or assumed harmful effects—whether or not supported by reference to firsthand experience. But sometimes such accusations lead to more general discussions of the principles that should underlie medical practice, and of medical method as a whole. Three treatises which can be used to illustrate both certain shared concerns and certain differences of approach are *On the Art, On Regimen in Acute Diseases*, and *On Ancient Medicine.* I shall begin with the most theoretical of these, or rather with the one that is least secure in its practical medical knowledge, the treatise *On the Art.*

As is well known, this is a sophistic type of demonstration-lecture (ἐπίδειξις) and it is possible that the author was no practitioner himself. But from our point of view, that cannot be held to detract from the value of the evidence the work provides on the type of question raised about medicine as a τέχνη, the kind of challenge that was offered, and the nature of some of the points used (whether or not from inside the 'profession') in its defence. The treatise opens with some general remarks about those who have 'turned the abuse of the τέχναι into a τέχνη in itself', and who are accused by the author of *On the Art* of being nasty, malicious, ignorant individuals and indeed of lacking τέχνη (i.e., of being ἄτεχνοι). He says it is up to others to defend other τέχναι, though chapter 9 seems to suggest a promise by the author himself to deal with some aspects of these. But his chief concern in this λόγος is the defence of ἰητρική.

The question this opening chapter immediately poses is whom the author has in mind. Who was in business to abuse the τέχναι or speak shamefully of them? We have plenty of evidence of Plato's eventually wanting to downgrade some of the τέχναι and he certainly had little good to say about those who practised them as full-time professionals. In *Gorgias* 464b–466a there is a contrast between certain genuine τέχναι (including ἰατρική) and their spurious 'flattering' counterparts, and again between a τέχνη and mere ἐμπειρία [465a]; *Phaedrus* 260e calls rhetoric no τέχνη but an ἄτεχνος τριβή and furthermore contrasts the practice of medicine as a τέχνη with its practice only as a knack and by experience [τριβῇ μόνον καὶ ἐμπειρίᾳ: 270b]. Moreover, in the passage of fundamental importance, *Philebus* 55e–58a grades the τέχναι according to their degree of exactness: medicine, in particular, comes in the lowest category (along with music, farming, navigation, and generalship) and below carpentry, which itself comes below the mathematical τέχναι (subdivided here into the impure, mundane applied branches and the purer, philosophical inquiries).

But apart from the problems of dating (and it is difficult to think that the author of *On the Art* 2, which concerns the relations between words and things, was familiar with Platonic dialogues such as the *Cratylus*), it must be thought unlikely for two main reasons that *On the Art* 1 was directed at Plato. First, the attackers envisaged in this treatise apparently reject the τέχναι wholesale; whereas, in the dialogues cited above at least, Plato is concerned to draw distinctions between superior and inferior τέχναι and indeed rates any τέχνη higher than a mere knack. Second, the author of *On the Art* refers to those who bring shame on the arts and their discoveries and abuse them, and this hardly fits the character of the mainly epistemological objections raised by Plato against the less exact τέχναι.

Is it, then, some group of sophists who 'turn the abuse of the τέχναι into a τέχνη'? So far as our evidence concerning that highly amorphous group goes, I do not believe that anyone can be made to fit the bill exactly. Of course there were those, including Gorgias, who privileged the art of rhetoric and wrote books called Τέχναι dealing with that subject; and in the Platonic dialogue, Gorgias is represented as claiming that rhetoric is more powerful than all the other arts: it is he, not his brother who was a doctor, who persuades people to take their medicine [*Gorg.* 456b]; and he claims that in any assembly or gathering set up to choose a doctor, he would win the contest, thanks to his skill in speaking, and defeat any actual doctor. That would obviously involve the discomfiture, the loss of face, the shame, of any doctor unfortunate enough to be so defeated. Many Hippocratic texts make clear just how upset their authors would be likely to be at the prospect of losing an argument with a layman on a medical matter. Nevertheless the identification breaks down. Although the effect of the Platonic Gorgias' claim would be shaming, the Platonic Gorgias himself says that rhetoric should not be used to detract from the δόξα (reputation) of doctors [*Gorg.* 457b]: rather his case is the positive one, that rhetoric, used justly, should be admired, not the destructive one that the other τέχναι should not be.

Is it, then, that the author of *On the Art* has simply invented the art-abusers he attacks? That too seems a difficult conclusion to draw, not least because of the huge gaps in our evidence for fifth- and fourth-century 'sophists' of one kind or another. That the Hippocratic author overstates his opponents' position, especially perhaps as regards their motivations, may well be the case. But we have enough independent grounds in the Hippocratic Corpus itself that radical challenges to the status of ἰητρική had been mounted, to allow us to say that, in this instance at least, *On the Art* has—not just Aunt Sallys—but genuine opponents, even if we cannot name them. We shall consider the evidence from *On Regimen in Acute Diseases* and *On Ancient Medicine* shortly; but in what are probably later works too, such as *Precepts* 9, there are signs of concern to defend medicine against detractors and to show that it *is* a τέχνη. That may be as far as it is wise for us to go: yet even if the idea of a group of sophists attacking the arts in general were a mere fiction, a figment of this author's imagination, it has one interesting implication nevertheless. Even if he had no such real opponents, he has himself implicitly raised the very issue he accuses them of pressing. They should not abuse the τέχναι nor disparage their discoveries. But this, of course, by implication poses the questions: What discoveries?, and How secure are the claims of the τέχναι to be τέχναι? These issues are most pertinent in a period when claims for new inventions, for founding new inquiries and for innovating in existing ones, were all the rage. Moreover, these issues once raised, this author's defence of the arts in general is both thin and indiscriminate: 'it seems to me that in general there is no τέχνη that does not exist', the gist of his argument being that what can be seen and recognised exists, and what cannot, does not.

What the author has to say on medicine in particular begins in chapter 3, and is marked by a certain self-consciousness in style. Thus, he will begin his demonstration or exposition (ἀπόδειξις) of the τέχνη of medicine with a definition (cf. ὁριεῦμαι). This definition comes in two main parts: (1) the complete removal of the sufferings of the sick, together with the alleviation of the violences of diseases; and (2) the refusal to treat cases 'where the disease has already won the mastery', realising that in such instances medicine is powerless. One cannot help admiring the cunning and bravura of this twofold statement. On the one hand, the first part claims the possibility of complete success and later chapters show that this is no mere opening gesture of the hand. Chapter 9 subdivides diseases into two main groups—broadly those with visible signs and those without—and affirms of the former group that 'in all cases the cures should be infallible, not because they are easy, but because they have been discovered'. (Chapter 10 goes on to say that in the latter group too the τέχνη should not be at a loss.) But the 'heads I win, tails you lose' character of his original definition comes out when he deals with cases where the τέχνη is not successful. There, the disease has already won the mastery and it would be quite unfair to expect the τέχνη to be able to achieve a cure: indeed it is perfectly right for doctors to refuse to treat such cases [cf. ch. 13]. So where there are successes, they are proof of the power of the τέχνη: but no failure, no inability to produce some alleviation, is allowed to count against the art, since he has made it constitutive of medicine, in his definition, that it refuses to deal with hopeless cases.

How far these extravagant claims, with their all-purpose fail-safe clause, were likely to carry conviction—even with an audience at a sophistic ἐπίδειξις—we cannot know. Certainly the actual medical knowledge the author displays is none too impressive. Yet his attempt to distinguish between the products of τέχνη and those of mere chance, τύχη, in chapters 4–6, does score some notable points. Some attribute the successes of medicine merely to chance, he says in chapter 4, and some point out that people have recovered even without calling a physician [ch. 5]. Now while he does not rule out chance—but claims rather that good luck follows good treatment, and bad luck, bad [ch. 4]—he mounts an argument in chapter 6 to suggest that the spontaneous (τὸ αὐτόματον) is a null category. Everything that happens, happens on account of something (διά τι). Even if the patient called in no doctor, his recovering from his complaint was due to *doing* something or again to his *not* doing something [ch. 5], that is, to precisely the means the doctor *would* have used, had he been called in. So the medical art is at work whenever there is a cure, whether or not the doctors are in attendance—though chapter 6 ends with the further point that medicine is real not just because it acts διά τι but also because its results can be foretold.

The author of my second main text, *On Regimen in Acute Diseases*, is altogether more knowledgeable about actual medical practice; but he too shares some of the worries of *On the Art*. The art as a whole, he says in chapter 3, has a very bad name among laymen, to the point that there is thought to be no ἰητρική at all. But here the problems arise not—as in *On the Art*—because of malicious slanderers, but rather because of the incompetence of, and especially the disagreements among, doctors themselves. In chapter 3 he identifies a state of some chaos in the profession on the question of the diet to be prescribed in cases of acute diseases; and he follows this up with the remark that such disagreements are likely to make laymen object that the τέχνη is like divination, where diviners disagree about which birds on the right or left are propitious and on what the signs in the entrails portend.

Now that comparison is interesting for several reasons and chiefly because divination (μαντική) was in most quarters a highly respectable and respected τέχνη. When Prometheus boasts of the benefits he has brought mankind, μαντική figures prominently; and *On Regimen* even cites it is as a prime example of a τέχνη. Clearly, however, for the author of *On Regimen in Acute Diseases*, if medicine were no better than divination, that would not be enough. Yet we may note that one requirement he places upon the doctor is to know features of the patient's condition without being told, that is, to practise 'prognosis', not here predicting the future, but inferring the present, a practice medicine shared with divination (and the terms in which both *Prognostic* 1 and *Epidemics* i 5 speak of the doctor 'telling in advance the present, the past and the future' indicate pretty clearly that they were aware of the parallelism with prophecy).

Once again, as in *On the Art*, the author of *On Regimen in Acute Diseases* is led to explore topics to do with the nature of causation, particularly in relation to the classification of diseases. One of the criticisms he has of the revisers of the *Cnidian Sentences* is that though they were keen to enumerate the types of each kind of disease, their account was incorrect; and he goes on to complain that it is impossible to proceed on the assumption that *any* difference in the symptoms constitutes a different disease—or again that a mere variety in the *name* does so. Again the problems posed by the fact that the same condition may be due to different causes are mentioned in chapter 11. It is a very serious error, for instance, to mistake weakness that is due to the pain of an acute disease for weakness caused by lack of nourishment and to feed the patient. Conversely, it is also an error not to see that the weakness *is* due to such a lack. That is not so dangerous to the patient, but it does lay the doctor open to ridicule far more. Another doctor or even a layman seeing what is the matter would immediately be able to help the patient: that is the type of mistake that leads to practitioners being despised by ordinary folk, while the doctor or layman who comes in and saves the patient is like someone who raises him from the dead.

On Regimen in Acute Diseases does recognize some insecurity in the claims of medicine to be a τέχνη, and positively and constructively relies on some detailed advice about diagnosis and especially the characteristics of different treatments. The author of this treatise frequently criticises his fellow-practitioners, often claiming that current practice and procedures are mistaken: yet most of his recommendations take the form of contrary *assertions*, the plausibility or persuasiveness of which depend entirely on how convincing his appeals to his own experience sound.

On Ancient Medicine—my third text—has, however, much more to say about methodology. This author attacks those who try to base medicine on the new-fangled method of postulates (ὑποθέσεις). You may have to rely on these in dealing with 'meteorology', where it is not clear 'either to the speaker himself or to his audience whether what is said is true or not, since there is no criterion to which one should refer to obtain clear knowledge'. But that will not do in medicine: it would do there only if medicine were no τέχνη at all and just a matter of chance [ch. 1: cf. ch. 12]. But that is not the case, as is shown by the differences between good and bad practitioners. Such argument depends, to be sure, on the successes being agreed and being attributable to the τέχνη of those who achieved them. But on that his chief point is that medicine has a principle (ἀρχή) and a method (ὁδός) which are tried and tested. It has been discovered (cf. εὕρημα, ζήτημα) as the result of systematic inquiry: indeed he believes that one day the whole of medicine will be discovered, using the same method [ch. 8].

The claim is, then, that medicine has a methodology, delivers results, and is not a matter of chance. But the particular argument the author advances, connecting medicine with cookery and dietetics, in part to support a claim that this was how the τέχνη started and to suggest its antiquity, has a potentially embarrassing feature. It is perfectly reasonable, he says in chapter 4, that dietetics is not considered an art, because no one is a layman in it, but all are knowledgeable since all have to use it. But if medicine starts from dietetics, trial and error procedures with the food you eat and what you drink, at what point, precisely, it *becomes* a τέχνη and is not just what everyone knows, ought to be a question for the author to take up: yet it is not one for which he has a very clear answer.

In other respects, however, he not only indicates the type of semiology he suggests the doctor should practise in diagnosis and provides a number of specific analyses of the effects of particular regimens, he also has further points to add on the key question of isolating the causal factors that can be held responsible for a condition. Chapter 21 identifies as a common mistake among doctors as well as laymen the assumption that anything unusual done near the beginning of a complaint was its cause—when it may not be at all—and chapter 19 specifies that 'we must consider the causes of each condition to be those things which are such that, when they are present, the condition necessarily occurs, but when they change to another combination, it ceases'. It is striking that this statement of causal factors that are, as we might say, necessary and sufficient conditions of a disease was propounded in the context of particular problems posed by medical diagnosis.

But this treatise makes an even more important contribution towards the definition of the medical τέχνη in its remarks on the degree of exactness (τὸ ἀτρεκές, τὸ ἀκριβές, and their cognates) which it can attain and which it should be expected to attain. Chapter 9 begins by noting that substituting weaker for stronger foods is far from effective treatment in all cases. The matter is more complex and requires greater exactness. 'One should aim at some measure. But as a measure you will find neither number nor weight by referring to which you will know what is exact, and no other measure than the feeling of the body.' Exactness is difficult to achieve and small errors are bound to occur. Up to a point the subject can be, and has been, made exact; but perfect exactness is unattainable.

> But I assert that the ancient art of medicine should not be rejected as non-existent or not well investigated because it has not attained exactness in every item. Much rather, since, as I think, it has been able to come close to perfect exactness by means of reasoning where before there was great ignorance, its discoveries should be a matter of admiration, as well and truly the result of discovery and not of chance.

While the medical τέχνη has achieved exactness in some areas, in others it has not and it could never be expected to be perfectly exact: in particular, it is positively inappropriate in medicine to attempt to use the measures produced by numbering and weighing. To appreciate the full force of this set of statements we must refer both to medical and to non-medical texts.

As is well known, the date of *On Ancient Medicine* is controversial, as also is the question of who exactly are those whom this treatise criticises for importing ὑποθέσεις into medicine. But it is less controversial, even if not universally agreed, to identify mathematics as one domain in which a method of ὑπόθεσις was developed before Plato: it is, after all, by reference to how geometers proceed that Plato himself explains the method when it is first introduced in *Meno* 86e, even though the term itself is not directly attributed to them. It is possible, though far from certain, that the medical writers attacked in *On Ancient Medicine* 1 were influenced by some mathematical usage; and I conjectured, some time ago, that Philolaus might provide an appropriate bridge between medicine and mathematics. (We know he was interested in both subjects and indeed the type of medical and physiological theories attributed to him, which are based on the hot, correspond broadly to those particularly criticised in *On Ancient Medicine.*)

But whatever we may think about a conjectured mathematical context to the method of ὑπόθεσις attacked in *On Ancient Medicine*, the author's remarks about exactness obviously resist some notion or tendency to turn medicine into an exact inquiry. We need not look far for examples in the Hippocratic Corpus itself of just such a notion or tendency. The whole topic of the periodicities of diseases is an extremely complex one. Different texts set out different schemata, some based on odd/even contrasts (where Pythagorean ideas are often seen in the background) and others using other systems. Some such schemata are explicitly qualified as holding only 'for the most part', even while others carry no such reservation. *Prognostic*, in particular, after proposing an intricate theory of the periodicities of fevers, goes on to remark [ch. 20] that the periods cannot be numbered in whole days exactly—rather they are like the lunar month and the solar year. But the evidence of a substantial group of texts points clearly in the direction of their authors attempting precise, dogmatic theories in this area. Moreover a similar controversy can be traced in regard to the question of exactness in drug prescription, where some Hippocratic texts specify determinate quantities, while others take a line closer to that in *On Ancient Medicine* and insist that the ingredients and dosages have always to be adjusted to particular patients. With its rejection of numbering and weighing, *On Ancient Medicine* takes a stand on an issue about which the medical writers were deeply divided: for some—whether or not they had the model of pure or applied mathematics explicitly in mind—were all for turning medicine into an exact inquiry, while others, as the author of this treatise, resisted that ambition, though insisting nevertheless that medicine is a τέχνη based on experience and reasoning.

Much more could be said, and many more texts adduced from the Hippocratic Corpus, to illustrate and elaborate these themes. But it is time to draw some of the threads of this selective discussion together. Medicine exists in one form or another in every society, whether healing is entrusted to specialists or not, to one group or to more. In the context of fifth- and fourth-century Greece, where so much was being called into question, the foundations of medicine and its status were made the subject of explicit debate, maybe not earlier than other subjects, but among the earliest; and that spirit of challenge itself raises a whole series of problems in a comparative perspective. In the process, as aspects of the pluralism of Greek medicine and its relations with other inquiries became the topic of explicit analysis, there was a heightening of self-consciousness and the construction of not one but several overlapping and competing models of what medicine is or should be.

Many of the issues were sharpened considerably with Plato's insistence on the distinctions among various ἐπιστῆμαι, on the contrast between τέχνη and experience (ἐμπειρία) or mere knack (τριβή), and on the differences between τέχναι that were more or less exact, more or less tied to procedures of measurement. But some of those questions are anticipated in earlier, or made independently in contemporary, medical texts. Could medicine be called a τέχνη at all? Several Hippocratic writers were emphatic that it could, though their ideas differ on how. Its successes are not the results of chance but of reason based on experience, the effects of causal factors that *could* be used by the τέχνη even when they were at work independently; above all, predictions could be made on the basis of reliable signs, and that was a test of the doctor's skill. Some, whether or not aware of the problems that remained if the τέχνη depended solely on the pragmatic criterion, went further and insisted that medicine is an inquiry based on research conducted according to its own methodological principles: its theories should not be based on arbitrary assumptions but should, in principle, be testable. Above all, while some might suppose, and some did suppose, that there was nothing to stop medicine being turned into an exact inquiry, others resisted that ambition and insisted that though not perfectly exact, it was a τέχνη nevertheless.

As remarked, many of these themes become appreciably clearer and more explicit with Plato—not that Plato was happy, in the final analysis, to concede to medicine, as a conjectural art, a very high status at all. With Plato and with Aristotle of the *Posterior Analytics*, the requirement of certainty for knowledge and the securing of incontrovertible conclusions by deduction from self-evident axioms, come into the foreground in philosophical analysis—as they were implicitly to dominate (and to some extent were already dominating) the exact sciences throughout antiquity. But Aristotle's practice in the inquiry concerning nature accommodates more easily than some of his theoretical pronouncements the category of what is true 'for the most part' as well as what is true always; and of course he explicitly allowed differences in the degree of exactness of different disciplines, not just between mathematics and moral philosophy but also between mathematics and physics. Some of these Aristotelian themes may be represented as picking up (though they do not do so explicitly) points already made by Hippocratic writers, while in the next generations the continuing Hellenistic debates on the status and methods of the medical τέχνη were to be one context in which the issues between scepticism and dogmatism were to be fought out. Sophisticated developments were to come: but whatever we may think of the many different ways the name of Hippocrates was conjured with in the Hellenistic medical sects, at least the first beginnings of important topics are to be found in some of the treatises that came to be ascribed to him.

12

Between Data and Demonstration: The *Analytics* and the *Historia animalium*

JAMES G. LENNOX

There is a subset of Aristotle's treatises which we usually refer to as his biology or zoology. Aristotle himself occasionally mentions the investigation of animals and plants, although seldom in a way that marks it off decisively from the study of coming to be and passing away in general [*Meteor.* 339a5–8, 390b19–22; *De part. an.* 644b22–645a10]. When we turn to these works individually and as a group, a number of simple but important questions arise regarding the manner in which the study of animals is partitioned and about the ways in which the different works are related to each other. Of course, we may have ready responses to these questions, but such responses are in part influenced by recent developments in the biological sciences that have little to do with Aristotle or his conception of nature or of science [cf. Balme 1987a, 9–11]. So, to answer these questions in a way that will increase our understanding of Aristotle's science, we need to understand better his aims and methods.

Take the case of the *Historia animalium.* This treatise stands apart from those aimed at offering various explanations for the parts, development, motions, and so on, of animals. Much of the information it contains is duplicated in these other treatises,[1] and there are numerous casual references to it in them. Further, unlike many of Aristotle's works, the title of the *Historia animalium* is used to refer to it within the corpus itself.[2]

---

[1] Le Blond [1945, 19] and Balme [1987a, 13–17] draw precisely opposed conclusions from this fact.

[2] Actually, Aristotle sometimes refers simply to 'the histories' [*De part. an.* 646a11; *De iuv.* 478b1; *De gen. an.* 719a10, 740a23, 746a15] or to 'the natural histories' [*De part. an.* 639a12, 650a31–2; *De incess. an.* 704b10]: the majority

Why, then, are the 'researches concerning animals' distinguished from these other studies? Obviously, there is danger that in answering this question we will obscure Aristotle's own purposes. Accordingly, the aim of this paper is to answer the 'what-is-it' question about the documents Aristotle refers to as 'the animal histories' or sometimes simply as 'natural histories'. I shall argue that attending to Aristotle's wider logical and epistemological vision can help us to understand better his approach to the systematic study of living things.[3]

1. Modern views of the *Historia animalium*

From a post seventeenth-century perspective, there are two standard ways to conceive the function of the *Historia animalium* within a systematic study of the animal kingdom. The first takes it as the classificatory ground work of Aristotelian zoology. Yet so viewed the treatise is hopelessly inadequate [cf. Balme 1987b, 80–85; Pellegrin 1986, 1–12]. The taxonomic vocabulary is restricted to the two terms γένος and εἶδος and these terms refer to groups of animals at all levels of the taxonomic hierarchy. There appears to be no concern for finding or consistently using certain features as classificatory markers in order to provide either a classification which is exhaustive or a hierarchy of *taxa* from widest to narrowest. The second treats the *Historia animalium* as a collection of 'natural histories', that is, as a series of more or less complete descriptive studies of each of the kinds of animals discussed. But from this standpoint the work is even more

of the references in *De part. an.*, however, are to specifically animal histories [660a9, 660b2, 668b30, 674b16–17, 680a1, 684b4-5, 689a18]. It is presumably from these references that some ancient editor (probably Andronicus: cf. Keaney 1963, 57–58) derived the title that has come down to us.

[3] I will focus on the extent to which the theory of finding middles relative to specific problems and the theory of inquiry in *An. post.* ii can give us a purchase on Aristotle's concept of an ἱστορία. In a companion study [Lennox 1987a], I concentrated on the remarks in the *An. post.* about the move from incidental to unqualified understanding, about how locating predications at the commensurately universal level is a crucial part of this move, and about how the *Hist. an.* is consistently concerned to locate such predications. Some of the results of this companion study are directly relevant to this paper's theme, and (in a somewhat developed and modified form) will be presented in section 3 of this paper. I intend to leave open questions of the chronological order of the research reported in, and of the composition of, the various 'zoological' treatises and the *Analytics*. That the *Analytics* may serve to shed light on the biological works, or *vice versa*, is consistent both with the view that the *Analytics* is a result of reflecting on scientific research and explanation done or in progress, and with the view that it is a sort of model for the presentation of such research and explanation.

disappointing [cf. Balme 1987a, 9; 1987b, 85–88]. As David Balme [1987b, 88] has put it, 'To any reader looking for information about given genera or species, the HA seems an incoherent jumble.'

The *Historia animalium* is clearly not organized according to a rigorous taxonomic scheme, nor as a reference work on the various kinds of animals discussed. To expect such an organization is to rely in part on the etymological tie between the name given to modern works of this character and Aristotle's [Pratt 1982]. But since the *Historia animalium* is so disappointing when looked at in these two ways, we must either dismiss it (with an excuse, perhaps) or ask in charity whether we have misunderstood it. It is one of the lasting achievements of David Balme's scholarship to have provided us with the framework for a serious re-evaluation of the *Historia animalium*.

> Yet Aristotle does state his purpose: 'first, to grasp the *differentiae* and attributes that belong to all animals; then to discover their causes' (HA I. 491a9). The HA is a collection and preliminary analysis of the differences between animals. The animals are called in as witnesses to *differentiae*, not in order to be described as animals. [Balme 1987b, 88]

The thesis developed in this paper is fully in the spirit of this reassessment: like Balme [1987b, 80], I shall insist that the *Historia animalium* is a work directed 'toward a methodical apodeixis of living nature'.

2. Pre-demonstrative science

In this section I wish to accomplish two interrelated tasks: to present evidence that Aristotle distinguished a pre-demonstrative yet theoretical scientific inquiry in his philosophy of science; and to show that he was inclined to refer to this pre-demonstrative inquiry as ἱστορία.[4]

[4] Previous theoretical uses of the term do not take one very far. Herodotus [*Hist.* i 1] announces his work as an ἐπίδειξις ἱστορίης, but the context suggests that he means little more than a presentation of reliable information. There is a consistent implication that ἱστορία is a basis for knowledge [cf. *Hist.* i 44, ii 118, 119], though often this basis is a reliable report rather than something Herodotus has directly observed. In *De vet. med.* 21, the author tells us that by ἱστορία he means knowledge of what man is and of the causes of his coming to be. In *Phaedrus* 244c8, Plato conjoins ἱστορία with νοῦς, where it seems to have the force of information acquired on the basis of signs. And at *Phaedo* 96a8–10, Socrates tells us that those who have investigated natural coming-to-be and passing away describe their wisdom as περὶ φύσεως ἱστορία. The wider context suggests that they meant their wisdom to include an understanding of the causes

The *Posterior Analytics* is well advertised by scholars nowadays as the first attempt in the history of philosophy to provide a rigorous theory of explanatory proof. Its first six chapters do characterize scientific understanding of a fact in terms of deductive proof from premises which are true, unmediated, and primary, and which state facts more familiar than, prior to, and causative of the fact stated in the conclusion [71b19–23]. But this advertising has been so successful that the work is often discussed now as if its second book—which announces itself as an extended account of different sorts of inquiry (ζήτησις) and their interrelations—did not exist.[5] The result is a general interpretation of the *Posterior Analytics* which makes it seem oddly out of touch with Aristotle's substantive scientific and philosophical works.[6] The subject of this section, then, is *An. post.* ii and, specifically, its concept of a stage of inquiry aimed at establishing that a predication holds, an inquiry preliminary to investigation of the reason why it holds.

'The things we seek (τὰ ζητούμενα) are equal in number to the things which we understand' [89b23-4]. So opens book 2 of the *Posterior Analytics.* Aristotle claims [89b24–35] to be able to reduce the objects of scientific investigation to four: the fact (τὸ ὅτι), the reason why (τὸ διότι), whether something exists (εἰ ἔστι), and what it is (τί ἐστι). These inquiries are paired, in the following way:

(1) Is it the case that S is P? → Why is it the case that S is P?
(2) Are there Ss (or Ps)? → What are Ss (or Ps)?

where the arrows indicate that the first question in each pair must be answered before the second, as Aristotle's remarks in *An. post.* ii suggest. Like so many of Aristotle's introductory sentences, this apparently straightforward, sensible division of investigations opens up a Pandora's box of difficulties which the rest of the book aims to resolve. In *An. post.* ii, perhaps the most important concern the ways in which the two pairs

of natural things. I draw attention to three points. First, while it is common to translate ἱστορία as 'inquiry' or 'investigation', the term more often designates the report or result of inquiry. Second, if Aristotle is restricting it to the report of a pre-causal inquiry, he is legislating this usage, which is not reflected in the above passages. Third, Herodotus *does* use the term to refer to reports and information which serve as the basis for knowledge, rather than to the state of knowing or the report of knowledge itself. Cf. Louis 1955.

[5] There are a number of recent correctives to this: cf. Ackrill 1981, Bolton 1987, Ferejohn 1982, Lennox 1987a.

[6] Jonathan Barnes' two papers on this subject [1975b, 1981], for example, are limited almost entirely to the theory of demonstration in the *An. post.*

of inquiries, as well as their respective results, mesh with one another. I will, however, overlook most of these problems, since I am here primarily interested in whether this broad picture of types of inquiry has any parallels in Aristotle's various works which record the results of his investigation of animals.

De incessu animalium presents itself as a work concerned with why each of the parts involved in animal locomotion is as it is, and lists a large number of specific causal questions it aims to answer. Toward the end of this list of why-questions, which constitutes most of the first chapter, Aristotle states:

> For that (ὅτι μέν) these things are in fact thus is clear from our inquiry into nature (τῆς ἱστορίας τῆς φυσικῆς), but why (διότι δέ) they are thus we must now examine. [*De inc. an.* 704b9–10: cf. *De part. an.* 646a8–12, *Hist. an.* 491a7–14][7]

It seems probable that the wording of this remark intentionally reflects the distinction between the two stages of the first pair of inquiries (1) listed above. If so, it indicates that the 'natural histories' mentioned are supposed to establish that certain predicative relationships hold true and, thus, that they are a necessary preliminary to the inquiry aimed at establishing the causal basis for these predications.

A similiar distinction is defended as a matter of principle in *De part. an.* i, which is sometimes referred to as 'Aristotle's philosophy of zoology' [Balme 1972, 69; Le Blond 1945, 51–72]. This book begins by distinguishing two sorts of 'proficiency' relevant to a given study: a first order proficiency in understanding the subject-matter, and a second order proficiency in judging whether the study is well presented. The rest of the book is then organized around a series of questions bearing on the second type of proficiency, since

> ... the inquiry about nature (τῆς περὶ φύσιν ἱστορίας), too, must possess certain principles of the kind to which one will refer in appraising the method of demonstration (ἀποδέξεται τὸν τρόπον τῶν

[7] Theophrastus introduces *De causis plantarum* by way of a similar contrast:

> That (ὅτι μέν) the modes of generation of the plants are numerous and how many they are and of what sort was said previously in the histories; but since not all modes of generation occur in all the plants, it is appropriate to distinguish which belong to each plant and through which causes, making use of principles in accord with their proper being. . . .

On the overall relationship between the methods of Aristotle's *Hist. an.* and the *Hist. plant.* by Theophrastus, cf. Gotthelf 1987b in Fortenbaugh 1987.

> δεικνυμένων), apart from the question of how the truth has it, whether thus or otherwise. [*De part. an.* 639a12–15][8]

The second question[9] in the series is,

> Should the natural philosopher, like the mathematicians when they demonstrate (δεικνῦναι) astronomy, first survey the appearances (τὰ φαινόμενα) in regard to the animals and their parts in each case, and only then go on to state the because-of-what (i.e., the causes), or should he proceed in some other way? [*De part. an.* 639b7–10]

The question is answered in the affirmative at 640a14–15: '[natural philosophers are] first to take the appearances in respect of each kind, and only then go on to speak of their causes.' Now the stage of natural inquiry in which one surveys (θεωρεῖν) the appearances regarding a kind before stating their causal explanations is not here described as ἱστορία. Still, the connection between the use of this term in referring to the *Historia animalium* and the sort of survey of the appearances that is discussed in *De part. an.* i 1 can be made more secure in two steps.

First, Aristotle introduces the distinction as familiar from the domain of astronomy. *An. post.* i 13 records that astronomy is one of the sciences which have mathematical and physical aspects, where the latter are called τὰ φαινόμενα [78b39].[10] The general point Aristotle makes about such sciences is that to establish the facts one attends to the appearances, whereas one considers the appropriate mathematical principles in order to demonstrate the reasons why the facts are as they are [79a2–6]. That is, he sees this difference in aspects as an instance of the more general distinction between the two sorts of inquiry given as pair (1) above. Second, *De part. an.* ii 1 opens by noting that 'in the histories' it was made clear from which parts each of the animals is constituted, while the present work will investigate the causes through which each of the animals is so constituted [646a8–12].[11] This would appear to be the same distinction as that found in

[8] All translations from *De part. an.* i are from Balme 1972 unless otherwise indicated.

[9] The first question Aristotle raises is discussed in Lennox 1987a, 114–115.

[10] It is of historical interest that the treatise by Euclid which comes to us under this title is highly mathematical in character.

[11] Though, as Allan Gotthelf has reminded me, after making the distinction Aristotle goes on to say that the investigation will proceed χωρίσαντας καθ' αὑτὰ τῶν ἐν ταῖς ἱστορίαις εἰρημένων [*De part. an.* 646a11–12]. Since *De part. an.* gives us all the data to be explained, and since at least some of it is inconsistent with *Hist. an.*, Gotthelf suggests that this passage instructs us to 'put aside the information reported in HA'. I prefer the sense of Ogle's translation [1912, *ad* 646a12 n1]

De part. an. i 1, with the *Historia animalium* serving as the treatise which reports on the first investigation.

I need not rely entirely on indirect evidence of this sort, however. For there are two passages, closely allied in language, one in the *Prior Analytics* and one in the *Historia animalium* itself, which explicitly describe the pre-demonstrative stage of inquiry as ἱστορία. The first passage insists that, just as demonstrations in astronomy were discovered only after the principles were supplied by astronomical observation, any craft or science has its principles supplied by experience [*An. prior.* 46a17–22].

> So that if the predicates (τὰ ὑπάρχοντα) about each thing have been grasped, we will be well prepared to exhibit their demonstrations (ἀποδείξεις). For if none of the predicates which truly belong to the subjects have been left aside by our inquiry (ἱστορία), we will be able, with respect to everything for which there is a demonstration, to discover the demonstration and carry it out; but of that which in the nature of things has no demonstration, we will be able to make this apparent. [*An. prior.* 46a22–27]

The function of the ἱστορία is to enable one to 'grasp' the predicates which hold of each item in the general subject being investigated. This is apparently intended to explicate the way in which experience with the phenomena of a subject sets the stage for demonstration.

Treating this passage in isolation does not, however, give one a sense of how detailed Aristotle's recommendations in fact are. For this one must turn to the beginning of chapter 27, regarding the proper method to be used in 'picking out' or 'selecting' (ἐκλαμβάνειν, ἐκλέγειν) premises appropriate to the deduction of a given predication [*An. prior.* i 27–29]. Very briefly, the method involves taking as given the subject and predicate of the predication at issue, and developing a list of everything that the predicate belongs to universally as well as a list of all the things that belong universally to the subject.

> For those wishing to establish something of some whole, they must look to the subjects of what is established, that is, the subjects of which it happens to be predicated, and to whatever follows that of which it is to be predicated. For if any of these are the same, the one must belong to the other. [*An. prior.* 43b39–43]

---

which takes Aristotle to distinguish these causal inquiries from the sort of report found in *Hist. an.*

Suppose the predication we wish to prove deductively is that A belongs to every C (i.e., AaC). Aristotle recommends that we generate lists of propositions of the form,

Predicate: A	Subject: C
AaD	**FaC**
AaE	GaC
AaF	HaC

in the hope of finding, as here, a middle term which can 'unite' the terms [cf. *An. prior.* 41a11–13]. For '... no syllogism can establish the attribution of one thing to another unless some middle is taken, which is somehow related to each by predication' [41a2–4]. Chapter 27 is careful to state that this rather algorithmic procedure is only relevant to demonstration in so far as the lists identify other true predications, and it provides a set of rules for identifying predications at the appropriate levels of generality and specificity as well as distinguishing what is in the essence, what is predicated as a property, and what is predicated as an accident [cf. esp. 43b1–32]. That is to say, this is a recipe for organizing information in such a way as to identify middles: it is not a description of how the credentials of the information are established.

Since nothing is said here about how one is to establish the truth of a predication, or about how one is to determine which among a set of universal predicates are predicated in the essence and which are not, this is clearly not a method which will simply allow us to read off demonstrations. When Aristotle concludes his account of selecting premises and division by remarking that 'it is apparent from what things and in what way demonstrations come about and to what sorts of things we should look concerning each problem' [46b38–39], we must take him to mean, I think, that the foregoing method is a necessary condition for the production of demonstrations. Suppose, for example, that in addition to the statements, AaF and FaC (which yield AaC), one's selection also provides FaA; in other words, that according to our divisions, F and A are commensurately universal terms. Thus, we have two syllogisms in Barbara:

$$\begin{array}{c} AaF \\ FaC \\ \hline AaC \end{array} \qquad \begin{array}{c} FaA \\ AaC \\ \hline FaC \end{array}$$

There is nothing here that will help us determine which of two commensurate predicates of a subject is explanatory of which, or even whether there is any explanatory relationship between them at all. This is, of course, the problem which Aristotle raises in *An. post.* i 13 when he distinguishes demonstration that and demonstration why; and it is the problem that he returns to and discusses in detail in *An. post.* 98a35–b24. There, we are told that the predicate which is in the account of the other is the explanatory middle, though nothing is said about how one acquires this knowledge.

The description of ἱστορία as an inquiry establishing which predicates truly belong to which things is consistent with Aristotle's occasional claims that this algorithmic procedure for selecting premises is relevant to demonstration—which one would expect, since demonstration is a species of deduction. The organization of true propositions in this way is presented as facilitating the development of a demonstrative science. And one can see why: it aids in using information imbedded in divisions to identify commensurate universal predications and gives us a 'short list' of candidates for demonstrative middles.

The *Historia animalium* characterizes ἱστορία in terms very similar to those used in *An. prior.* 46a22–27.

> These things have now been said thus in outline to give a taste of the number of things that one must study and how far—we will speak in detail later—so that first we may grasp the *differentiae* present and the attributes in every case. After this, we must attempt to discover the causes of these. For it is natural that the study (μέθοδος) be carried out in this way, when there is an inquiry (ἱστορία) concerning each thing. For about which things (περὶ ὧν) and from which things (ἐξ ὧν) the demonstration (ἀπόδειξις) should be becomes apparent from these. [*Hist. an.* 491a7–14][12]

The technical language of the theory of demonstration in this passage is hard to deny; and equally clear is the distinction between an investigation aimed at establishing the *differentiae* as well as the incidental features of each kind of animal and a search for causes based on this. It is as a result of the first investigation that the elements of demonstrations become apparent. As in the *Prior Analytics* this investigation is called a ἱστορία.

It is time to take stock of our progress thus far. A number of passages from the biology and the *Analytics* agree in detail that there is a distinction to be made between an investigation aimed at establishing that p is the case and one aimed at establishing why p is the case. The former is,

[12] For detailed discussion of this passage, see also Kullmann 1974, 196–202; Lloyd 1979, 137–138 (and n64), 212; Lennox 1987a, 101–102; Gotthelf 1987b; Balme 1987b, 79–80.

apparently, a pre-demonstrative inquiry, that is, an inquiry devoted to organizing empirical information in such a way that the identification of middle terms is facilitated. This suggests that the causal inquiry may not actually be a search for new, more basic entities so much as an inquiry into the causal relationships which hold among the predicates established during the initial inquiry.

Aristotle, as we have seen, is inclined to restrict the range of the term ἱστορία to the first stage of natural inquiry, that is, to a particular sort of pre-demonstrative investigation. The second stage of inquiry is directed toward scientific demonstration.[13] This suggests that the way to understand the distinction between the *Historia animalium* and such works as the *Parts of Animals* or *Generation of Animals* is in terms of Aristotle's own distinction between two stages of inquiry into predications, one involved in grasping that the predication is the case, another involved in establishing the reason why. The reasons for dividing up the investigation of living things as Aristotle does are to be found in his theories of explanation and inquiry in the *Posterior Analytics*.

3. The *Analytics* on 'problems'

These passages, however much they may cohere, are all theoretical in nature, even those in the biology. In order to show that we *must* understand the distinction between the *Historia animalium* and the other biological works in terms of the distinction between factual and causal inquiries, we need to look carefully at the *Historia animalium* to test the claim that it does in fact respect these theoretical ideals. But to perform such a test, we need to know what to look for. Now, to determine whether the *Historia animalium* is an inquiry of the sort described in *An. post.* ii as a ὅτι-investigation, we require some idea of what the report of the results of such an investigation would look like. The key here, as I have argued elsewhere, is to work back from Aristotle's concept of demonstrative understanding. For that concept places constraints on how empirical information is to be organized if it is to be converted by demonstration into science [see also Lennox 1987a].

---

[13] For other passages indicating that biological explanation is to be demonstrative in character, see *De gen. an.* 742b18–743a1; *De part. an.* 639a14, 640a2–9, 645a1–2; *De incess. an.* 704b12–705a2. Gotthelf [1987a, 170-172, 197–198] makes a strong case for a technical sense of ἀπόδειξις in these passages.

What, then, are these constraints? First of all, we must recognize that not just any universally true predication can be the subject of demonstration. Two sorts of predications are distinguished at the end of *An. post.* i 4.

> If, then, a chance case is proved primitively to have two right angles or whatever else, it belongs universally (καθόλου) to this primitively, and the demonstration of this [universal primitive predication] holds universally in itself (καθ' αὑτό... καθόλου); but it holds of the others in some fashion not in itself, nor does it hold universally of the isosceles but further than it. [*An. post.* 73b39–74a3][14]

The use of καθόλου is restrictive and based on a stipulation made in this chapter, viz. a predication is universal if the predicate belongs to the subject 'in every case, and in itself and as such' [73b26-7]. That is, the subject and predicate of the proposition to be proved must be coextensive, and the predicate must belong to the subject as that subject, not incidentally. Notice, for example, that when Aristotle says that the property of having interior angles equal to two right angles does not hold of isosceles universally, he does not mean merely universally, because all isosceles triangles do in fact have this property. His point is that the property is true of other sorts of triangle as well ('extends beyond isosceles'), and so does not belong to the isosceles *qua* isosceles. Rather it belongs to the isosceles triangle *qua* triangle. Thus, it is the proof showing why this property belongs to triangles as such that is basic.

Aristotle does, however, allow for demonstration of the weaker predication, though he insists that the demonstration holds in some weaker fashion. In later passages, it becomes clear that Aristotle has in mind a special class of non-coextensive universal predications, namely, those cases where a predicate belongs coextensively to a kind and, consequently, belongs to all the differentiated forms of that kind. To describe such predications he will occasionally remark that a predicate 'extends beyond [this form], but not beyond its γένος' [see *An. post.* 85b7–15, 96a24–31, 99a18–21 and 24]. In such cases Aristotle admits partial demonstrations [see *An. post.* i 24], meaning, I take it, demonstrations covering a part of the kind.

Thus we come to the first constraint on ἱστορία imposed by Aristotle's theory of demonstration: it must aim for predications in which the predicate is coextensive with its subject. Furthermore, the subject kind must be

[14] Translations of the *An. post.* are by Barnes [1975a], unless otherwise noted. The expansions in square brackets are my own.

differentiated into its immediate sub-kinds or forms if there are to be 'partial demonstrations' asserting that the sub-kind has the feature in question because it is of the kind that has the feature primitively.

An. post. i 5 discusses extensively the types of ignorance that can prevent one from having unqualified rather than sophistical understanding, and each type turns on failure to recognize the primitive level at which a predication holds [cf. 74a25–32]. But in *An. post.* ii 13–18, there is a marked concern with the manner of acquiring predications at the primitive level, and it is to this problem that we shall now turn.

An. post. ii 14 opens with a cryptic statement of method.

> Relative to grasping problems one should select (ἐκλέγειν) from both the dissections and the divisions, and do so by positing the kind that is common to all of them. For example, if the objects of study are animals, select what belongs to every animal; and, having grasped these, once again select what follows (ἕπεσθαι) all the first of the remaining kinds. For example, if this is Bird, select what follows every bird; and in this way always select what follows the proximate kind. For it is clear that we will immediately be able to say why (τὸ διὰ τί) the things which follow belong to those kinds under the common one, for example, why they belong to Human Being or Horse. Let *A* stand for Animal; *B* for the things which follow every animal; and *C*, *D*, and *E* for certain animals. It is quite clear why *B* belongs to *D*, for it is because of *A*; similarly with *C* and *E*. And the same account always applies in the case of subordinate kinds. [*An. post.* 98a1–12; my trans.]

The recurrence of the process of singling out certain features that 'follow' indicates that *An. prior.* i 27–30 is the formal background for this chapter [cf. Barnes 1975a, 239–240; Lennox 1987a, 97–99]. Further, that one must select from divisions suggests (as does *An. prior.* i 31) that division is at best a preliminary stage of the method described here.[15] Clearly, it is from divisions already made that one selects, at each level of generality, what belongs universally; and equally clear is the fact that this division presupposes a division of the kind, animal. So it looks as if *An. post.* ii 14 describes a procedure for using information imbedded in divisions to produce propositions of the sort required for a demonstrative science.

[15] Cf. Alexander, *In an. prior.* i 333.19ff. It is worth noting that Diogenes Laertius, *Vitae* v 25 lists a *Dissections* in eight books and one book of *Selections from the Dissections*; regrettably neither work survives, and to my knowledge there is nothing in the doxographical tradition which even hints at the form they might have taken.

This procedure directs attention to predications at the level of commensurate universality. If, among the things which follow, that is, which belong to every S, one finds a feature also belonging to T, the natural question to ask is whether S and T are both forms of some kind K which has that property primitively. (Or, if one is already aware that S is a K, one would note that the property in question belongs not just to S but to K in general.) So if, for example, one finds that having a heart belongs to birds in virtue of the fact that hearts belong to all blooded animals, the next step is to ask why hearts belong to all blooded animals. In fact, one would not really understand why birds have hearts until this more basic question is answered: saying that birds have hearts because they are blooded animals means that they have hearts for the same reason all blooded animals do.[16] Notice that the grasp of the problem one ends up with in the above

---

16 My remarks on this problem owe a great deal to Allan Gotthelf's contribution to the Symposium on Classification and Explanation in Aristotle's Biology at the APA Pacific Division Meetings in 1986.

Partial demonstrations raise two central questions, one having to do with their form and the other with their explanatory force. I suggest that the form of such explanations is that the middle term will refer to the kind to which the referent of the minor term belongs as a sub-kind [see also Lennox 1987a]. For example,

Having a heart belongs to all blooded animals
Being blooded belongs to all birds
―――――――――――――――――――
Having a heart belongs to all birds

The virtues of this model are three in number. It appears to underlie numerous passages in the *An. post.* [e.g., 73a16–20, 74a1–3, 74a25–32, 85b4–15]; it is a pattern of reasoning that is found regularly in the biology [cf. Lennox 1987a, 108-110]; and the major premise is the sort of proposition one would expect to find as the conclusion of the more basic explanations of commensurately universal predications, thereby giving a means of logical transition between universal and partial demonstrations. This is clearly the way in which Philoponus [*In an. post.* 417.26–28] reads the passage: 'Since these things follow animal, you will prove that perception or motion belong to humans and the rest through the middle, animal.'

But this reading also has certain drawbacks. The discussion of demonstration and its relationship to definition in *An. post.* ii 8–10 generally treats the middle term as giving an account of the minor term, and that seems quite unlike the model just suggested. There, progress in understanding comes through acquiring more basic middle terms of the same logical sort, thus giving us better accounts of the minor term. It is not clear where this leaves us, however, for that discussion is not concerned with the move from knowing that a sub-kind has a feature to knowing that it has the feature in virtue of being the kind of thing that has the feature in itself.

This brings us to the issue of the explanatory force of partial demonstrations. Professor Gotthelf has urged that the appearance of the γένος in the position of the middle term may simply be shorthand for saying 'the sub-kind has the

passage has the same form as that achieved by the move from incidental to unqualified understanding in *An. post.* i 4, 5. Aristotle's method, then, is intended to identify the widest kind to which a predicate selected from a division belongs. Once this has been done, that predicate will immediately show itself as belonging to immediate forms of that kind: the subject in question will have this feature just because it is (a form of) the kind to which the feature belongs universally.

The continuation of *An. post.* ii 14 also echoes concerns of i 5. In the latter it was noted that while there once were distinct proofs that proportionals alternate in the case of numbers, lines, solids, and times, now it is proved universally in a single demonstration. The original failure to see the universal demonstration was because 'all these things... do not constitute a single named item and differ in sort (εἶδος) from one another'.[17] The lack of a name to signify the universal contributed to the mathematicians' failing to see that 'it did not belong to things as lines or as numbers, but as this which they suppose to belong universally' [*An. post.* 74a23–25].

An. post. ii 14 says more concerning what it is to 'grasp problems'.

> Now at present we argue in terms of the common names that have been handed down; but we must not only inquire in these cases, but also if anything else has been seen to belong in common, we must extract that and then inquire what it follows and what follows it... [98a13–16]

Searching for 'what follows that which belongs in common and what follows it' is the method recommended in *An. prior.* i 28 for finding a middle term that is relative to a problem, and the example which follows, an example familiar from *De part. an.* 674a23–b18,[18] clarifies the strategy:

> having a third stomach and not having incisors <follow> having horns; again, <we should inquire> what having horns follows. For

feature for the very reason that the kind does' [cf. Ackrill 1981, 380], and there are passages which do suggest this gloss [e.g., *An. post.* 91a2–5].

At the very least, Gotthelf has convinced me that formulations in Lennox 1987a and earlier drafts of the present work which suggest that partial demonstrations (which I term A-type explanations in Lennox 1987a) could be demonstrative prior to understanding why the more primitive predication holds are wrong, for the reasons given in the text above.

[17] A scholium to Euclid, *Elements* v attributes the discovery of the general theory of proportion to Eudoxus [cf. Heiberg and Stamatis 1977, i 213.1–12; Heath 1956, ii 112–113].

[18] See also the careful analysis of this passage in Gotthelf 1987a, 179–185.

> it is clear why what we have mentioned will belong to them, for it will belong because they have horns. [*An. post.* 98a16–19]

Here two *differentiae* are noted (one negative, incidentally) which follow 'the possession of horns'. Thus, given

(1) having a third stomach belongs to every horned animal
(2) lacking upper incisors belongs to every horned animal,

we are asked to inquire, to what does 'the possession of horns' belong universally?, that is, to find a value for P such that

(3) horns belong to every P.

For, given this, we may then infer that

(4) having a third stomach belongs to every P.

As Aristotle says, having a third stomach belongs due to the possession of horns, the possession of horns being the middle through which the link between the possession of a third stomach or the absence of a second row of teeth and the third item, P, is established.

The choice of this example in *An. post.* ii 14 is interesting. Aristotle needs a case that clearly goes beyond the common nomenclature, for that is the point he is making. Thus, his example requires the use of specialized descriptive phrases[19] to refer to items predicated of one another. Apparently, Aristotle wishes to emphasize that this predication is established by realizing that the possession of these other two features follows from the possession of horns, and that they will thus belong to anything which possesses horns because that thing is horned.[20] The final lines of *An. post.* ii 14 extend the method even further, and likewise use an example familiar from the biological works: 'Again, another way is by excerpting in virtue of analogy; for you cannot get one identical thing which both pounce and spine

[19] Not true names, as Allan Gotthelf reminds me [cf. Balme 1962, 90]. The *An. post.* 93b29–32 allows for definitions of 'name-like phrases', as does *Top.* 102a1–5. While I suspect that terms in the biology such as τὰ κερατοφόρα, τὰ ὀστρακόδερμα, τὰ ζωοτόκα τῶν τετραπόδων, are among the sort of name-like phrase he has in mind, we need to know more about the significance of 'name-like' here, specifically, about how name-like terms differ from terms which cannot be defined [see Bolton 1985, for some suggestions].

[20] See *De part. an.* 663b35–664a3, 674b7–17, where the production of horns (for self-defense) leaves less earthy material for teeth (accounting for the lack of two rows of teeth) and, thus, indirectly necessitates the possession of more stomachs (for the digestion of the poorly chewed nutrients). In this way, an animal's possession of horns accounts for these features.

and bone should be called; but there will be things that follow them too, as though there were some single nature of this sort [*An. post.* 98a20–23].'

Let us try to reconstruct the steps in the process here sketched. The three sorts of 'skeletal' parts referred to are related by analogy. But there may be predicates within the divisions being used which belong to all of them (as if they had a single nature). Thus, 'excerpting in virtue of analogy' means searching 'the dissections and divisions' for *differentiae* common to subjects related by analogy. At *De part. an.* 653b33–36, Aristotle says that 'among those animals having bones, the nature of the bones, being hard, has been devised for the sake of the preservation of the soft parts; and in those not having bones the analogue <has been devised for this>, for example among some of the fish, fish-spine, among others, cartiladge.' Accordingly, *An. post.* 98a20–23 may propose that certain passive capacities, e.g., hardness or brittleness, belong to each of these analogous parts in virtue of a common function that each plays in the life of its respective kind, or in virtue of a common material nature (since all these parts are earthen). Again, it might also mean that all three analogous hard parts are associated with soft flesh and viscera, an association which would naturally suggest the idea that the analogous parts play an identical functional role in their respective kind's life [so Barnes 1975a, 240].

Here, then, is a characterization of methods for achieving predications of the sort that prepare the investigator for acquiring understanding through causal explanation. The methodology is clearly an application of the more formal methods of *An. prior.* i 27–31 to the specific goals of the scientific investigator, which should be no surprise given that the *Analytics* is introduced as an investigation of demonstration and demonstrative understanding [*An. prior.* 24a10–11, 25b26–31].

An. post. ii 15–17 explore the complexities involved in finding a causal account relative to a pre-established problem, where a problem is here essentially a why-question asked of accepted facts: That P belongs to all the Ss is clear; why then does it belong? A botanical example will allow us to tie the concerns of *An. post.* ii 14 to the second stages of each pair of the investigations with which we began, the whether-it-is/what-it-is pair and the that-it-is/why-it-is pair.

In *An. post.* ii 8–10, Aristotle finally come to grips with the question of the relation between definition and demonstration within a science. In the process, he has much to say about the way in which inquiries διὰ τί and inquiries τί ἐστι are related to one another. The interpretation of these chapters is controversial [see Bolton 1976, 1985, 1987; Sorabji 1980; Ackrill 1981]. But certain features of the debate may be lifted from the fray for present purposes. At least in cases relevantly similar to the examples in these chapters, the middle term in a demonstration of why some predicate belongs to some subject will also serve to account for what the predicate is [*An. post.* 93a3–5, 93b3–14, 94a1–10, 95a16–21, 99a3–4, 99a25–9]. An investigation of why those occasional noises in the clouds occur—an investigation based on our awareness that they do—is completed when we have grasped the most fundamental causal explanation of those noises. Aristotle claims that not only is 'quenching of fire' a candidate for the middle which accounts for the occurrence of thunder in the clouds, but that it is also a possible answer to the question, What is thunder? Thus, the familiar account of thunder as a certain characteristic noise in the clouds is underwritten by a more basic account—more basic in that it serves to explain the perceptually familiar features by which we became acquainted with thunder initially [see Bolton 1976].

Can one take this view of the way the results of these two inquiries converge in the case of the sorts of facts biologists wish to explain? Aristotle seems to think so.

> The middle term is an account of the first extreme, and thus all the sciences come about through definition. For example, loss of leaves follows the vine while exceeding it, and follows the fig while exceeding it; but it does not exceed all, but is equal in extent with them. Now if one takes the primary middle term, it is an account of shedding leaves. For there will be a first middle in the other direction, that all are such; then a middle of this, that sap coagulates or some other such thing. But what is shedding of leaves? It is the coagulation of the sap at the connection of the seed pod. [*An. post.* 99a21–29; my trans.]

At the outset we see once more the language of 'following' (i.e., belonging to all), and the idea of a predicate which both follows and exceeds two kinds of plant while being coextensive with (one must suppose) all the kinds that shed their leaves. Aristotle is not making the less than startling point that all the kinds which shed their leaves shed their leaves; rather, he is indicating the necessity of identifying the group of plants with *differentiae* coextensive with this one if we are going to account for it scientifically. As a matter of fact, in the previous chapter at 98b4–21, Aristotle identified just such a group by noting a feature common to them all and coextensive with the shedding of leaves, namely, being broadleafed (πλατύφυλλον). The fact that these *differentiae* (being broadleafed, shedding leaves) are coextensive is noted [98a35–b3] within the context of the question, Does the causal basis of something have to be coextensive with that of which it is the cause? The question is then raised (as it has been ever since of attempts to describe explanation in purely extensional terms) as to whether either of the coextensive terms can be used to prove the other—there being no question that one can construct a valid and sound syllogism in Barbara with either term in the middle position.[21]

Throughout this discussion Aristotle assumes that being broadleafed is the cause of any plant's losing its leaves, and the sample explanation we are given 'demonstrates' that vines lose their leaves because they are broadleafed [*An. post.* 98b5-16]. In the language used elsewhere, this is a 'partial' or 'incidental' demonstration, given that the 'problem' being explained predicates loss of leaves to one of the sub-kinds which loses its leaves. *An. post.* ii 16 closes by correcting the impression that this is a primitive scientific explanation.

> Or if problems are always universal, must the explanation (τὸ αἴτιον) be some whole and what it is explanatory of universal? E.g. shedding leaves is determined to some whole, even if it has sorts, and <it belongs> to these universally (either plants or plants of such and such a sort); hence in these cases the middle term and what it is explanatory of must be equal and convert. E.g. why do trees shed their leaves? Well, if it is because of solidification of their moisture, then if a tree sheds its leaves solidification must belong to it, and if solidification belongs—not to anything whatever but to a tree—<it must> shed its leaves. [*An. post.* 98b32–38]

Problems need to be universalized, in the sense, as Ross notes, of *An. post.* i 4. But when this is done, the kind that took the middle position in the partial demonstration is now that to which shedding leaves belongs, the subject of the predication to be explained.

We are now in a position to make sense of *An. post.* ii 17. Aristotle tells us that if one takes the primary middle term, it is an account of shedding leaves: 'for first there will be a middle in the other direction, that all are such; then a middle of this, that sap coagulates or some other such thing' [99a25–28]. Two middle terms are mentioned here, only one of which is identified as an account of shedding leaves. The other is called a 'middle in the other direction'. Suppose that this is the property of being broadleafed. Being broadleafed serves as a middle in the direction of the various forms of plants which shed their leaves—'Shedding leaves belongs to all the vines because they are broadleafed.' But there will now be a primary middle for this, where 'this' indicates the proposition which predicates shedding leaves

---

21 Cf. *An. post.* 78a28–b13. These are, of course, the passages at the basis of the distinction between *demonstratio quia* and *demonstratio propter quid.*

of being broadleafed.[22] Thus, only when one has elevated problems (or why-questions) to the level of the commensurate or primitive universal does the middle term also become an account of what the predicated property is. Here is one further way in which the methods outlined by Aristotle are important in setting the stage for a demonstrative understanding of a subject. These methods move us to the stage where further exploration can aim for the primitive definitions that may serve as the starting-points of our explanations.

4. The *Historia animalium* as pre-demonstrative science

In section 2, I reviewed the evidence that the *Historia animalium* was offered by its author neither as a report of a systematic taxonomy of the animal kingdom nor as a series of natural histories of them, but as a rendering of the true propositions currently known about animals for the purpose of causal demonstration. The theory of problems and the methodology for working with them that are presented in the later chapters of *An. post.* ii and discussed in section 3, indicate that Aristotle would have something quite specific in mind, when he came to organize information into propositional form for scientific purposes. He would, for example, make use of information imbedded in divisions. This would mean that the terms he was working with would refer to *differentiae* which were ordered so as to reveal how a general feature could be specified or determined (e.g., wing→feathered wing *v.* membranous wing *v.* dermatous wing). Thus, Aristotle would seek to identify the coextensive relationships among *differentiae* from different divisions, that is, to identify groups all of which and only which had certain *differentiae*.[23] This process would have to begin by identifying universal predications of a given subject, and what the subject was itself universally predicated of. But doing this would lead to the recognition of coextensive or 'primitive' universal predications; for example, that wing follows bird but not *vice versa*, and that feathered-wing follows bird and bird follows feathered-wing. Aristotle would not be concerned to stick

[22] Broadly speaking, this interpretation of the passage has defenders from Philoponus [*In an. post.* 429.32–430.7] to Ross [1949, 671].

[23] On the developments in Aristotle's theory of division and its role in the biology, see Balme 1987b, 74–89.

with the popularly designated kinds.[24] If he identified a feature belonging invariably to all the animals with some other feature, he would want to see what these features followed and what followed them—even to the extent of seeking features predicated of all of a group of analogically related features. In doing this, he would prepare the ground for the sort of causal understanding that he regarded as the goal of science.

We have seen that the *Historia animalium* aims to 'grasp the *differentiae* and the attributes which belong to all the animals', since, after this is done, one can try to discover their causes. In this treatise, Aristotle maintains that this is the appropriate way to proceed on the grounds that, if the ἱστορία has been carried out properly, one should be able to distinguish the things from which demonstration proceeds from the things about which we

[24] I am here passing over a set of very difficult questions about what Aristotle would call a γένος and why. In a classic article, Balme [1962] first distinguished *differentia*-classes and true kinds. The former lack actual nominal identifications and collect animals for convenience on the basis of some shared feature or other. Pellegrin [1982; 1985, 103–106, 112; 1987, 334–337] argues that this is in fact typical of the way Aristotle identifies kinds—indeed Pellegrin goes further and argues that typically γένη are *parts* in Aristotle's biology, that his biology is more properly said to be a *moriology* than a zoology. (For my reservations on this score, see Lennox 1984, which is a review of Pellegrin 1982.) And Allan Gotthelf [1985,1987a] has raised a number of questions related to this topic in the process of his work on the concepts of substance, essence, and definition in *De part. an.* Two points from our discussion of the *Analytics* are relevant to this issue. First, Aristotle clearly allows for definitions of things which have name-like phrases along with those having actual names. Of course, this merely pushes the question back one step to the question, What will count as a name-like phrase, i.e., How does Aristotle go about screening out those 'names' of merely accidental unities? Second, at least part of *An. post.* ii 14 presses us to admit into our scientific vocabulary phrases which are certainly not names (or even name-like), and provides us with a rationale for doing so. These phrases are similar in character to those Aristotle uses to designate previously unnamed μέγιστα γένη in the biological works: translations usually mask this fact, 'oviparous quadruped' typically rendering a phrase which might be translated literally as 'the ones among the four-footed that lay eggs'. Indeed, Aristotle will list the two *differentiae* in either order (though I do not mean to imply that he does so randomly).

This is clearly an issue that requires fresh and thorough re-investigation. At this stage, I am prepared to say that Aristotle is aware of the need to use certain mechanisms to extend the vocabulary of science in the direction of identifying unnamed groups that are unified in some way or other, that a syllogistic model of the logic underlying science requires that most of its terms be expressible as either subjects or predicates and be from diverse categories, and that Aristotle's zoological terminology is dominated by phrases that identify groups of animals as 'the ones that are (or have, or do) X, where X is a peculiar feature of those animals. These are among the 'phenomena' which an account of Aristotle's theory of scientific γένη will have to explain.

want demonstrations. I have argued that Aristotle's *An. post.* ii and a number of related texts provide us with the theoretical background for viewing ἱστορία as a pre-demonstrative preparation for causal explanation—just the sort of study that the *Historia animalium* introduces itself as.[25]

We are now in a position to ask whether the *Historia animalium* is a work which aims to organize information found in divisions in a way that is preliminary to demonstration as Aristotle understood it. Research carried out in collaboration with Professor Gotthelf and with just this question in mind indicates that in fact it does, though this should not be taken to imply that this is all it does or that it reflects the workings of a mind mechanically following a set of formal rules.[26] But before exploring one passage in detail, I should like to draw attention to the support for this contention that is provided by the work's overall organization.

As the important transition near the end of *Hist. an.* i 6 indicates, the first six chapters of book 1 are in some sense introductory. At least five theoretical preliminaries are addressed. (1) We are introduced to the distinction between parts that are uniform (flesh, bone), simple and non-uniform (eye, finger), and complex and non-uniform (head, limb). (2) Aristotle then distinguishes sameness in form, sameness in kind, and sameness by analogy, as they apply to animals, to the parts of animals, and to the degree of sameness and difference exhibited by animals and their parts. Special attention is paid to the ways in which things differ when they are the same in kind but not in form [see Lennox 1987b, 352–353; Pellegrin 1987, 331–336]. (3) Having introduced these distinctions in the context of parts alone, Aristotle next says that animals are differentiated according to their lives, activities, dispositions, and parts.[27] (4) The use of these ideas in the study of animals is then clarified by a series of examples organized under the categories mentioned in (3), differences of the first three kinds being discussed down to 488b28 and differences among parts from there to the beginning of chapter 6. (5) Finally, Aristotle establishes a number of extensive kinds (μέγιστα γένη), that is, kinds embracing a significant variety of forms sufficiently alike to be treated together.

---

[25] The extent to which elements of these ideas are reflected in Aristotle's zoology is explored in Kullmann 1974; Gotthelf 1987a, 1987b; Lennox 1987a; Bolton 1987.

[26] For other products of this collaboration, see Gotthelf 1987b and Lennox 1987a. Gotthelf 1987b points out ways in which this research is consistent with the late David Balme's most recent work on the *Hist. an.*, work that was in progress as part of the preparation of a new edition of this text as well as a translation and commentary.

[27] See Gotthelf 1987a, 192–193.

Aristotle introduces (3) and (4) by writing, 'the differences among the animals are with respect to their lives, their activities, their dispositions and their parts, about which let us first speak in outline (τύπῳ), while later we will speak attending (ἐπιστήσαντες) to each kind' [*Hist. an.* 487a11–13]. As we saw, he later refers back to this outline as a preliminary sketch that is intended merely to give us a flavor of what is to come. One difference between this outline and the remainder of *Historia animalium* is that the later account will be about each kind (περὶ ἕκαστον γένος). We will see the force of this contrast shortly.

Now the preliminary material that Aristotle mentions in *Hist. an.* 487a11–13 clearly draws on divisions. Let me simply quote two brief passages which can stand for dozens of a similar character in these chapters.

> ... some of these animals are water dwellers, others are land dwellers; water dwellers are of two sorts: some live and feed in water, take in and expel water, and are unable to live if deprived of it (e.g., many of the fishes); others take nourishment and pass time in the water, yet do not take in or expel water and give birth out of water. Many of these are also footed, such as the otter, beaver, and crocodile; others are winged, such as the diver and the grebe; and still others are without feet, such as the watersnake. [487a15–23]

> Among fliers, some are feather-winged (for example, the eagle and hawk), some membranous-winged (e.g., the bee and the cockchafer), and some are dermatous-winged (for example, the flying fox and the bat). [490a5–8]

A number of features of these passages are relevant to our earlier discussion. First, even at the most abstract level, we begin with the assumption of four broad categories of *differentiae*. No kind of animal can be adequately characterized without a study of the life it leads in its environment, the activities it performs (locomotive, generative, perceptive, nutritive), its dispositional differences (Is it gregarious or a loner, timid or brazen, predator or prey?), and its parts. Further divisions are indicated under each category: thus, under manner of life, water-dweller/land-dweller; under water-dweller, those which do/do not take water in or generate in water;[28] under winged, feathery, dermatous, membranous. Specific kinds of animal are presented to illustrate the differences mentioned: these kinds are not

[28] This contrast is dealt with in much more detail and more systematically at the beginning of Aristotle's discussion of differences in manner of life: see *Hist. an.* 589a10–590a18.

themselves the subjects of the division. This does not mean that the universe of division is *differentiae*, however. It is often said explictly, and it is nearly always implicit in the method used, that we are first to identify *animals* by a common feature (e.g., all the ones with wings rather than all the wings) and then to divide according to the way in which that common feature is differentiated [see n23 above]. Notice what strange bedfellows this method produces: at the common level, birds, bats, and bees are united as winged. Moreover, depending on which general *differentia* is chosen, animals will be grouped and re-grouped, to use David Balme's phrase. Such a methodology will indeed perplex the reader bent on taxonomy.

One must be careful about what one views as a division in these texts. In the first of the pair of passages just translated, Aristotle notes that some of the 'partial' water-dwellers are footed, whereas others are winged and others footless. Is this a further subdivision of this group? If so, since he appears to insert a division by locomotive organs into a division according to mode of life, Aristotle would seem to violate a basic rule of division, namely, that one should never divide by something accidental to the axis of division.

In fact, however, sketching out divisions is only a part of what one finds in these pages. Aristotle is also concerned to correlate animals grouped and divided according to one sort of *differentia* with those grouped according to others. He indicates that animals sharing one mode of life are diverse when viewed from the standpoint of locomotion. In a similar vein, he adds that among land-animals all those with lungs inhale and exhale air [487a29–31]; that all insects live and find their food on land [487a31–32]; that no creature which inhales and lives in water finds its food on land, though some that inhale and live on land find their food in the water [487b1–2]; that all animals have a mouth, stomach, the sense of touch (and an unnamed organ for same), and a life-sustaining liquid (with container) [488b29–31, 489a17–19, 489a20–24]; that all animals with stomachs have bladders, though not every animal with a bladder has a stomach [489a3–6]—and so on. The correlations are occasionally disjunctive (animals with feathered or membranous wings have either two feet or none [490a10–12]), and occasionally conjunctive (the feathered and the dermatous winged flyers are all blooded [490a9–12]). Such material, then, combines a sketch of how one should lay out divisions under various broad categories of animal differences with sample identifications of positive and negative correlations between groups in different divisions.

Aristotle himself constantly points out that he is here merely giving us a sense of the method and the sorts of differences that occur under the four categories he is discussing, and that the more systematic study to follow will need to center these methods on each kind (περὶ ἕκαστον γένος). Thus, the last step (5) in this introductory stretch of text, the articulation of nine 'extensive kinds' (which are themselves broadly grouped according to whether they have blood or its analogue) is very important.

None of this preliminary material is mere window dressing. The overall structure of *Historia animalium* owes much to the principles articulated and discussed in these chapters. At the broadest level, the entire work is organized around the four categories of *differentiae*: a study of parts [i 7–iv 7], activities and lives [v–viii], and dispositions [ix]. Within the study of the parts, the investigation of the blooded animals [i 7–iii 22] is distinguished from that of bloodless animals [iv 1–7]. And the investigation of the parts of the blooded animals is divided into an account of the external non-uniform parts [i 7–ii 14], the internal non-uniform parts [ii 15–17], the genitalia which are not always clearly internal or external [iii 1], and, finally, the uniform parts [iii 2–22]. In the bloodless animals, parts external in one group are often internal in another; and this may be at least part of the reason why Aristotle investigates the internal and external parts together in each kind before moving on to the next.

We have seen that part of the more systematic nature of this work will involve its studying animal *differentiae* 'concerning each kind'. What role do the μέγιστα γένη play in the way in which the information in the *Historia animalium* is presented? The study of the external non-uniform parts of the blooded animals moves from man to viviparous quadrupeds, through those which are biped in one respect and quadruped in another (apes and baboons) to the oviparous quadrupeds, birds, fish, and serpents.[29] Yet parts which extend (at some level of description) across more than one extensive kind are said to do so when they are first introduced; consequently, the later in the discussion a kind comes the less tends to be written about it. The study of the bloodless kinds is organized similarly. In both cases, numerous groups are noted which either do not fit into these extensive kinds at all or fit into one of them in one respect but not in another. On the other hand, the investigation of the internal non-uniform parts and the uniform parts of the blooded animals is organized, not kind by kind, but part by part.

[29] Serpents are denied the status of μέγιστον γένος, but they are discussed at length. The cetacea, however, are listed as one of the blooded extensive kinds [*Hist. an.* 490b9] and yet are not discussed in the review of external parts. Given their extreme peculiarity, which Aristotle stresses elsewhere [*Hist. an.* 588a31–b2], this seems doubly odd: cetacea are mentioned in books 2 and 3, but only by way of contrast. The most extensive account of the cetacea occurs in the sections of *Hist. an.* dealing with *differentiae* of activity [reproduction, 566b2–27; respiration, 589a27ff.] and with dispositions [631a9–b4].

This may reflect Aristotle's belief that viscera do not differ as radically as external features from kind to kind, and so may be considered across the entire blooded clan.

The method of these passages can be seen clearly in the following discussion of the lungs and related organs.

> As many animals as are quadruped and viviparous, all have an esophagus and windpipe,[30] placed in the same way as in humans; the placement is similar among the quadrupeds which are oviparous, and among the birds; but these kinds differ in the forms of these parts. Generally, all those which taking in air inhale and exhale, have a lung, a windpipe, and an esophagus, and the position of the windpipe and esophagus is similar, but these organs are not the same in all, since the lung is neither alike in all nor similar in position. [*Hist. an.* 505b32–506a5]

Aristotle goes on to note that not all blooded animals have lungs, and identifies those that do not (e.g., fish and any other animal with gills) [506a11–12].[31] Differences in these three organs are regularly referred to during the discussion of the other groups mentioned here [507a11–12, a24–27; 508a17–21, 32–33; 508b30–509a16].

This passage first establishes correlations between three distinct extensive kinds and three organic parts, and goes on to record a more general correlation between the animals that breathe and these three organs.[32] This is a move to the identification of the animals that have these organs as such, that is, to the 'primitive universal'. Such a move is achieved by uniting the kinds with these features by means of another common *differentia*, their breathing. Aristotle then discusses the differentiation of these organs under two headings, position and 'similarity'. Throughout the entire group the

---

30 The *Hist. an.* makes frequent use of 'doubly quantified' expressions of the form ὅσα ἐστὶ X, πάντα ἔχει Y. Gotthelf [1987b] suggests that the preponderance of this otherwise rare form of expression may signal Aristotle's concern in *Hist. an.* to identify primitively universal predications.

31 Notice that this provides a means of identifying animals without lungs while leaving the extension of the group open-ended. No reason is here given as to why animals with gills will not possess lungs, though one is provided in *De iuv.* 476a6–15.

32 Earlier, in his review of the parts found in humans, Aristotle [*Hist. an.* 495a18–22] states that 'the so-called esophagus (so named due to its length and narrowness) and the windpipe are within the neck; but the windpipe is positioned in front of the esophagus in all those animals which have it—and all have it which also have a lung.' Indeed, the entire passage, 495a18–495b23, is taken for granted by the discussion in book 2.

windpipe and esophagus are alike in position, though both differ both in their 'affective' qualities and their quantitative dimensions from kind to kind. The lung, on the other hand, differs in all these respects from kind to kind.

What is not said is of equal interest—for instance, the functional relationships among these organs, the reasons why all breathers have all three, the reasons for the differences in their character and position, and the reason why animals with gills have none of them. But the level of generality at which the search for such explanations should proceed is made progressively clearer, as the coextensive *differentiae* at that level (including the activities of inhaling and exhaling) are noted.

Aristotle's procedure in discussing the lungs has a number of features which recur to a greater or lesser degree throughout his description of the viscera of the blooded animals:

(1) the specification of an organ's pervasiveness among blooded kinds,

(2) the identification of coextensive organic structures,

(3) the attempt to identify the entire class with the feature in a unified rather than a conjunctive manner,

(4) in combination with (3) an emphasis on the diversity in the forms of the parts in the variety of kinds which share them,

(5) the distinction between the qualitative and the positional differences among the groups with regard to these organs,[33]

(6) the correlation of those differences with identified kinds of animals, and

(7) the identification of features coextensive with the differences primarily under consideration.

These features call to mind the ideas discussed in section 3 above, regarding the organization of information in a manner suitable for demonstration. That this was Aristotle's intent is evident if one compares this discussion with those of the same structures in the *Parts of Animals* ii–iv. Does Aristotle there offer explanations at the level of primitive universality that

[33] In his discussion of the ways in which the parts can differ, Aristotle [*Hist. an.* 486a25- 487a1] makes a broad distinction between the ways in which parts can vary by 'excess and defect' (i.e., in degree, which includes variations in the quality, size or number of a structure), by analogy, and by the position of the part. This passage uses that distinction carefully, though the study of the relative positions of the windpipe and esophagus and of the differences in form of the lung at *Hist. an.* i 16 [see n30] is taken for granted and not elaborated.

he has here identified, and by reference to the activities and parts at the same level? It seems clear that he does. For he begins by noting that only animals which have windpipes and esophaguses have a neck, since the neck is simply a device for their protection [*De part. an.* 664a12–17: cf. *Hist. an.* 495a18–20]. He also remarks that the windpipe exists for the sake of breathing, since it is through this that the air passes on its way to and from the lung [*De part. an.* 664a16–20: compare the unexplained universal that all animals that have a lung have a windpipe at *Hist an.* 495a20–22; and the equally unexplained claim that all animals which breathe have all three organs at 506a1–5]. Moreover, he asserts that the esophagus is not required for nutritional reasons (witness that fish get along without one). It is rather a by-product of the fact that those animals with lungs must have a windpipe of some length. The presence of a windpipe in turn produces a certain distance between mouth and stomach; consequently, there must be an organ to connect them. This explains why all and only breathing animals have an esophagus, an organ that seems to have little to do with breathing. [Compare *De part. an.* 664a20–31 with the purely descriptive discussion at *Hist. an.* 495a22–30]. Next in this chapter Aristotle reviews the relative placement of the windpipe and esophagus [cf. *Hist. an.* i 16, ii 15]. He points out that having the windpipe in front of the esophagus seems less than optimal, since food must pass over the windpipe when such animals eat. Such organisms must have a means of closing the windpipe when eating: in vivipara, this is accomplished by the epiglottis; in ovipara, by a windpipe that can open and close at the top. By contrast with this rich explanatory discussion, the *Historia animalium* is content simply to describe these organs and their locations: it never says that they are necessary or what they are for.

The discussion of lungs in *Parts of Animals* iii 6 presents an array of interesting difficulties. I shall focus only on its conclusion.

> Generally, then, the lung is for the sake of breathing, while it is also bloodless for the sake of a certain kind of animal. But what is common to animals with lungs is without a name, that is, unlike 'bird' which names things in a certain kind. Wherefore, just as the being for a bird (τὸ ὀρνίθῳ εἶναι) is constituted from something, the possession of a lung likewise belongs in the being (οὐσία) of these. [*De part. an.* 669b8–12][34]

[34] See Gotthelf 1985, 31 and Balme 1962, 90. As Gotthelf points out, while *De part. an.* iii 6 explains why certain animals possess a lung, it concludes by saying that having a lung is in their being. This raises a host of questions about the

Aristotle begins here by alluding to a teleological explanation for the possession of a lung, and for its possession in a different form in a sub-kind.[35] Not only do lungs belong to all the animals which breathe; they belong for the sake of breathing. The discussion leading up to this passage, in fact, shows concretely how closely tied together are the explanation of why certain animals have a certain organ and the account of what that organ is. Not only does understanding why animals breathe explain why they must have a lung; it also provides us with an account of what the lung is.[36]

The remainder of these concluding lines is puzzling at first. But, reminding ourselves of the following points may help to remove some of the problems. First, recall that the lung is neither limited to one of the extensive kinds identified by Aristotle, nor does it extend to all the blooded kinds. And yet, as we have seen, there is a complex network of anatomy and physiology related to breathing and the possession of lungs. Apparently, the 'universal' common to all these animals has not been named (unlike 'bird', which picks out those feather-winged, beaked, two-legged creatures). But that, we must remember, should not stop us from seeking scientific understanding: 'we must not only inquire in cases where there is a common name, but also if anything else has been seen to belong in common, we must extract that and then inquire what it follows and what follows it' [*An. post.* 98a14–16]. The basic account for lung, windpipe, esophagus, and neck must show why all the animals which have them do in fact have them. The studies of these interconnected organs and the animals that possess them in *Historia animalium* and *Parts of Animals* and the relationships between these studies, display the methodological concerns of the *An. post.* ii. There appears to be a common conception of the activity and aims of scientific inquiry underlying Aristotle's science and his theory of science.

5. *Hist. an.* iv 1–7: A case study

In order to display the structure and aims of the *Historia animalium* concretely, I will conclude with a study of Aristotle's account of the parts of

sorts of feature to be specified in the account of an animal's being, issues which lie beyond my present concerns.

[35] I emphasize that this is merely an allusion to such an explanation, since all of *De part. an.* iii 6 is devoted to this task, a task which must involve a physiological account of breathing that can handle the fact that not all blooded animals breathe (by which Aristotle meant 'take in and expel environmental air'), and that some water animals do: see Lennox 1987a, 110–111 for a brief discussion.

[36] See Gotthelf 1987b and Lennox 1987a, 109–114.

the bloodless organisms. Aristotle identifies four 'extensive kinds' among bloodless animals: τὰ μαλάκια, τὰ μαλακόστρακα, τὰ ὀστρακόδερμα, and τὰ ἔντομα. These groups correspond roughly to our cephalopods, crustacea, testacea, and insects, and in due course I shall use these terms. But it is important to emphasize a number of points about these Greek names. First, when they appear alone they are always in the plural.[37] Moreover, all are derived from vividly descriptive adjectives—literally, these kinds are the softies, the soft earthenwares, the earthenware-skinned, the divisible; and Aristotle's initial differentiation of them remains close to this basic sense. Aristotle distinguishes these kinds on the basis of whether their hard parts are inside or outside (or all the way through!), and the nature of that hardness. The cephalopods, if they have a hard part, have it inside; the crustacea have a hard but crushable exterior, a soft interior; testacea have a hard fragmentable exterior and soft interior; insects are uniformly hard throughout. This descriptive terminology is introduced in *Hist. an.* 490b10–16, with an occasional remark that suggests there are no common names for these groups as such. But throughout the *Historia animalium* these terms consistently identify kinds with a variety of forms, forms possessing the general features of the kind differentiated 'in degree'.

The parts of the bloodless kinds are discussed in *Hist. an.* iv 1–7, beginning with cephalopods. The discussion opens in a manner typical of the entire work with an account of the external parts.

> The following are the external parts of the so-called softies: first, the so-called feet; second, the head which is continuous with these; third, the sac, which contains the internal parts and which some erroneously call the head; and again the fin which encircles the sac. But it so happens that the head is between the feet and the belly in all the cephalopods. Now all have eight feet, and all have a double row of suckers, except in one kind (γένος) of the octopuses. But, distinctively (ἰδίᾳ), the cuttlefish and the small and large calamary have two long tentacles, with rough tips and two rows of suckers.... [*Hist. an.* 523b21–31]

Those external *differentiae* that can be predicated in general of the cephalopods are given first. Aristotle then remarks on a peculiar feature common to the cuttlefish and calamary, and goes on [524a3–19] to discuss features peculiar to octopuses as a group. Next, he describes differentiations among

[37] The importance of this point was brought home to me in discussion with Allan Gotthelf.

cuttlefish and calamary, and then among octopuses [524a20–32]. After describing the head, the eyes, and the mouth (with its two teeth and tongue-analogue), he moves down the esophagus and discusses the internal parts: features common to the cephalopods [524b1–22], a hard part peculiar to the cuttlefish and calamary [524b22–23] which is nonetheless differentiated (διαφέρει δέ), the sepia of the cuttlefish being harder, bonier, and flatter than the calamary's firmer, thinner, more cartilaginous 'pen' [524b22–28]. The octopuses as a group have no hard external part. Aristotle then discusses sexually related differences at various levels and, finally, certain features which distinguish a number of kinds of octopus [524a14–28].

Likewise, Aristotle begins his account of crustacea with 'Now common to all these is, first...' [*Hist. an.* 526b21], and adds 'But now the distinctive *differentiae* must be studied with respect to each kind' [526b34–527a1: τὰς δ' ἰδίας διαφορὰς καθ' ἕκαστον δεῖ θεωρεῖν]. The insects are introduced as a kind with many forms [531b21], and two groups are identified which have numerous forms closely akin to one another but which are not bound together by a common name [531b22–23: οὐκ ἐπέζευκται κοινὸν ὄνομα οὐδέν]: bees, hornets, wasps, and the like [cf. 623b23], and those insects with encased wings, which Aristotle refers to as κολεόπτερα [cf. 552b30, 601a3]. Then, as in his study of the cephalopods, Aristotle turns to features common to all insects: to the articulation of the body into head, thorax, and abdomen [531b26–28]; that they all live when divided [531b30–532a5]; and that all have eyes [532a5]. Yet only some have stings [532a14–17] and wings [532a19–22]; and among these latter some have encased wings, while others do not; but all their wings are membranous, lacking the stock and divisions of feathers. In sum, as he says, '... the parts of all the animals, both external and internal, both those peculiar to and those common to each kind, belong in this manner' [532a27–29].

In this passage we see a mind striving to identify the widest group of animals to which a feature belongs universally. But 'feature' is ambiguous. How widely one predicates a feature often depends on how generally or specifically it is described. No one for whom division was a scientific tool will forget this—all the calamaries and cuttlefish have a hard structure; so if one wishes to understand why they do (and the octopus does not), this is the predication that is crucial. But if one wishes to understand why the cuttlefish has a sepium rather than a pen, that hard feature must be described and identified more specifically. Aristotle's method throughout the texts we have surveyed is tailor-made to achieve these explanatory goals, that is, to provide propositional descriptions of the animal world that meet his strictures on proper explanation.

There are, however, important differences between the chapters devoted to the parts of the bloodless animals and the earlier discussion of the parts of the blooded ones. The differences related to the order in which the internal and external parts are surveyed has already been mentioned. In addition, there are very few attempts to identify features which extend beyond a given extensive kind. And the universal predications are primarily of the form which correlate a *differentia* of some sort with a named group, rather than with another *differentia*. Thus, these chapters are virtually devoid of the 'as-many-as-are-X,-all-(most, some, or none)-have Y' form of proposition that is found with such regularity in the earlier books. Attending to these sorts of differences among the various discussions that make up the whole of *Historia animalium* is not likely to invalidate the claims I have been making, but it will in all likelihood lead to a yet richer and more complex picture of this great work.

An independent test of the soundness of this view of the *Historia animalium* is to consider its ability to treat naturally those features which are anomalies on other accounts. As David Balme [1987a, 9] has stressed, one such anomaly for anyone who reads the *Historia animalium* either as systematics or as natural history, is that various animals are mentioned regularly, but only in order to point out some oddity. Balme's favorite example is the blind mole-rat: we are told on a number of occasions of its peculiar, rudimentary, subcutaneous eyes, though, for all else we are told it could also have wings, scales, gills, and ten feet [cf. 491b27–34]. Similarly, there is a variety of octopus, the ἐλεδώνη, which Aristotle mentions only once and only to tell us that it has a single (rather than a double) row of suckers on its narrow tentacles. Such selectivity in Aristotle's treatment is to be expected if the majority of the features of such animals are in fact more appropriately discussed as features of the wider kind of which they are one example. As we have seen, the features common to all octopuses are predicated of octopus, while those shared by all cephalopods are discussed at this more general level. Only that peculiarity of the ἐλεδώνη, the single row of suckers on its narrow tentacles, is termed an ἴδιον [525a16–19]. From the perspective of the theory of explanation in the *Posterior Analytics*, this feature will be explained, if at all, in terms of other features peculiar to the ἐλεδώνη. And as a matter of fact, the discussion of the cephalopods in *De part. an.* iv does just that.

> The other cephalopods have two rows of suckers, but one kind (γένος) of octopus has only one. The cause (αἴτιον) of this is the length and slimness of their nature; for their being narrow necessitates a single row. Now they have these things arranged thus not because it is best (ὡς βέλτιστον) but because it is necessary due to

> the peculiar account of their being (ὡς ἀναγκαῖον διὰ τὸν ἴδιον λόγον τῆς οὐσίας). [*De part. an.* 685b12–16]

Both the property identified as the cause and the property identified as the effect in the *De partibus animalium* are said in the *Historia animalium* to belong to the ἐλεδώνη alone; but there is not the slightest hint of a causal relationship between these features in the latter. The ἐλεδώνη has a long and slim nature and, thus, long and narrow tentacles. Accordingly, it must have a single row of suckers. It is not that having one row of suckers is better than having two. If, however, a case could be made out that either one or two rows were equally possible and that one of these possibilities were better, Aristotle might be inclined to say of the ἐλεδώνη that its possessing one row is better than its having two rows, on the ground that 'we see that nature does nothing pointless, but always the best for each being among the possibilities'. But, in this case, he takes the possession of one row as necessitated by the antecedently given nature of animal.[38]

6. Conclusion

The methodological unity of the chapters in the *Historia animalium* which are devoted to the parts of the bloodless animals is clearly not one imposed by the aim of a hierarchical classification. There is, for example, no effort to introduce a vocabulary for *taxon*-categories of different extension. Γένος is the all-purpose word for animal-kinds at any level of generality. The cephalopods as a whole are a 'kind' [523b3], but so are the large calamaries [524a29], and there are many 'kinds' of octopus [525a13]. The crustacea are a 'kind', but so are the crabs and carids; and there are many kinds of each of these [525a33–b1]. The testacea are a 'kind', but so are the snails, oysters [528b11–12], and the sea-urchins [528a2]. Finally, animals which do not fall into any of the extensive kinds, such as sea-anemones, are kinds as well [531a31]. The inclusion of this last 'kind' indicates another fact about the *Historia animalium* which points to its lack of interest in systematics—the untroubled recognition of kinds which do not fit into the wider kinds Aristotle has identified.

---

[38] See Gotthelf 1985, 41–42 for a discussion of this passage and its relationship to the patterns of explanation outlined in *De part. an.* 640a33–b1. Gotthelf stresses the apparent difficulty for strong functionalist readings of Aristotle's biology posed by Aristotle's inclusion of such features as the dimensions of a part in the account of the being of an animal.

Finally, at the most general level, this is a work organized as a study of the *differentiae* which belong to animals; it is not presented as a classification of animals. This explains the fact that information regarding any particular kind of animal is sprinkled, in a seemingly haphazard manner, throughout the nine books of this treatise. But the appearance of happenstance is in the eyes of the modern biological beholder. It is in general a well-ordered treatise—given its (declared) demonstrative aims, aims that are revealed in a variety ways. For there is a persistent concern to identify groups which share a number of features as a group and yet have not been identified as a single unit, i.e., that have not been united by a common name. Identifying such groups, I have suggested, is an activity fostered by the scientific ideals of the *Posterior Analytics*. Moreover, there is a recurrent effort to find the widest group to which a given organ or tissue belongs universally; to note how these organs or tissues are differentiated qualitatively, quantitatively, and positionally, in different subgroups; and to identify the widest group which possesses the various differentiated forms of the general type of organ or tissue being discussed. Again, this is what *An. post.* ii 14–18 would lead us to expect. (It would be interesting to pursue the philosophical issue of whether there are compelling reasons for preferring the (Linnean) methodology of establishing a hierarchy of kinds on the basis of a single diagnostic character, and the historical question of whether certain individuals—Baron Cuvier comes to mind—were so impressed with Aristotle because of their own tendency to approach the biological realm with an eye to understanding differentiation rather than with the aim of organizing its species hierarchically.)

Aristotle's guiding question in his zoology seems to be, Why do all and only these animals have this feature? His answer seems to require starting with the *differentiae* and asking how widely a given *differentia* extends in relation to others—that is, he seeks to identify groups relative to some difference and not to identify the difference relative to a pre-established group. This method succeeds in identifying animals with commensurately universal *differentiae*, the first step toward causal accounts in the explanatory model proposed in the *Posterior Analytics*.

The distinction between ὅτι- and διότι-inquiries in *An. post.* ii 1 and 2 is general and theoretical. I hope that my study of Aristotle's method in the *Historia animalium* and of its relationships to its companion studies of animals has given reasons for thinking that his zoological treatises took this distinction seriously. At the same time, I also hope that my remarks have deepened our understanding of ὅτι-investigations beyond the facile notion that they 'collect the facts'.

Acknowledgement. Earlier drafts of this paper benefited from comments on three occasions. At the conference on Aristotle's Biology held by Cambridge and Trenton State Universities (July 1985), the comments of David Balme, Alan Code, Geoffrey Lloyd, Allan Gotthelf, Martha Nussbaum, and Malcolm Schofield were especially useful. At the APA Pacific Division's Symposium, 'Classification and Explanation in Aristotle's Biology' (March 1986), the contributions of David Charles and Allan Gotthelf helped to clarify many issues relevant to the topic of this paper, and pointed out a number of weaknesses in its predecessor. During the conference, 'The Interaction of Science and Philosophy in Fifth and Fourth Century Greece', held by the Institute for Research in Classical Philosophy and Science (May 30–June 1, 1986), Charles Kahn, the late Joan Kung, Geoffrey Lloyd (yet again), Father Joseph Owens, and Robert Turnbull offered helpful suggestions. In addition Michael Ferejohn and Geoffrey Lloyd provided helpful written comments. This paper, as with all my work on Aristotle's biology, is pervasively indebted to discussions on the subject with Allan Gotthelf.

BIBLIOGRAPHY

Aaboe, A. and Price, D. J. de Solla. 1964. 'Qualitative Measurement in Antiquity: The Derivation of Accurate Parameters from Crude but Crucial Observations'. Pp. 1–20 in *Mélanges Alexandre Koyré. Tome 2: L'Aventure de l'esprit*. Paris: Hermann.

Achilles. *Isagoga excerpta*. See Maass 1898, 25–75.

Ackrill, J. L. 1981. 'Aristotle's Theory of Definition: Some Questions on *Posterior Analytics* ii 8–10'. See Berti 1981, 359–384.

Adler, M. J. 1978. 'Aristotle's Conception of Practical Truth and the Consequences of that Conception'. In *Paideia*. Second Special Issue. Brockport: State University of New York Press.

Aëtius. *De placitis reliquiae*. See Diels 1879.

Alexander of Aphrodisias. *In Aristotelis analyticorum priorum librum i commentarium*. M. Wallies 1883. ed. Commentaria in Aristotelem Graeca 2.1. Berlin: Reimer.

Annas, J. 1976. *Aristotle's Metaphysics Books* M *and* N. Oxford: Clarendon Press.

Anton, J. P. 1980. ed. *Science and the Sciences in Plato*. Delmar, NY: Caravan Books.

Apollonius of Perga. *Opera*. See Heiberg 1891–1893, Heath 1896.

Archytas. Fragment 1. See Bowen 1982, 79–83.

Archimedes. *Opera*. See Heiberg 1910–1915, Heath 1897, Dijksterhuis 1956.

Aristarchus. *De magnitudinibus et distantiis solis et lunae*. See Heath 1913, 352–410.

Aristides Quintilianus. *De musica*. See Winnington-Ingram 1963.

Aristotle. *Analytica priora et posteriora*. See Ross 1964.

__________ *De anima*. See Ross 1965.

__________ *De sensu*. See Hett 1964.

__________ *Metaphysica*. See Jaeger 1957.

__________ *Physica*. See Ross 1936, Wicksteed and Cornford 1929–1934.

__________ *Analytica posteriora*. See Ross 1949.

__________ *De generatione animalium*. See Balme 1972.

__________ *De partibus animalium*. See Balme 1972.

__________ *Ethica Eudemia*. See Woods 1982.

__________ *Ethica Nicomachea*. See Ross 1908–1952, ix.

__________ *Physica*. See Hardie and Gaye 1930, Wicksteed and Cornford 1929–1934.

__________ *Topica*. See Ross 1908–1952, i.

[Aristotle] *De audibilibus*. See Düring 1932, 67.24–77.18; Hett 1955.

[Aristotle] *Problemata*. See Hett 1953–1957, Barker 1984–1989.

Aristoxenus. *Harmonica elementa*. See Meibom 1652, Macran 1902, Da Rios 1954.

Artmann, B. 1985. 'Über voreuklidische "Elemente," deren Autor Proportionen vermied' *Archive for History of Exact Sciences* 33:291–306.

Aujac, G. 1979. ed. and trans. *Autolycos de Pitane: La Sphère en mouvement, Levers et couchers héliaques, Testimonia*. Paris: Les Belles Lettres.

__________ 1984. 'La Langue formulaire dans la géométrie grecque' *Revue d'histoire des sciences et leurs applications* 37.2:97–109.

Autolycus. *De sphaera quae movetur liber*. See Mogenet 1950, Aujac 1979.

Baillet, J. 1892. ed. *Le Papyrus mathématique d'Akhmîm*. Mémoires . . . de la mission archéologique française au Caire 9.1. Paris.

Balme, D. M. 1962. 'GENOS and EIDOS in Aristotle's Biology' *Classical Quarterly* NS 12:81–98.

__________ 1972. *Aristotle's* De Partibus Animalium I *and* De Generatione Animalium I. Clarendon Aristotle Series. Oxford: Clarendon Press.

__________ 1987a. 'The Place of Biology in Aristotle's Philosophy'. See Gotthelf and Lennox 1987, 9–20.

__________ 1987b. 'Aristotle's Use of Division and Differentia'. See Gotthelf and Lennox 1987, 69–89.

Barbera, A. 1984. 'Placing *Sectio Canonis* in Historical and Philosophical Contexts' *Journal of Hellenic Studies* 104:157–161.

Barker, A. D. 1977. 'Music and Mathematics: Theophrastus against the Number-Theorists' *Proceedings of the Cambridge Philological Society* 203 (NS 23): 1–15.

__________ 1978a. 'Οἱ καλούμενοι ἁρμονικοί: The Predecesors of Aristoxenus' *Proceedings of the Cambridge Philological Society* 204(NS 24):1–21.

__________ 1978b. 'Σύμφωνοι ἀριθμοί: A Note on *Republic* 531C1–4' *Classical Philology* 78:337–342.

__________ 1981. 'The Euclidean *Sectio Canonis*' *Journal of Hellenic Studies* 101: 1–16.

__________ 1984. 'Aristoxenus' Theorems and the Foundations of Harmonic Science' *Ancient Philosophy* 4:23–64.

__________ 1984–1989. ed.. and trans. *Greek Musical Writings*. 2 vols. Cambridge: Cambridge University Press.

———— 1989 (forthcoming). 'Archito di Taranto e l'armonica pitagorica'.

Barnes, J. 1969. 'Aristotle's Theory of Demonstration' *Phronesis* 14:123–152. Revised in Barnes, Schofield, and Sorabji 1975.

———— 1975a. *Aristotle's Posterior Analytics.* Clarendon Aristotle Series. Oxford: Clarendon Press.

———— 1975b. 'Aristotle's Theory of Demonstration'. See Barnes, Schofield, and Sorabji 1975, 65–87.

———— 1976. 'Aristotle, Menaechmus, and Circular Proof' *Classical Quarterly* NS 26:278–292.

———— 1981. 'Proof and the Syllogism'. See Berti 1981, 17–59.

Barnes, J., Schofield, M., and Sorabji, R. R. K. 1975. edd. *Articles on Aristotle. 1. Science.* London: Duckworth.

Becker, O. 1959. 'Die Archai in der griechischen Mathematik' *Archiv für Begriffsgeschichte* 4:210–226.

Bekker, I. 1831–1870. ed. *Aristotelis opera.* 5 vols. Berlin: Reimer.

Bélis, A. 1982. 'Les "Nuances" dans le traité d'harmonique d'Aristoxène de Tarente' *Revue des études grecques* 450–451:54–73.

———— 1986. *Aristoxène de Tarente et Aristote: Le traité d'harmonique.* Paris: Klincksieck.

Berggren, J. L. 1976–1977. 'Spurious Theorems in Archimedes' *Equilibria of Planes*, Book I' *Archive for History of Exact Sciences* 16:87–103.

———— 1984. 'History of Greek Mathematics: A Survey of Recent Research' *Historia Mathematica* 11:394–410.

Berggren, J. L. and Goldstein, B. R. 1987. edd. *From Ancient Omens to Statistical Mechanics: Essays on the Exact Sciences Presented to Asger Aaboe.* Acta historica scientiarum naturalium et medicinalium edidit bibliotheca universitatis Hauniensis 39. Copenhagen: Munkegaard.

Berggren, J. L. and Thomas, R. S. D. 1992 (in press). *Euclid's* Phaenomena*: A Translation and Study of a Hellenistic Treatise in Spherical Astronomy.* IRCPS: Sources in the History and Philosophy of Classical Science 2. London/New York: Garland Publishing.

Berka, K. 1980. 'Was There an Eleatic Background to Pre-Euclidean Mathematics?' See Hintikka *et al.* 1981, 125–131.

Berti, E. 1981. ed. *Aristotle on Science: The* Posterior Analytics. Padova: Editrice Antenore.

Bishir, J. W. and Drewes, D. W. 1970. *Mathematics in the Behavioral and Social Sciences.* New York: Harcourt, Brace & World.

Boethius. *De institutione musica.* See Friedlein 1867.

Bolton, R. 1976. 'Essentialism and Semantic Theory in Aristotle: *Posterior Analytics* ii.7–10' *Philosophical Review* 85:514–544.

———— 1985. 'Aristotle on the Signification of Names'. Pp. 153–162 in *Language and Reality in Greek Philosophy.* Athens: Greek Philosophical Society.

__________1987. 'Definition and Scientific Method in Aristotle's *Posterior Analytics* and *Generation of Animals*'. See Gotthelf and Lennox 1987, 120–166.

Bonitz, H. 1870. *Index Aristotelicus.* See Bekker 1831–1870, v.

Bourgey, L. 1953. *Observation et experiénce chez les médecins de la collection hippocratique.* Paris: Vrin.

Bourgey, L. and Jouanna, J. 1975. edd. *La Collection hippocratique et son rôle dans l'histoire de la médecine.* Leiden: Brill.

Bowen, A. C. 1982. 'The Foundation of Early Pythagorean Harmonic Science: Archytas, Fragment 1' *Ancient Philosophy* 2:79–104.

__________1983. 'Menaechmus *versus* the Platonists: Two Theories of Science in the Early Academy' *Ancient Philosophy* 3:12–29.

__________1984. rev. Szabó 1978. *Historia Mathematica* 11:335–345.

__________1985. rev. Neugebauer 1983. *Archaeoastronomy* 9:187–194.

__________1992 (forthcoming). *The Ancient Greek Science of Music: A Preface to Platonic Dialectic.* Toronto/Buffalo: University of Toronto Press.

Bowen, A. C. and Bowen, W. R. 1991 (forthcoming). 'The Translator as Interpreter: Euclid's *Sectio canonis* and Ptolemy's *Harmonica* in the Latin Tradition'. See Maniates 1991.

Bowen, A. C. and Goldstein, B. R. 1983. See Goldstein and Bowen 1983.

__________1991. 'Hipparchus' Treatment of Early Greek Astronomy: The Case of Eudoxus and the Length of Daytime' *Proceedings of the American Philosophical Society* 135:233–254.

Brown, N. O. 1953. trans. *Hesiod, Theogony.* Indianapolis: Bobbs Merrill Press.

Bruins, E. M. 1964. ed. *Codex Constantinopolitanus.* 3 vols. Leiden: Brill.

Brumbaugh, R. S. 1976. 'Plato's Relation to the Arts and Crafts'. Pp. 40–52 in W. H. Werkmeister 1976. ed. *Facets of Plato's Philosophy. Phronesis* Supplemental Volume 2. Assen: Van Gorcum.

Burkert, W. 1959. 'ΣΤΟΙΧΕΙΟΝ: Eine semasiologische Studie' *Philologus* 103:167–197.

__________1972. *Lore and Science in Ancient Pythagoreanism.* E. L. Minar, Jr. trans. Cambridge, MA: Harvard University Press.

Burnet, J. 1900–1907. ed. *Platonis opera.* 5 vols. Oxford: Clarendon Press.

Burnyeat, M. F. 1981. 'Aristotle on Understanding Knowledge'. See Berti 1981, 97–139.

Chace, A. B. and Manning, H. P. 1927–1929. edd. *The Rhind Mathematical Papyrus.* 2 vols. Oberlin, OH: Mathematical Association of America.

Code, A. 1976. 'The Persistence of Aristotelian Matter' *Philosophical Studies* 19:357–367.

Cohen, S. M. 1981. 'Proper Differentiae, the Unity of Definition, and Aristotle's Essentialism' *The New Scholasticism* 55:229–240.

Corcoran, J. 1972. 'Completeness of an Ancient Logic' *Journal of Symbolic Logic* 37:696–702.

———— 1974. ed. *Ancient Logic and Its Modern Interpretations.* Dordrecht/Boston: Reidel.

Cornford, F. M. 1951. *The Republic of Plato.* Oxford: Clarendon Press.

Corpus Hippocraticum

De aeribus aquis locis/On Airs, Waters, Places. See Heiberg 1927; Jones 1923–1931, i.

De arte/On the Art. See Heiberg 1927; Jones 1923–1931, ii.

De decenti habitu/Decorum. See Heiberg 1927; Jones 1923–1931, ii.

De morbo sacro/On the Sacred Disease. See Grensemann 1968; Jones 1923–1931, ii.

De mulierum morbis i/*On the Diseases of Women* i. See Littré 1839–1961, viii.

De vetere medicina/On Ancient Medicine. See Heiberg 1927; Jones 1923–1931, i.

De victu/On Regimen. See Littré 1839–1961, vi; Jones 1923–1931, iv.

De victus ratione in morbis acutis/On Regimen in Acute Diseases. See Littré 1839–1961, ii; Jones 1923–1931, ii.

Epidemiae i/*Epidemics* i. See Littré 1839–1961, ii; Jones 1923–1931, i.

Jusjurandum/The Oath. See Heiberg 1927; Jones 1923–1931, i.

Praeceptiones/Precepts. See Heiberg 1927; Jones 1923–1931, i.

Prognosticum/Prognostic. See Littré 1839–1961, ii; Jones 1923–1931, ii.

Da Rios, R. 1954. *Aristoxeni elementa harmonica.* Rome: Academia Lynceorum.

De Falco, V. and Krause, M. 1966. edd. and trans. *Hypsicles: Die Aufgangzeite der Gestirne.* Abhandlungen der Akademie der Wissenschaft in Göttingen, Philologisch-historische Klasse. Ser. 3:№ 62. Göttingen: Vandenhoeck und Ruprecht.

Dedekind, R. 1901. *Essays on the Theory of Numbers. A Translation of* Stetigkeit und die Irrationale Zahlen *and* Was sind und was sollen die Zahlen. W. W. Beman trans. Chicago: Open Court. Repr. New York: Dover, 1963.

Defares, J. G. and Sneddon, I. N. 1964. *An Introduction to the Mathematics of Medicine and Biology.* Amsterdam: North Holland Publishing Company.

Demosthenes. *In Neaeram.* See Rennie 1931.

Dicks, D. R. 1959. 'Thales' *Classical Quarterly* NS 9:294–309.

———— 1966. 'Solstices, Equinoxes, and the Presocratics' *Journal of Hellenic Studies* 86:26–40.

Diels, H. 1879. ed. *Doxographi graeci.* Berlin: Reimer.

Diels, H. and Kranz, W. 1951–1952. edd. *Die Fragmente der Vorsokratiker.* 3 vols. 6th edn. Zürich: Weidmann.

Dijksterhuis, E. J. 1929–1930. *De Elementen van Euclides.* 2 vols. Groningen: Noordhoff.

__________ 1956. *Archimedes*. Copenhagen: Munksgaard. Repr. Princeton, NJ: Princeton University Press, 1987.

Diller, H. 1952. 'Hippokratische Medizin und attische Philosophie' *Hermes* 80:385–409.

Diogenes Laertius. *Vitae philosophorum*. See Hicks 1925.

Diophantus of Alexandria. *Opera*. See Tannery 1893–1895, Heath 1910, Sesiano 1982, Rashed 1984.

Düring, I. 1930. *Die Harmonielehre des Klaudios Ptolemaios*. Göteborgs Högskolas Årsskrift 36.1. Göteborg: Elanders Boktryckeri Aktiebolag.

__________ 1932. *Porphyrios Kommentar zur Harmonielehre des Ptolemaios*. Göteborgs Högskolas Årsskrift 38.2. Göteborg: Elanders Boktryckeri Aktiebolag.

Düring, I. and Owen, G. E. L. 1960. edd. *Aristotle and Plato in the Mid-Fourth Century*. Studia Graeca et Latina Gotoborgensia 11. Göteborg: Elanders Boktryckeri Aktiebolag.

Edelstein, L. 1967. *Ancient Medicine*. O. and C. L. Temkin edd. Baltimore: Johns Hopkins.

Einarson, B. and De Lacy, P. H. 1967. edd. *Plutarch's Moralia*. xiv. Loeb Classical Library. London: Heinemann.

Ernout, A. and Meillet, A. 1951. *Dictionnaire étymologique de la langue latine, histoire des mots*. 3rd edn. Paris: Klinckseick.

Euclid. *Catoptrica*. See Heiberg 1895.

__________ *Elementa*. See Heiberg and Stamatis 1969–1977, Heath 1956, Dijksterhuis 1929–1930, Frajese and Maccioni 1970.

__________ *Optica*. See Heiberg 1895.

__________ *Phaenomena*. See Menge 1916.

__________ *Sectio canonis*. See Jan 1895, Menge 1916.

Euripides. *Fragmenta*. See Nauck 1964.

Eutocius. *Commentaria in Archimedem*. See Heiberg 1910–1915 iii.

__________ *Commentaria in Apollonium*. See Heiberg 1891–1893, ii.

Evans, J. D. G. 1977. *Aristotle's Conception of Dialectic*. Cambridge/London: Cambridge University Press.

Ferejohn, M. 1982–1983. 'Definition and the Two Stages of Aristotelian Demonstration' *Review of Metaphysics* 36:375–395.

Fortenbaugh, W. W. and Sharples, R. W. 1988. edd. *Theophrastean Studies*. Rutgers Studies in Classical Humanities 3. New Brunswick: Transaction Books.

Fowler, D. H. 1979. 'Ratio in Early Greek Mathematics' *Bulletin of the American Mathematical Society* NS 1:807–846.

__________ 1980–1982. 'Book II of Euclid's Elements and a Pre-Eudoxan Theory of Ratio' *Archive for History of Exact Sciences* 22:5–36, 26:193–209.

———— 1983. 'Eratosthenes' Ratio for the Obliquity of the Ecliptic' *Isis* 74:556–562.

———— 1985a. '400 Years of Decimal Fractions' *Mathematics Teaching* 110: 20–21.

———— 1985b. '400.25 Years of Decimal Fractions' *Mathematics Teaching* 111:30–31.

———— 1987. *The Mathematics of Plato's Academy: A New Reconstruction.* Oxford: Clarendon Press.

Fowler, D. H. and Turner, E. G. 1983. 'Hibeh Papyrus, 27: An Early Example of Greek Arithmetical Notation' *Historia Mathematica* 10:344–359.

Fowler, H. N. 1921. trans. *Plato, The Sophist.* Loeb Classical Library. Rev. edn. London: Heinemann.

———— 1925. trans. *Plato, The Statesman.* Loeb Classical Library. Rev. edn. London: Heinemann.

Frajese, A. and Maccioni, L. 1970. *Gli elementi di Euclide.* Turin: Unione Tipografico-Editrice Torinese.

Frede, D. 1974. 'Comment on Hintikka's Paper "On the Ingredients of an Aristotelian Science"' *Synthese* 28:79–89.

Friedländer, P. 1958. *Plato: An Introduction.* Hans Meyerhoff trans. New York, NY: Pantheon Books.

Friedlein, G. 1867. ed. *Anicii Manlii Torquati Severini Boetii de institutione arithmetica libri duo et de institutione musica libri quinque.* Leipzig: Teubner.

———— 1873. ed. *Procli Diadochi in primum Euclidis elementorum librum commentarii.* Leipzig: Teubner.

Galen. *Institutio logica.* See Kalbfleisch 1896.

Gigon, O. 1945. *Der Ursprung der griechischen Philosophie.* Basel: Schwabe.

Gillings, R. J. 1972. *Mathematics in the Time of the Pharaohs.* Cambridge, MA: MIT Press.

Gold, J. B. 1978. 'The Ambiguity of "Name" in Plato's Cratylus' *Philosophical Studies* 34:223–251.

Goldstein, B. R. and Bowen, A. C. 1983. 'A New View of Early Greek Astronomy' *Isis* 74:330–340.

Gómez-Lobo, A. 1981. 'Definitions in Aristotle's *Posterior Analytics*'. See O'Meara 1981, 25–46.

Gotthelf, A. 1985a. ed. *Aristotle on Nature and Living Things: Philosophical and Historical Studies.* Pittsburgh: Mathesis Publications/Bristol: Bristol Classical Press.

———— 1985b. 'Notes Towards a Study of Substance and Essence in Aristotle's Parts of Animals ii–iv'. See Gotthelf 1985a, 27–54.

———— 1987a. 'First Principles in Aristotle's *Parts of Animals*'. See Gotthelf and Lennox 1987, 167–198.

———— 1987b. '*Historiae I: Plantarum et Animalium*'. See Fortenbaugh and Sharples 1988, 100–135.

Gotthelf, A. and Lennox, J. G. 1987. edd. *Philosophical Issues in Aristotle's Biology*. Cambridge: Cambridge University Press.

Granger, F. 1962. trans. *Vitruvius on Architecture*. 2 vols. Loeb Classical Library. London: Heinemann.

Grensemann, H. 1968. ed. *Ars medica*. Abt. 2.1. Berlin: de Gruyter.

Griffin, J. 1977. 'The Epic Cycle and the Uniqueness of Homer' *Journal of Hellenic Studies* 97:39–53.

Grmek, M. D. 1980. ed. *Hippocratica*. Paris: Centre national de la recherche scientifique.

Gulley, N. 1958. 'Greek Geometrical Analysis' *Phronesis* 33:1–14.

Hardie, R. P. and Gaye, R. K. 1950. trans. *Aristotelis physica*. In Ross 1908–1952, ii.

Harig, G. and Kollesch, J. 1977. 'Neue Tendenzen in der Forschung zur Geschichte der antiken Medizin und Wissenschaft' *Philologus* 121:114–136.

Heath, T. L. 1896. *Apollonius of Perga: Treatise on Conic Sections*. Cambridge: Cambridge University Press. Repr. Cambridge: Heffer and Sons, 1961.

———— 1897. *The Works of Archimedes*. Cambridge: Cambridge University Press. Repr. (with the 1912 supplement, *The Method of Archimedes*) New York: Dover, n.d.

———— 1910. *Diophantus of Alexandria: A Study in the History of Greek Algebra*. 2nd edn. Cambridge: Cambridge University Press.

———— 1913. *Aristarchus of Samos, the Ancient Copernicus*. Oxford: Clarendon Press.

———— 1921. *A History of Greek Mathematics*. 2 vols. Oxford: Oxford University Press.

———— 1949. *Mathematics in Aristotle*. Oxford: Clarendon Press.

———— 1956. *The Thirteen Books of Euclid's Elements*. 2nd edn. 3 vols. New York: Dover Publications. Repr. Cambridge: Cambridge University Press, 1926.

Heiberg, J. L. 1883–1888. ed. *Euclidis opera omnia*. i–iv. Leipzig: Teubner.

———— 1891–1893 ed. *Apollonii Pergaei quae graece exstant*. 2 vols. Leipzig: Teubner. Repr. Stuttgart: Teubner, 1974.

———— 1895. ed. *Euclidis opera omnia*. vii. Leipzig: Teubner.

———— 1898–1903. ed. *Claudii Ptolemaei opera quae exstant omnia*. 2 vols. Leipzig: Teubner.

———— 1910–1915. *Archimedis opera omnia cum commentariis Eutocii*. 2nd edn. 3 vols. Leipzig: Teubner. Rev. repr. E. S. Stamatis 1972. ed. Stuttgart: Teubner.

———— 1927. ed. *Corpus medicorum graecorum* i.1. Leipzig: Teubner.

Heiberg, J. L. and Stamatis, E. 1969–1977. edd. *Euclidis elementa cum scholiis*. 5 vols. Leipzig: Teubner. Rev. edn of Heiberg 1883–1888.

Heiberg, J. L. and Zeuthen, H. G. 1912. edd. *Paul Tannery: Mémoires scientifiques* iii. Paris: Gauthier-Villars.

Hendley, B. P. 1987. ed. *Plato, Time, and Education: Essays in Honor of Robert S. Brumbaugh*. Albany, NY: State University of New York Press.

Hero of Alexandria. *Opera*. See Schöne and Heiberg 1903–1914, Bruins 1964.

Hett, W. S. 1953–1957. trans. *Aristotle: Problems*. 2 vols. Loeb Classical Library. London: Heinemann.

__________1955. trans. *Aristotle, Minor Works*. Loeb Classical Library. London: Heinemann.

__________1964. trans. *Aristotle: On the Soul, Parva Naturalia, On Breath*. Loeb Classical Library. London: Heinemann.

Hicks, R. D. 1925. ed. and trans. *Diogenes Laertius, Lives of Eminent Philosophers*. Loeb Classical Library. London: Heinemann.

Hilbert, D. 1959. *Foundations of Geometry*. E. J. Townsend trans. La Salle, IL: Open Court.

Hiller, E. 1878. ed. *Expositio rerum mathematicarum ad legendum Platonem utilium*. Leipzig: Teubner.

Hintikka, J. 1972. 'On the Ingredients of an Aristotelian Science' *Noûs* 6:55–69.

__________1974. 'Reply to Dorothea Frede' *Synthese* 28:91–96.

__________1981. 'Aristotelian Axiomatics and Geometrical Axiomatics'. See Hintikka *et al*. 1981, 133–144.

Hintikka, J. *et al*. 1981. edd. *Proceedings of the 1978 Pisa Conference*. i. Dordrecht: Reidel.

Hintikka, J. and Remes, U. 1974. *The Method of Analysis*. Dordrecht: Reidel.

Hirsch, E. D., Jr. 1967. *Validity in Interpretation*. New Haven, CT: Yale University Press.

Hough, G. 1966. *An Essay on Criticism*. New York: Norton.

Huffman, C. A. 1985. 'The Authenticity of Archytas Fr. 1' *Classical Quarterly* NS 35:344–348.

Hultsch, F. 1876–1878. ed. *Pappi Alexandrini collectionis quae supersunt*. 3 vols. Berlin: Weidmann. Repr. Amsterdam: Hakkert, 1965.

__________1886. 'Autolykos und Euklid'. *Berichte über die Verhandlungen der königlichen sächsischen Gesellschaft der Wissenschaften zu Leipzig, Philologisch-historische Klasse* 38:128–155.

__________1906. 'Euklid'. See Pauly and Wissowa 1894–, vi. cols. 1003–1052.

Hunt, J. 1707. *Theodosii sphaericorum libri tres*. Oxford: Sheldonian Theatre.

Husserl, E. 1965. 'Philosophy as Rigorous Science'. See Lauer 1965, 71–147.

Hypsicles. *Anaphoricus*. See De Falco and Krause 1966.

Irwin, T. 1977. *Plato's Moral Theory: The Early and Middle Dialogues.* Oxford: Clarendon Press.

——— 1977–1978. 'Aristotle's Discovery of Metaphysics' *Review of Metaphysics* 31:210–229.

Itard, J. 1961. *Les Livres arithmétiques d'Euclide.* Paris: Hermann.

Jackowski, A. J. and Sbraga, J. B. 1970. *Fundamentals of Modern Mathematics.* Englewood Cliffs, NJ: Prentice Hall.

Jaeger, W. 1957. ed. *Aristotelis metaphysica.* Oxford Classical Texts. Oxford: Clarendon Press.

Jan, K. von. 1895. ed. *Musici scriptores graeci.* Leipzig: Teubner.

Joly, R. 1966. *Le Niveau de la science hippocratique.* Paris: Les Belles Lettres.

——— 1977. ed. *Corpus Hippocraticum.* Mons: Université de Mons.

Jones, A. 1986. ed. and trans. *Pappus of Alexandria: Book 7 of the* Collection. New York: Springer-Verlag.

Jones, W. H. S. 1923–1931. ed. and trans. *Hippocrates* i–ii, iv. Loeb Classical Library. London: Heinemann.

——— 1946. *Philosophy and Medicine in Ancient Greece.* Supplement to the Bulletin of the History of Medicine 8. Baltimore: Johns Hopkins University Press.

Kahn, C. H. 1970. 'On Early Greek Astronomy' *Journal of Hellenic Studies* 90:99–116.

——— 1981. 'The Role of *Nous* in the Cognition of First Principles in *Posterior Analytics* II 19'. See Berti 1981, 385–414.

——— 1985. *Anaximander and the Origins of Greek Cosmology.* 2nd edn. Centrum: Philadelphia.

Kalbfleisch, K. 1896. ed. *Galeni institutio logica.* Leipzig: Teubner.

Keaney, J. J. 1963. 'Two Notes on the Tradition of Aristotle's Writings' *American Journal of Philology* 84:52–63.

King, D. A. 1987. 'Universal Solutions in Medieval Islamic Astronomy'. See Berggren and Goldstein 1987, 121–132.

Kirk, G. S., Raven, J. E., and Schofield, M. 1983. *The Presocratic Philosophers.* 2nd edn. Cambridge: Cambridge University Press.

Knorr, W. R. 1975. *The Evolution of the Euclidean Elements: A Study of the Theory of Incommensurable Magnitudes and Its Significance for Early Greek Geometry.* Dordrecht: Reidel.

——— 1978a. 'Archimedes' Lost Treatise on Centers of Gravity of Solids' *Mathematical Intelligencer* 1:102–108.

——— 1978b. 'Archimedes and the Pre-Euclidean Proportion Theory' *Archives internationales d'histoire des sciences* 28:183–244.

——— 1981a. rev. Szabó 1978. *Isis* 72:135–136.

——— 1981b. 'On the Early History of Axiomatics: The Interaction of Mathematics and Philosophy in Greek Antiquity'. See Hintikka *et al.* 1981, 145–186.

__________ 1982a. 'Infinity and Continuity: The Interaction of Mathematics and Philosophy in Antiquity'. See Kretzmann 1982, 112–145.

__________ 1982b. 'Techniques of Fractions in Ancient Egypt and Greece' *Historia Mathematica* 9:133–171.

__________ 1983. 'Construction as Existence Proof in Ancient Geometry' *Ancient Philosophy* 3:125–148.

__________ 1985a. 'Euclid's Tenth Book: An Analytic Survey' *Historia Scientiarum* 29:17–35.

__________ 1985b. 'Archimedes and the Pseudo-Euclidean *Catoptrics*: Early Stages in the Ancient Geometric Theory of Mirrors' *Archives internationales d'histoire des sciences* 35:28–105.

__________ 1986a. *The Ancient Tradition of Geometric Problems*. Boston/Basel: Birkhäuser.

__________ 1986b. 'Archimedes' *Dimension of the Circle*: A View of the Genesis of the Extant Text' *Archive for History of Exact Sciences* 35:281–324.

__________ 1989a. *Textual Studies in Ancient and Medieval Geometry*. Boston/ Basel: Birkhäuser.

__________ 1989b. 'The Practical Element in Ancient Exact Sciences' *Synthese* 81:313–328.

__________ 1992 (forthcoming). 'Pseudo-Euclidean Reflections in Ancient Optics'.

Koelbing, H. M. 1977. *Arzt und Patient in der antiken Welt*. Zürich: Artemis.

Kretzmann, N. 1982. ed. *Infinity and Continuity in Ancient and Medieval Thought*. Ithaca, NY: Cornell University Press.

Kucharski, P. 1959. 'Le *Philèbe* et les éléments harmoniques d'Aristoxène' *Revue philosophique* 149:41–72.

Kudlien, F. 1967. *Der Begin des medizinischen Denkens bei den Griechen*. Zürich/ Stuttgart: Artemis.

Kuhrt, A. 1987. 'Berossus' *Babyloniaka* and Seleucid Rule in Babylonia'. Pp. 32–56 in A. Kuhrt and S. Sherwin-White 1987. edd. *Hellenism in the East*. Berkeley/Los Angeles: University of California Press.

Kullmann, W. 1974. *Wissenschaft und Methode: Interpretationen zur aristotelischen Theorie der Naturwissenschaft*. Berlin/New York: Gruyter.

__________ 1981. 'Die Funktion der mathematischen Beispiele in Aristoteles' *Analytica Posteriora*'. See Berti 1981, 245–270.

Kung, J. 1985. 'Tetrahedra, Motion, and Virtue' *Noûs* 18:17–27.

Lanczos, C. 1964. *Applied Analysis*. Englewood Cliffs, NJ: Prentice Hall.

Landor, B. 1981. 'Definitions and Hypotheses in *Posterior Analytics* 72a19–25 and 76b35–77a4' *Phronesis* 26:308–318.

Lasserre, F. and Mudry, P. 1983. edd. *Formes de pensée dans la collection hippocratique*. Geneva: Droz.

Lauer, Q. 1965. *Phenomenology and the Crisis of Philosophy*. New York: Harper and Row.

Le Blond, J.-M. 1945. *Aristote, Philosophe de la vie: Le livre premier du traité sur les parties des animaux.* Aubier: Editions Montaigne.

Lear, J. 1982. 'Aristotle's Philosophy of Mathematics' *Philosophical Review* 91:161–191.

Lee, H. D. P. 1935. 'Geometrical Method and Aristotle's Account of First Principles' *Classical Quarterly* 29:113–124.

——— 1955. trans. *Plato: The Republic.* Harmondsworth: Penguin Books.

Lennox, J. G. 1984. 'Recent Philosophical Studies of Aristotle's Biology' *Ancient Philosophy* 4:73–82.

——— 1986. 'Aristotle, Galileo and "Mixed Sciences"'. Pp. 29–51 in W. A. Wallace 1986. ed. *Reinterpreting Galileo. Studies in Philosophy and the History of Philosophy.* Washington, DC: Catholic University of America Press.

——— 1987a. 'Divide and Explain: The *Posterior Analytics* in Practice'. See Gotthelf and Lennox 1987, 90–119.

——— 1987b. 'Kinds, Forms of Kinds, and "the more and the less"'. See Gotthelf and Lennox 1987, 339–359.

Leszl, W. 1980. 'Unity and Diversity of the Sciences: The Methodology of the Mathematical and of the Physical Sciences and the Role of Nominal Definition' *Revue internationale de philosophie* 34:384–421.

Lichtenthaeler, C. 1963. 'Le Logos mathématique de la première clinique hippocratique'. Pp. 109–135 in *Quatrième série d'études hippocratiques* 7–10. Geneva/Paris: Lichtenthaeler.

Littré, E. 1839–1961. ed. *Oeuvres complètes d'Hippocrate.* 10 vols. Paris: Baillière.

Lloyd, G. E. R. 1963. 'Who is Attacked in *On Ancient Medicine*?' *Phronesis* 8:108–126.

——— 1979. *Magic, Reason and Experience.* Cambridge: Cambridge University Press.

——— 1983a. *Science, Folklore and Ideology.* Cambridge: Cambridge University Press.

——— 1983b. 'Plato on Mathematics and Nature, Myth and Science'. International Christian University Publication IV–B, Humanities, Christianity and Culture 17. Mitaka, Tokyo: Institute for the Study of Christianity and Culture, International Christian University.

——— 1987. *The Revolutions of Wisdom.* Berkeley: University of California Press.

Lonergan, B. 1972. *Method in Theology.* New York: Seabury Press.

Louis, P. 1955. 'Le Mot HISTORIA chez Aristote' *Revue de philologie, de litterature et d'histoire anciennes* 39:39–44.

Maass, E. 1898. *Commentariorum in Aratum reliquiae.* Berlin: Weidmann.

Machamer, P. K. and Turnbull, R. G. 1976. edd. *Motion and Time, Space and Matter.* Columbus: Ohio State University Press.

MacIntyre, A. 1984. 'Does Applied Ethics Rest on a Mistake?' *The Monist* 67:498–513.

Macran, H. S. 1902. *The Harmonics of Aristoxenus*. Oxford: Clarendon Press.

Mahoney, M. S. 1968–1969. 'Another Look at Greek Geometrical Analysis' *Archive for History of Exact Sciences* 5:319–348.

Maniates, M. R. 1991 (in press). ed. *Music Discourse from Classical to Early Modern Times: Editing and Translating Texts*. Papers read at the 26th Annual Conference on Editing Problems, 19–20 October 1990, University of Toronto. New York: AMS Press.

Mariétan, J. 1901. *Problème de la classification des sciences d'Aristote à St-Thomas*. Paris: Alcan.

Mathiesen, T. J. 1975. 'An Annotated Translation of Euclid's Division of a Monochord' *Journal of Music Theory* 19:236–258.

McKirahan, R. 1983. 'Aristotelian *Epagoge* in *Prior Analytics* 2.21 and *Posterior Analytics* 1.1' *Journal of the History of Philosophy* 21:1–14.

McRae, R. 1961. *The Problem of the Unity of the Sciences: Bacon to Kant*. Toronto: University of Toronto Press.

Meibom, M. 1652. ed. *Antiquae musicae auctores septem graece et latine*. 2 vols. Amsterdam: Elzevir.

Mendell, H. R. 1984. 'Two Geometrical Examples from Aristotle's *Metaphysics*' *Classical Quarterly* NS 34:359–372.

__________ 1986. *Aristotle and the Mathematicians: Some Cross-Currents in the Fourth Century*. Ph.D. dissertation, Stanford University.

Menge, H. 1916. ed. *Euclidis opera omnia* viii. Leipzig: Teubner.

Mignucci, M. 1975. *L'argomentazione dimostrativa in Aristotele*. Padova: Antenore.

Mogenet, J. 1950. *Autolycus de Pitane: Histoire du texte, suivie de l'édition critique des traités de la sphère en mouvement et des levers et couchers*. Louvain: Publications universitaires de Louvain.

Moravcsik, J. 1979. 'Forms, Nature, and the Good in *Philebus*' *Phronesis* 24:81–104.

Morrow, G. R. 1970. trans. *Proclus: A Commentary on the First Book of Euclid's* Elements. Princeton, NJ: Princeton University Press.

Mourelatos, A. P. D. 1980. 'Plato's "Real Astronomy": *Republic* 527d–531d'. See Anton 1980, 33–73.

__________ 1981. 'Astronomy and Kinematics in Plato's Project of Rationalist Explanation' *Studies in History and Philosophy of Science* 12:1–32.

__________ 1987. 'Knowledge, Speculation, and Myth in Plato's Accounts of the Order and the Distances of Celestial Bodies'. See Hendley 1987, 83–105.

Mueller, I. 1970. 'Homogeneity in Eudoxus's Theory of Proportion' *Archive for History of Exact Sciences* 7:1–6.

———— 1974. 'Greek Mathematics and Greek Logic'. See Corcoran 1974, 35–70.

———— 1980. 'Ascending to Problems: Astronomy and Harmonics in *Republic* VII'. See Anton 1980, 103–122.

———— 1981. *Philosophy of Mathematics and Deductive Structure in Euclid's Elements*. Cambridge, MA: MIT Press.

Nauck, A. 1964. ed. *Euripidis tragoediae*. iii. 2nd edn. Leipzig: Teubner.

Neuenschwander, E. 1972–1973. 'Die ersten vier Bücher der Elemente Euklids' *Archive for History of Exact Sciences* 9:325–380.

Neugebauer, O. 1941. 'Exact Science in Antiquity'. Pp. 22–31 in *Studies in Civilization*. Philadelphia, PA: University of Pennsylvania Press. Repr. in Neugebauer 1983, 23-31.

———— 1946. The History of Ancient Astronomy: Problems and Methods' *Publications of the Astronomical Society of the Pacific* 58:17–43, 104–142. Repr. in Neugebauer 1983, 33–98.

———— 1957. *The Exact Sciences in Antiquity*. 2nd edn. Providence, RI: Brown University Press.

———— 1963. 'The Survival of Babylonian Methods in the Exact Sciences of Antiquity and Middle Ages' *Proceedings of the American Philosophical Society* 107.6:528–535.

———— 1975. *A History of Ancient Mathematical Astronomy*. New York: Springer-Verlag.

———— 1983. ed. *Astronomy and History: Selected Essays*. New York/Berlin: Springer-Verlag. See Bowen 1985.

Nicomachus of Gerasa. *Harmonices manuale*. See Jan 1895.

Ogle, W. 1912. trans. *Aristotle: De Partibus Animalium*. See Ross 1908–1952, v.

O'Meara, D. J. 1981. ed. *Studies in Aristotle. Studies in Philosophy and the History of Philosophy*. ix. Washington: Catholic University of America Press.

Owen, G. E. L. 1960. 'Logic and Mathematics in Some Earlier Works of Aristotle'. See Düring and Owen 1960, 163–190.

Pappus. *Collectio*. See Hultsch 1876–1878, Jones 1986.

Parker, H. 1984. *Flawed Texts and Verbal Icons*. Evanston, IL: Northwestern University Press.

Parker, R. A. 1972. *Demotic Mathematical Papyri*. Providence, RI: Brown University Press.

Patterson, G. N. 1985. *Message from Infinity*. 2nd edn. Toronto: University of Toronto Press.

Pauly, A.; Wissowa, G.; and Kroll, W; *et al.* 1894–. edd. Paulys Real-Encyclopädie der classischen Altertumswissenschaft. Stuttgart.

Pedersen, O. 1974. *A Survey of the Almagest*. Odense: Odense University Press.

Peet, T. E. 1923. *The Rhind Mathematical Papyrus, British Museum 10057 and 10058*. London: Hodder and Stoughton.

Pellegrin, P. 1982. *La Classification des animaux chez Aristote: Statut de la biologie et unité de l'aristotelisme*. Paris: Les Belles Lettres.

__________ 1985. 'Aristotle: A Zoology without Species'. See Gotthelf 1985a, 95–116.

__________ 1986. *Aristotle's Classification of Animals*. A. Preus trans. Berkeley/Los Angeles: University of California Press.

__________ 1987. 'Logical Difference and Biological Difference: The Unity of Aristotle's Thought'. See Gotthelf and Lennox 1987, 313–338.

Philippson, P. 1936. 'Genealogie als mythische Form' *Symbolae Osloenses* Fasc. Suppl. 6.

Philoponus, Ioannes. *In Aristotelis analytica posteriora*. M. Wallies 1909. ed. Commentaria in Aristotelem Graeca 13.3 Berlin: Georg Reimer.

Plato. *Opera*. See Burnet 1900–1907.

__________ *Politicus*. See H. N. Fowler 1925.

__________ *Sophistes*. See H. N. Fowler 1921.

[Plutarch] *De musica*. See Einarson and De Lacy 1967.

Porphyry. *In Ptolemaei harmonica commentarium*. See Düring 1932.

Pratt, V. 1982. 'Aristotle and the Essence of Natural History' *History and Philosophy of the Life Sciences* 4.2:202–223.

Proclus. *In primum Euclidis elementorum librum commentarii*. See Friedlein 1873, Morrow 1970.

Ptolemy. *Almagest*. See Heiberg 1898–1903, Toomer 1984.

__________ *Harmonica*. See Düring 1930.

Randall, J. H. Jr. 1960. *Aristotle*. New York: Columbia University Press.

Rashed, R. 1984. ed. *Diophante: Les Arithmétiques, Livres* IV–VII. 2 vols. Paris: Les Belles Lettres.

Rennie, W. 1931. ed. *Demosthenes, Orationes* iii. Oxford Classical Texts. Oxford: Clarendon Press.

Reynolds, L. D. and Wilson, N. G. 1968. *Scribes and Scholars: A Guide to the Transmission of Greek and Latin Literature*. Oxford: Clarendon Press.

Rhind Mathematical Papyrus. See Peet 1923, Chace and Manning 1927–1929, Robins and Shute 1987.

Richards, I. A. 1929. *Practical Criticism: A Study of Literary Judgment*. London: Kegan Paul/Trench/Trubner.

Robins, G. and Shute, C. 1987. *The Rhind Mathematical Papyrus: An Ancient Egyptian Text*. London: British Museum Publications.

Robinson, R. 1936. 'Analysis in Greek Geometry' *Mind* NS 45:464–473.

Rome, A. 1936. ed. *Commentaires de Pappus et de Théon d'Alexandrie sur l'Almageste. Tome II: Théon d'Alexandrie, Commentaire sur les livres 1 et 2 de l'Almageste*. Studi e Testi 72. Vatican City: Biblioteca Apostolica Vaticana.

Roochnik, D. L. 1986. 'Socrates's Use of the Techne-Analogy' *Journal of the History of Philosophy* 24:295–310.

Rose, V. 1899. ed. *M. Vitruvii Pollionis de architectura libri decem.* Leipzig: Teubner.

Ross, W. D. 1908–1952. ed. *The Works of Aristotle Translated into English.* 12 vols. Oxford: Clarendon Press.

——— 1936. ed. *Aristotelis physica.* Oxford Classical Texts. Oxford: Clarendon Press.

——— 1949. *Aristotle's Prior and Posterior Analytics: A Revised Text with Introduction and Commentary.* Oxford: Clarendon Press.

——— 1953. *Aristotle.* 5th edn. London: Methuen.

——— 1964. ed. *Aristotelis analytica priora et posteriora.* Oxford Classical Texts. Oxford: Clarendon Press.

——— 1965. ed. *Aristotelis de anima. Oxford Classical Texts.* Oxford: Clarendon Press.

Ruelle, C. E. 1906. 'Sur l'inauthenticité probable de la division du canon musical attribuée à Euclide' *Revue des études grecques* 19:318–320.

Russell, D. A. 1981. *Criticism in Antiquity.* London: Duckworth.

Ryle, G. 1949. *The Concept of Mind.* London: Hutchinson.

Schmidt, O. 1943. *On the Relation Between Ancient Mathematics and Spherical Astronomy.* Ph.D. Dissertation, Brown University.

——— 1949. 'Some Critical Remarks about Autolycus' On Risings and Settings'. Pp. 202–209 in *Den 11. skandinaviske matematikerkongress i Trondheim 22–25 August 1949.*

Schöne, H. and Heiberg, J. L. 1903–1914. edd. *Heronis Alexandrini opera quae supersunt omnia* iii–v. Leipzig: Teubner.

Series, C. 1985. 'The Geometry of Markoff Numbers' *The Mathematical Intelligencer* 7:20–29.

Sesiano, J. 1982. ed. *Books IV to VII of Diophantus'* Arithmetica *in the Arabic Translation Attributed to Quṣṭâ ibn Lûqâ.* New York/Heidelberg: Springer.

Sidgwick, H. 1869. 'On a Passage in Plato, *Republic*, B. VI' *Journal of Philology* 2:96–103.

Smith, W. D. 1979. *The Hippocratic Tradition.* Ithaca, NY: Cornell University Press.

Solmsen, F. 1929. *Die Entwicklung der aristotelische Logik und Rhetorik.* Neue philologische Untersuchungen 4.

——— 1950. 'Chaos und "Apeiron"' *Studi italiani di filologia classica* 24:235–248. Repr. Solmsen 1968, 68–81.

——— 1968. *Kleine Schriften* i. Hildesheim: Olms.

Sorabji, R. R. K. 1980. *Necessity, Cause, and Blame.* Ithaca, NY: Cornell University Press.

Szabó, A. 1978. *The Beginnings of Greek Mathematics*. A. M. Ungar trans. Synthese Historical Library 17. Dordrecht/Boston: Reidel. See Bowen 1984.

Tannery, P. 1893–1895. ed. *Diophanti Alexandrini opera omnia*. 2 vols. Leipzig: Teubner.

__________ 1912. 'Inauthenticité de la division du canon attribuée à Euclide'. See Heiberg and Zeuthen 1912, 213–219. Originally published in *Comptes rendus des séances de l'Academie des inscriptions et belles-lettres* 4:439–445 (1904).

Tarán, L. 1981. *Speusippus of Athens*. Leiden: Brill.

Taylor, A. E. 1928. *A Commentary on Plato's* Timaeus. Oxford: Clarendon Press.

Tennyson, A. T. 'The Epic'. Pp. 81–83 in R. W. Hill, Jr. 1971. ed. *Tennyson's Poetry*. New York: Norton.

Themistius. *In Aristotelis analyticorum posteriorum commentaria*. See Wallies 1900, 1–66.

Theodosius. *Sphaerica*. See Hunt 1707.

Theon of Alexandria. *In Ptolemaei syntaxin mathematicam i–ii commentaria*. See Rome 1936.

__________ *Opticorum recensio*. See Heiberg 1895.

Theon of Smyrna. *Expositio rerum mathematicarum ad legendum Platonem utilium*. See Hiller 1878.

Thomas, I. (= Bulmer-Thomas, I.) 1939–1941. trans. *Selections Illustrating the History of Greek Mathematics*. 2 vols. Loeb Classical Library. London: Heinemann.

Toomer, G. J. 1984. trans. *Ptolemy's Almagest*. New York/Berlin: Springer.

Turnbull, R. G. 1976. '*Physics* I: Sense Universals, Principles, Multiplicity, and Motion'. See Machamer and Turnbull 1976, 28–56.

Unguru, S. 1979. 'History of Ancient Mathematics: Some Reflections on the State of the Art' *Isis* 70:555–565.

van der Waerden, B. L. 1954. *Science Awakening*. Groningen: Noordhoff.

__________ 1979. *Die Pythagoreer: Religiöse Bruderschaft und Schule der Wissenschaft*. Zurich/Munich: Artemis.

Van Helden, A. 1985. *Measuring the Universe: Cosmic Dimensions from Aristarchus to Halley*. Chicago: University of Chicago Press.

Vernant, J.-P. 1962. *Les Origines de la pensée grecque*. Paris: Presses universitaires de France. Translated by Vernant into English in 1982 as *The Origins of Greek Thought* (Ithaca, NY: Cornell University Press).

Vitruvius. *De architectura*. See Rose 1899, Granger 1962.

Vlastos, G. 1975. *Plato's Universe*. Seattle: University of Washington Press.

von Fritz, K. 1937. 'Oinopides'. See Pauly, Wissowa, and Kroll 1894–, xvii.2 cols. 2258–2272.

__________ 1969. *Platons Theaetet und die antiken Mathematik*. Darmstadt: Wissenschaftliche Buchgesellschaft.

———— 1971. *Grundprobleme der Geschichte der antiken Wissenschaft*. Berlin/ New York: de Gruyter.

Wade-Gery, H. T. 1949. 'Hesiod' *Phoenix* 3:81–93. Repr. Wade-Gery 1958, 1–13.

———— 1958. *Essays in Greek History*. Oxford: Blackwells.

Waismann, F. 1951. *Introduction to Mathematical Thinking: The Formation of Concepts in Modern Mathematics*. New York: Ungar.

Walker, D. P. 1978. *Studies in Musical Science in the Late Renaissance*. Studies of the Warburg Institute. Leiden: Brill.

Wallies, M. 1900. ed. *Themistii in Aristotelis analyticorm posteriorum paraphrasis*. Commentaria in Aristotelem Graeca 5.1. Berlin: Reimer.

Wehrli, F. 1969. *Eudemos von Rhodos*. 2nd edn. *Die Schule des Aristoteles* viii. Basel/Stuttgart: Schwabe.

Weisheipl, J. A. 1965. 'Classification of the Sciences in Medieval Thought' *Mediaeval Studies* 16:54–90.

Whitehead, A. N. 1933. *Adventures of Ideas*. New York: Macmillan. New American Library, 1955 [page references to the latter].

Wicksteed, P. and Cornford, F. M. 1929–1934. trans. *Aristotle, Physics*. 2 vols. Loeb Classical Library. London: Heinemann.

Williams, R. 1958. *Culture and Society* 1780–1950. London: Chatto and Windus.

Wimsatt, W. K., Jr. and Beardsley, M. C. 1946. 'The Intentional Fallacy' *Sewanee Review* 54:468–488.

Winnington-Ingram, R. P. 1963. ed. *Aristides Quintilianus, De musica*. Leipzig: Teubner.

Woods, M. J. 1982. trans. *Aristotle's* Eudemian Ethics, *Books I, II and VIII*. Clarendon Aristotle Series. Oxford: Clarendon Press.

Zeeman, E. C. 1986 'Gears from the Greeks' *Proceedings of the Royal Institution* 58:139–156.

Zeuthen, H. G. 1896. 'Die Geometrische Konstruction als "Existenzbeweis" in der antiken Geometrie' *Mathematische Annalen* 47:222–228.

INDEX OF NAMES

INDEX OF PASSAGES CITED